Methods in Molecular Biology

Series Editor
John M. Walker
School of Life and Medical Sciences
University of Hertfordshire
Hatfield, Hertfordshire, UK

For further volumes:
http://www.springer.com/series/7651

For over 35 years, biological scientists have come to rely on the research protocols and methodologies in the critically acclaimed *Methods in Molecular Biology* series. The series was the first to introduce the step-by-step protocols approach that has become the standard in all biomedical protocol publishing. Each protocol is provided in readily-reproducible step-by-step fashion, opening with an introductory overview, a list of the materials and reagents needed to complete the experiment, and followed by a detailed procedure that is supported with a helpful notes section offering tips and tricks of the trade as well as troubleshooting advice. These hallmark features were introduced by series editor Dr. John Walker and constitute the key ingredient in each and every volume of the *Methods in Molecular Biology* series. Tested and trusted, comprehensive and reliable, all protocols from the series are indexed in PubMed.

Autoimmunity

Methods and Protocols

Edited by

Rami Bechara

Université Paris-Saclay, Inserm, CEA, Center for Immunology of Viral, Autoimmune, Hematological and Bacterial Diseases, Le Kremlin-Bicêtre, France

Humana Press

Editor
Rami Bechara
Université Paris-Saclay, Inserm, CEA
Center for Immunology of Viral, Autoimmune, Hematological
and Bacterial Diseases
Le Kremlin-Bicêtre, France

ISSN 1064-3745 ISSN 1940-6029 (electronic)
Methods in Molecular Biology
ISBN 978-1-0716-4996-1 ISBN 978-1-0716-4997-8 (eBook)
https://doi.org/10.1007/978-1-0716-4997-8

This Humana imprint is published by the registered company Springer Science+Business Media, LLC, part of Springer Nature.
The registered company address is: 1 New York Plaza, New York, NY 10004, U.S.A.

Preface

Autoimmune diseases are complex and heterogeneous disorders driven by dysregulated immune responses targeting self-antigens. In recent years, major advances have been made in understanding the cellular and molecular mechanisms underlying autoimmunity, driven by advances in genomics, imaging, immunometabolism, and tissue modeling. However, translating these insights into reproducible experimental systems remains a major challenge for researchers entering or working within the field.

This volume of Methods in Molecular Biology brings together a comprehensive and up-to-date collection of protocols that reflect the diverse experimental strategies employed in the study of autoimmunity. The chapters span a broad range of model systems and techniques, from T-cell and macrophage characterization, neutrophil functional assays, and organoid culture methods, to cutting-edge approaches such as spatial transcriptomics, RNA FISH, microRNA profiling, and ribosome profiling. The inclusion of protocols for metabolic flux analysis and X chromosome inactivation studies reflects the expanding scope of autoimmune research into areas such as immunometabolism and sex-biased immune regulation.

Each chapter has been written by leading experts in the field and provides not only detailed step-by-step protocols but also critical insights into experimental design, troubleshooting tips, and relevant applications. This book is intended for immunologists, molecular biologists, and clinicians engaged in autoimmune disease research, as well as for trainees seeking a solid methodological foundation to explore disease-relevant immune mechanisms.

I would like to express my sincere gratitude to all the contributing authors for their generosity in sharing their protocols and expertise, and to the editorial team at Springer for their support in bringing this volume to fruition. It is my hope that this collection will serve as a valuable resource to both newcomers and experienced investigators dedicated to uncover the complexities of autoimmunity.

Le Kremlin-Bicêtre, France *Rami Bechara*

Contents

Contributors

DAOUDA ABBA MOUSSA • *IRMB, Université de Montpellier, INSERM, Montpellier, France; CHU Montpellier, Montpellier, France*
GEORGIA AFONSO • *IRMB, Université de Montpellier, Inserm, Montpellier, France*
SOPHIE ALLART • *Univ Toulouse, INSERM, CNRS, Infinity - Toulouse Institute for Infectious and Inflammatory Diseases, Toulouse, France*
CHRISTOS ARONIS • *Division of Rheumatology and Immunology, Department of Internal Medicine, Medical University of Graz, Graz, Austria*
SANYA ARSHAD • *Center for Systems Immunology and the Department of Immunology, University of Pittsburgh School of Medicine, Pittsburgh, PA, USA*
SAMUEL BITOUN • *Paris-Saclay University, INSERM UMR1184, Center for Immunology of Viral Infections and Autoimmune Diseases (IMVA-HB/IDMIT/CEA), Le Kremlin Bicêtre, France; Département de Rhumatologie, Hôpital Bicêtre, AP-HP, Fédération Hospitalo-Universitaire CARE (Cancer and Autoimmunity Relationship), Inserm U1184, Le Kremlin-Bicêtre, France; Rheumatology Department, Hôpital Bicêtre, Assistance Publique – Hôpitaux de Paris, Le Kremlin Bicêtre, France*
JULIE CHAUMEIL • *Université Paris Cité, CNRS, INSERM, Institut Cochin, Paris, France*
FANGFANG CHEN • *Department of Bioinformatics and Biochemistry, Braunschweig Integrated Centre of Systems Biology (BRICS), Technische Universität Braunschweig, Braunschweig, Germany; Research Group Cellular Metabolism in Infection, Helmholtz Centre for Infection Research, Braunschweig, Germany*
CHLOÉ CHIVÉ • *Université Paris Cité, Functional and Adaptive Biology Unit, UMR8251-CNRS, Paris, France; French Environment and Energy Management Agency, Angers, France*
ARMELLE CHOQUET • *IRCM, Univ Montpellier, ICM, INSERM, Montpellier, France*
THEKLA CORDES • *Department of Bioinformatics and Biochemistry, Braunschweig Integrated Centre of Systems Biology (BRICS), Technische Universität Braunschweig, Braunschweig, Germany; Research Group Cellular Metabolism in Infection, Helmholtz Centre for Infection Research, Braunschweig, Germany*
DIVI CORNEC • *LBAI, UMR1227, Univ Brest, Inserm, Brest, France; CHU de Brest, Brest, France*
ALEXANDRE DAVID • *IRCM, Univ Montpellier, ICM, INSERM, Montpellier, France; IRMB-PPC, Univ Montpellier, INSERM, CHU Montpellier, CNRS, Montpellier, France*
PATRICE DECKER • *University Sorbonne Paris Nord, Li2P, Bobigny, France; Inserm UMR 1125, Bobigny, France*
VALÉRIE DEVAUCHELLE-PENSEC • *LBAI, UMR1227, Univ Brest, Inserm, Brest, France; CHU de Brest, Brest, France*
ANA LUISA DIAN • *Institut Curie, PSL Research University, CNRS UMR 3348, INSERM U1278, Orsay, France; Université Paris-Saclay, CNRS UMR 3348, INSERM U1278, Orsay, France*
QUANG KHAI DOAN • *University Sorbonne Paris Nord, Li2P, Bobigny, France; Inserm UMR 1125, Bobigny, France*

Birte Dowerg • *Department of Bioinformatics and Biochemistry, Braunschweig Integrated Centre of Systems Biology (BRICS), Technische Universität Braunschweig, Braunschweig, Germany; Research Group Cellular Metabolism in Infection, Helmholtz Centre for Infection Research, Braunschweig, Germany*

Lucilla Fabbri • *Institut Curie, PSL Research University, CNRS UMR 3348, INSERM U1278, Orsay, France; Université Paris-Saclay, CNRS UMR 3348, INSERM U1278, Orsay, France*

Berenice Faz-Lopez • *Univ Toulouse, INSERM, CNRS, Infinity - Toulouse Institute for Infectious and Inflammatory Diseases, Toulouse, France*

Annemarie Fortuin • *IRMB, Université de Montpellier, Inserm, Montpellier, France*

Marie Frutoso • *LBAI, UMR1227, Univ Brest, Inserm, Brest, France*

Charlotte Gandubert • *LBAI, UMR1227, Univ Brest, Inserm, Brest, France; CHU de Brest, Brest, France*

Soizic Garaud • *LBAI, UMR1227, Univ Brest, Inserm, Brest, France*

Jean-Charles Guéry • *Univ Toulouse, INSERM, CNRS, Infinity - Toulouse Institute for Infectious and Inflammatory Diseases, Toulouse, France*

Rima Haddad • *Université Paris Cité, Inserm, CEA, Stabilité Génétique Cellules Souches et Radiations, LSHL/iRCM/IBFJ, Fontenay-aux-Roses, France; Université Paris-Saclay, Inserm, CEA, Stabilité Génétique Cellules Souches et Radiations, LSHL/iRCM/IBFJ, Fontenay-aux-Roses, France*

Imene Hani • *Université Paris-Saclay, Inserm, Physiologie et Physiopathologie Endocriniennes, Le Kremlin-Bicêtre, France*

Hélène Hérault • *IRCM, Univ Montpellier, ICM, INSERM, Montpellier, France*

Sophie Hillion • *LBAI, UMR1227, Univ Brest, Inserm, Brest, France; CHU de Brest, Brest, France*

Simon Höllerer • *Novartis Pharma AG, Basel, Switzerland*

Alok V. Joglekar • *Center for Systems Immunology and the Department of Immunology, University of Pittsburgh School of Medicine, Pittsburgh, PA, USA; Cancer Immunology and Immunotherapy Program, UPMC Hillman Cancer Center, Pittsburgh, PA, USA*

Quiterie Labaye • *IRMB, Université de Montpellier, INSERM, Montpellier, France*

Céline M. Labbé • *Institut Curie, PSL Research University, CNRS UMR 3348, INSERM U1278, Orsay, France; Université Paris-Saclay, CNRS UMR 3348, INSERM U1278, Orsay, France*

Daniela Latorre • *Institute of Microbiology, ETH Zurich, Zurich, Switzerland; Division of Neuroscience, IRCCS San Raffaele Hospital, Milan, Italy*

Christelle le Dantec • *LBAI, UMR1227, Univ Brest, Inserm, Brest, France*

Becky Lin • *Division of Infectious Diseases, Department of Medicine, University of Pittsburgh School of Medicine, Pittsburgh, PA, USA*

Kadi Lõhmussaar • *Novo Nordisk Foundation Center for Stem Cell Medicine, Faculty of Health and Medical Sciences, University of Copenhagen, Copenhagen, Denmark*

Pascale Louis-Plence • *IRMB, Université de Montpellier, Inserm, Montpellier, France*

Bineta Ly • *Paris-Saclay University, INSERM UMR1184, Center for Immunology of Viral Infections and Autoimmune Diseases (IMVA-HB/IDMIT/CEA), Le Kremlin Bicêtre, France*

Venkata Krishna Kanth Makani • *Center for Systems Immunology and the Department of Immunology, University of Pittsburgh School of Medicine, Pittsburgh, PA, USA; Cancer Immunology and Immunotherapy Program, UPMC Hillman Cancer Center, Pittsburgh, PA, USA*

Xavier Mariette • *Paris-Saclay University, INSERM UMR1184, Center for Immunology of Viral Infections and Autoimmune Diseases (IMVA-HB/IDMIT/CEA), Le Kremlin Bicêtre, France; Département de Rhumatologie, Hôpital Bicêtre, AP-HP, Fédération Hospitalo-Universitaire CARE (Cancer and Autoimmunity Relationship), Inserm U1184, Le Kremlin-Bicêtre, France; Rheumatology Department, Hôpital Bicêtre, Assistance Publique – Hôpitaux de Paris, Le Kremlin Bicêtre, France*

Laetitia Martinerie • *Université Paris-Saclay, Inserm, Physiologie et Physiopathologie Endocriniennes, Le Kremlin-Bicêtre, France; Endocrinologie Pédiatrique, Centre de Référence Maladies Endocriniennes Rares de la Croissance et du Développement, Hôpital Universitaire Robert-Debré GHU APHP Nord, Paris, France; Université Paris-Cité, Faculté de Santé, UFR de Médecine, Paris, France*

Luis E. Muñoz • *Department of Internal Medicine 3, Rheumatology and Immunology, Friedrich-Alexander-Universität Erlangen-Nürnberg (FAU) and Universitätsklinikum Erlangen, Erlangen, Germany; Deutsches Zentrum für Immuntherapie (DZI), Friedrich-Alexander-Universität Erlangen-Nürnberg (FAU) and Universitätsklinikum Erlangen, Erlangen, Germany*

Mar Naranjo-Gomez • *IRMB, Université de Montpellier, INSERM, Montpellier, France*

Mélissa Noack • *Immunogenomics and Inflammation Unit, Edouard Herriot Hospital, Pav. P, Hospices Civils de Lyon, Lyon, France*

Gaetane Nocturne • *Paris-Saclay University, INSERM UMR1184, Center for Immunology of Viral Infections and Autoimmune Diseases (IMVA-HB/IDMIT/CEA), Le Kremlin Bicêtre, France; Département de Rhumatologie, Hôpital Bicêtre, AP-HP, Fédération Hospitalo-Universitaire CARE (Cancer and Autoimmunity Relationship), Inserm U1184, Le Kremlin-Bicêtre, France; Rheumatology Department, Hôpital Bicêtre, Assistance Publique – Hôpitaux de Paris, Le Kremlin Bicêtre, France*

Audrey Paoletti • *Paris-Saclay University, INSERM UMR1184, Center for Immunology of Viral Infections and Autoimmune Diseases (IMVA-HB/IDMIT/CEA), Le Kremlin Bicêtre, France; Arthritis R&D, Neuilly Sur Seine, France; Département de Rhumatologie, Hôpital Bicêtre, AP-HP, Fédération Hospitalo-Universitaire CARE (Cancer and Autoimmunity Relationship), Inserm U1184, Le Kremlin-Bicêtre, France*

Julie Perrot • *Université Paris-Saclay, Inserm, Physiologie et Physiopathologie Endocriniennes, Le Kremlin-Bicêtre, France*

Françoise Pflumio • *Université Paris Cité, Inserm, CEA, Stabilité Génétique Cellules Souches et Radiations, LSHL/iRCM/IBFJ, Fontenay-aux-Roses, France; Université Paris-Saclay, Inserm, CEA, Stabilité Génétique Cellules Souches et Radiations, LSHL/iRCM/IBFJ, Fontenay-aux-Roses, France*

Cindy Qiu • *University Sorbonne Paris Nord, Li2P, Bobigny, France; Inserm UMR 1125, Bobigny, France; Arthritis R&D, Neuilly-sur-Seine, France*

Clara M. Reichardt • *Department of Internal Medicine 3, Rheumatology and Immunology, Friedrich-Alexander-Universität Erlangen-Nürnberg (FAU) and Universitätsklinikum Erlangen, Erlangen, Germany; Deutsches Zentrum für Immuntherapie (DZI), Friedrich-Alexander-Universität Erlangen-Nürnberg (FAU) and Universitätsklinikum Erlangen, Erlangen, Germany*

Lisa Sériau • *LBAI, UMR1227, Univ Brest, Inserm, Brest, France*

Nainisha Shah • *IRMB, Université de Montpellier, Inserm, Montpellier, France*

Lenka Súkeníková • *Institute of Microbiology, ETH Zurich, Zurich, Switzerland*

Arnaud Tête • *Université Paris Cité, HealthFex, Inserm UMR-S 1124, Paris, France*

Giraldina Trevejo-Nuñez • *Division of Infectious Diseases, Department of Medicine, University of Pittsburgh School of Medicine, Pittsburgh, PA, USA*

Stéphan Vagner • *Institut Curie, PSL Research University, CNRS UMR 3348, INSERM U1278, Orsay, France; Université Paris-Saclay, CNRS UMR 3348, INSERM U1278, Orsay, France*

Say Viengchareun • *Université Paris-Saclay, Inserm, Physiologie et Physiopathologie Endocriniennes, Le Kremlin-Bicêtre, France*

Matteo Villa • *Division of Rheumatology and Immunology, Department of Internal Medicine, Medical University of Graz, Graz, Austria*

Paul M. Zdinak • *Center for Systems Immunology and the Department of Immunology, University of Pittsburgh School of Medicine, Pittsburgh, PA, USA; Graduate Program in Microbiology and Immunology, University of Pittsburgh School of Medicine, Pittsburgh, PA, USA*

Chapter 1

Identification and Characterization of Human Autoreactive T Cells for Autoimmune Research

Lenka Súkeníková and Daniela Latorre

Abstract

T cells directed against self-antigens play a pivotal role in autoimmunity, yet their comprehensive characterization in human autoimmune disorders remains challenging. In this chapter, we present methods to isolate and profile autoreactive T cells from human biological samples, specifically peripheral blood, cerebrospinal fluid, and tissue biopsies. The workflow combines in vitro T cell screening, isolation of autoreactive T cells, and single T cell clone generation, followed by in depth analysis of T cell receptor (TCR) repertoire, HLA restriction, cytokine profile, and targeted self-epitope. These protocols aim to facilitate research of human autoimmune diseases by providing powerful tools for systematic analysis of autoreactive T cells in human biological samples.

Key words Autoimmunity, Autoreactive T cells, Single T cell clone, T cell receptor (TCR), Epitope specificity

1 Introduction

Human autoimmune diseases represent a heterogeneous group of disorders that affect up to 10% of the population, posing a significant public health challenge [1]. These conditions arise from a complex interplay of genetic predisposition, environmental factors, and immune dysregulation, resulting in loss of tolerance to self-antigens [2]. Immune dysregulation often involves pathogenic autoreactive T cells, including cytotoxic $CD8^+$ T cells that may directly destroy target cells, and $CD4^+$ helper T cells that may help B cells in producing pathogenic autoantibodies or release cytokines, driving chronic inflammation and tissue damage [3]. Despite their critical role, a detailed characterization of these cells remains elusive in most human autoimmune diseases. Advancing this knowledge is essential for deepening our understanding of the underlying immune processes and may contribute to the development of more effective biomedical interventions.

Rami Bechara (ed.), *Autoimmunity: Methods and Protocols*, Methods in Molecular Biology, vol. 2990,
https://doi.org/10.1007/978-1-0716-4997-8_1,

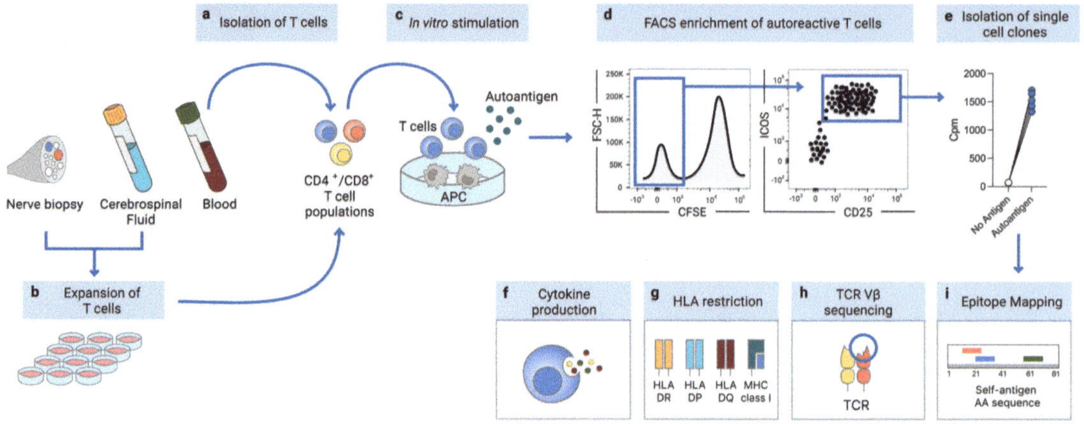

Fig. 1 Schematics showing steps involved in autoreactive T cell characterization. $CD4^+$ and $CD8^+$ T cell populations are enriched from freshly collected peripheral blood, CSF, and tissue biopsy samples: $CD4^+$ and $CD8^+$ memory T cells are sorted ex vivo from blood by FACS cytometry (**a**), while total $CD4^+$ and $CD8^+$ T cells are first polyclonally expanded in vitro from CSF or tissue biopsy before being enriched by FACS cytometry (**b**). On day 6–7 following in vitro stimulation (**c**), autoreactive T cells are identified by CFSE dilution and expression of the activation markers CD25 and ICOS by flow cytometry (**d**). After identification, autoreactive T cells are further analyzed at single-cell resolution through the generation of single T cell clones (**e**) and the characterization of their cytokine production (**f**), HLA restriction (**g**), TCR Vβ sequences (**h**), and targeted epitope (**i**)

In this chapter, we present a detailed experimental workflow for the identification, isolation, and characterization of rare autoreactive T cells from human biological samples, specifically peripheral blood, cerebrospinal fluid (CSF), and tissue biopsy (Fig. 1). The approach involves in vitro screenings of $CD4^+$ and $CD8^+$ T cells populations for reactivity against selected self-antigens presented by autologous antigen presenting cells (APCs) in co-culture assays. Autoreactive T cells are identified by proliferation using carboxyfluorescein succinimidyl ester (CFSE) dilution assay and the expression of the activation markers CD25 and ICOS by flow cytometry. $CFSE^{low}CD25^{high}ICOS^+$ T cells, enriched for autoreactive T cells, are then studied at the single-cell resolution by generating single T cell clones, which are further characterized for their recognized self-epitopes, TCR sequences, MHC restriction, and cytokine production (Fig. 1) and, optionally, by single-cell RNA sequencing (not described here).

The employment of this approach has led to the identification and characterization of rare autoreactive T cell populations targeting the hypocretin antigen in narcolepsy, a rare sleep disorder of the central nervous system (CNS), and myelin proteins of the peripheral nerves in a subset of patients with Guillain-Barré syndrome, an inflammatory disorder of the peripheral nervous system (PNS), providing critical insights into the immune mechanisms underlying these conditions [4–8].

2 Materials

2.1 Reagents

1. Ficoll-Paque Plus density gradient media.
2. Isopropanol.
3. PHA Purified (phytohemagglutinin).
4. CD14 MicroBeads, human.
5. LS/MS Columns.
6. 5(6)-Carboxyfluorescein diacetate N-succinimidyl ester (CFSE).
7. LIVE/DEAD™ Fixable Aqua Dead Cell Stain Kit.
8. Oligo dT(25) Primers.
9. Maxima H Minus Reverse Transcriptase.
10. 0.2% Triton.
11. 100 mM dNTP mix.
12. RiboLock RNase inhibitor.
13. Q5 Hot Start High-Fidelity DNA Polymerase.
14. TCR primers—*see* Table 4.
15. UltraPure Agarose.
16. TAE buffer 50×.
17. GelRed® Nucleic Acid Gel Stain.
18. 6× TriTrack DNA Loading Dye.
19. Quick-Load® 1 kb DNA Ladder.
20. GeneRuler 100 bp DNA Ladder.
21. Ultima Gold XR—Liquid Scintillation Counting.
22. Conical Propylene Centrifuge Tubes (50 and 15 mL).
23. T75 Culture Flask.
24. Serological Pipettes (2, 5, 10, and 25 mL).
25. Cryogenic Vials.
26. 40 μm Strainer.
27. 96 U-bottom Microplates.
28. 96-well PCR Microplates/Microtubes.

2.2 Culture Media and Buffers

1. Complete Medium (CM): RPMI 1640 w/o HEPES supplemented with 5% human serum, 1 mM sodium pyruvate, 1% non-essential amino acids, 2 mM L-Glutamine, 1% penicillin/streptomycin solution, 50 μM β-mercaptoethanol, and, optionally, 500 UI/mL IL-2 (CM + IL-2).
2. Wash Buffer: RPMI 1640 with HEPES supplemented with 0.5% human serum.

3. Freezing Medium (cryopreservation media): 90% fetal bovine serum with 10% DMSO (*see* **Note 1**).
4. Flow cytometry buffer (FACS buffer): PBS supplemented with 2 mM EDTA pH 8.0, 1% human serum (*see* **Note 2**).

2.3 Antibodies

1. Staining Antibodies: CD4–PE/Dazzle 594 (RPA-T4); CD45RA–BV650 (HI100); CD8–APC Fire750 (RPA-T8); CD8–FITC (B9.11); CCR7–BV421 (G043H7); CD14–PE–Cy5 (RMO52); CD25–PE–Cy5 (B1.49.9); CD56–PE–Cy5 (N901); CD19–FITC (HIB19); CD25–PE (M-A251); ICOS–Pacific Blue (H4A3); CD3–BV785 (UCHT1).
2. Anti-MHC neutralizing antibodies: anti-HLA-DR (L243), anti-HLA-DQ (SPVL3), anti-HLA-DP (clone B7/21), anti-panHLA-II (HB145), anti-panHLA-I (W6/32).

2.4 Peptides

1. Peptides: Peptides are commercially synthesized as 15-mers, each overlapping by 10 amino acids to span the full sequence of the target antigen. They are supplied as crude peptides in lyophilized form. Examples of peptide sequences used in previous studies, representing self-antigens derived from peripheral nerve myelin [7], are shown in Table 1. Peptide pools are prepared as described in Subheading 3.4.

2.5 Equipment

1. Multi-channel pipettes.
2. Tabletop centrifuge.
3. Biosafety cabinet.
4. CO_2 incubator.
5. Water bath.
6. X-ray Irradiator System.
7. FACS sorter.
8. PCR cycler.
9. Gel electrophoresis.
10. Quantum Gel Documentation Imaging (Vilber).
11. Microbeta Microplate Counter.
12. FilterMate Harvester.

3 Methods

3.1 Peripheral Blood Mononuclear Cells (PBMCs) Isolation and Storage

1. In a biosafety cabinet, transfer the blood from the lithium heparin collection tubes into a tissue culture T-75 flask and avoid any contamination (*see* **Note 3**).
2. Dilute the blood at a 1:1 ratio with RPMI medium with HEPES.

Table 1
List of 15-mers peptides overlapping by 10 spanning the entire sequences of P0 (UniProtKB: P25189-1, A1-D12, $n = 48$), P2 (UniProtKB: P02689, E1-G1, $n = 25$) and PMP22 (UniProtKB: Q6FH25, G2-I7, $n = 30$) myelin proteins

P0 myelin protein		P2 myelin protein		PMP22 myelin protein	
Peptide ID	Sequence	Peptide ID	Sequence	Peptide ID	Sequence
A1	MAPGAPSSSPSPILA	E1	MSNKFLGTWKLVSSE	G2	MLLLLLSIIVLHVAV
A2	PSSSPSPILAVLLFS	E2	LGTWKLVSSENFDDY	G3	LSIIVLHVAVLVLLF
A3	SPILAVLLFSSLVLS	E3	LVSSENFDDYMKALG	G4	LHVAVLVLLFVSTIV
A4	VLLFSSLVLSPAQAI	E4	NFDDYMKALGVGLAT	G5	LVLLFVSTIVSQWIV
A5	SLVLSPAQAIVVYTD	E5	MKALGVGLATRKLGN	G6	VSTIVSQWIVGNGHA
A6	PAQAIVVYTDREVHG	E6	VGLATRKLGNLAKPT	G7	SQWIVGNGHATDLWQ
A7	VVYTDREVHGAVGSR	E7	RKLGNLAKPTVIISK	G8	GNGHATDLWQNCSTS
A8	REVHGAVGSRVTLHC	E8	LAKPTVIISKKGDII	G9	TDLWQNCSTSSSGNV
A9	AVGSRVTLHCSFWSS	E9	VIISKKGDIITIRTE	G10	NCSTSSSGNVHHCFS
A10	VTLHCSFWSSEWVSD	E10	KGDIITIRTESTFKN	G11	SSGNVHHCFSSSPNE
A11	SFWSSEWVSDDISFT	E11	TIRTESTFKNTEISF	G12	HHCFSSSPNEWLQSV
A12	EWVSDDISFTWRYQP	E12	STFKNTEISFKLGQE	H1	SSPNEWLQSVQATMI
B1	DISFTWRYQPEGGRD	F1	TEISFKLGQEFEETT	H2	WLQSVQATMILSIIF
B2	WRYQPEGGRDAISIF	F2	KLGQEFEETTADNRK	H3	QATMILSIIFSILSL
B3	EGGRDAISIFHYAKG	F3	FEETTADNRKTKSIV	H4	LSIIFSILSLFLFFC
B4	AISIFHYAKGQPYID	F4	ADNRKTKSIVTLQRG	H5	SILSLFLFFCQLFTL
B5	HYAKGQPYIDEVGTF	F5	TKSIVTLQRGSLNQV	H6	FLFFCQLFTLTKGGR

(continued)

Table 1
(continued)

P0 myelin protein		P2 myelin protein		PMP22 myelin protein	
Peptide ID	Sequence	Peptide ID	Sequence	Peptide ID	Sequence
B6	QPYIDEVGTFKERIQ	F6	TLQRGSLNQVQRWDG	H7	QLFTLTKGGRFYITG
B7	EVGTFKERIQWVGDP	F7	SLNQVQRWDGKETTI	H8	TKGGRFYITGIFQIL
B8	KERIQWVGDPRWKDG	F8	QRWDGKETTIKRKLV	H9	FYITGIFQILAGLCV
B9	WVGDPRWKDGSIVIH	F9	KETTIKRKLVNGKMV	H10	IFQILAGLCVMSAAA
B10	RWKDGSIVIHNLDYS	F10	KRKLVNGKMVAECKM	H11	AGLCVMSAAAIYTVR
B11	SIVIHNLDYSDNGTF	F11	NGKMVAECKMKGVVC	H12	MSAAAIYTVRHPEWH
B12	NLDYSDNGTFTCDVK	F12	AECKMKGVVCTRIYE	I1	IYTVRHPEWHLNSDY
C1	DNGTFTCDVKNPPDI	G1	KGVVCTRIYEKV	I2	HPEWHLNSDYSYGFA
C2	TCDVKNPPDIVGKTS			I3	LNSDYSYGFAYILAW
C3	NPPDIVGKTSQVTLY			I4	SYGFAYILAWVAFPL
C4	VGKTSQVTLYVFEKV			I5	YILAWVAFPLALLSG
C5	QVTLYVFEKVPTRYG			I6	VAFPLALLSGVIYVI
C6	VFEKVPTRYGVVLGA			I7	ALLSGVIYVILRKRE
C7	PTRYGVVLGAVIGGV				
C8	VVLGAVIGGVLGVVL				
C9	VIGGVLGVVLLLLLL				
C10	LGVVLLLLLLFYVVR				
C11	LLLLLFYVVRYCWLR				
C12	FYVVRYCWLRRQAAL				
D1	YCWLRRQAALQRRLS				

(continued)

Table 1 (continued)

P0 myelin protein		P2 myelin protein		PMP22 myelin protein	
Peptide ID	**Sequence**	**Peptide ID**	**Sequence**	**Peptide ID**	**Sequence**
D2	RQAALQRRL SAMEKG				
D3	QRRL SAMEKGKLHKP				
D4	AMEKGKLHKPGKDA S				
D5	KLHKPGKDASKRG RQ				
D6	GKDASKRGRQTPVL Y				
D7	KRGRQTPVL YAMLDH				
D8	TPVLYAMLDH SRSTK				
D9	AMLDHSRSTKA VSEK				
D10	SRSTKAVSEKKAKGL				
D11	AVSEKKAKGLGESRK				
D12	KAKGLGESRKDKK				

3. Gently layer 24 mL of the diluted blood onto 12 mL of Ficoll-Paque Plus (*see* **Note 4**).
4. Spin the blood at 400 g for 30 min at room temperature with brakes off.
5. Using a sterile Pasteur pipet attached to a vacuum suction in a biosafety cabinet, aspirate the upper layer that contains RPMI and platelets.
6. Using a sterile 10 mL pipet, transfer the PBMCs layer to one new 50 mL tube.
7. Wash cells by adding excess volume of RPMI with HEPES and centrifuge at 350 g for 15 min at room temperature.
8. Discard the supernatant carefully without disturbing the cell pellet, resuspend cells in RPMI with HEPES, and centrifuge at 300 g for 10 min at room temperature.

9. Discard the supernatant carefully, resuspend cells in wash buffer, and filter the cell suspension through a 40 μm strainer. Centrifuge at 130 g for 10 min at room temperature.
10. Count cells and determine viability by trypan blue exclusion or another suitable method.
11. Label the cryovial tubes and place them on ice or in the freezer until use. Prepare pre-cooled to 4 °C Freezing Containers filled with 100% isopropanol.
12. Resuspend the PBMCs at a concentration of $2.0\text{–}3 \times 10^7$ viable cells/mL in pre-cooled to 4 °C freezing media.
13. Dispense 1 mL of the cell suspension into each cryovial. Place the cryovials in the freezing container and transfer it to −80 ° C. Leave undisturbed overnight.
14. Transfer frozen vials from the freezing container to a liquid nitrogen storage tank for long-term storage.

3.2 Enrichment of Immune Cell Populations from PBMCs

1. Submerge the cryovials containing frozen PBMCs into a 37 °C water bath (*see* **Note 5**).
2. Transfer the cells into 15 mL conical propylene tubes.
3. Slowly add pre-warmed wash buffer to the cells to prevent cellular shock.
4. Centrifuge the cells at 350 g for 5 min at room temperature.
5. Carefully discard the supernatant without disturbing the cell pellet. Wash the cells by adding excess volume of wash buffer and centrifuge at 350 g for 5 min at room temperature.
6. Repeat the washing **step 5**.
7. Resuspend the cells in FACS buffer.
8. Count the cells and determine their viability using the trypan blue exclusion method or another suitable method.
9. Wash cells by adding excess volume of FACS buffer and centrifuge at 350 g for 5 min at 4 °C.
10. Resuspend the pellet in FACS buffer in 15 mL conical propylene tube and incubate with CD14-coated microbeads. Obtain $CD14^+$ and $CD14^-$ cell fractions according to manufacturer's instructions.
11. Wash both the positively enriched $CD14^+$ cells and negatively separated $CD14^-$ cells by adding excess volume of FACS buffer and centrifuge at 350 g for 5 min at 4 °C.
12. Discard the supernatant, resuspend $CD14^+$ cells in 10 mL of wash buffer, and store them at 4 °C until use (*see* Subheading 3.5.1).
13. Discard the supernatant and resuspend $CD14^-$ cells by gently tapping the tube.

14. Add 100 μL of surface staining mix (Table 2) and incubate for 15–20 min at 4 °C, protected from light (*see* **Note 6**).
15. Wash cells by adding excess volume of FACS buffer and centrifuge at 350 g for 5 min at 4 °C.
16. Resuspend the stained $CD14^-$ cells in 300 μL of FACS buffer and filter them through a 40 μm strainer to remove clumps.
17. Sort memory $CD4^+$ and $CD8^+$ T cell populations by flow cytometry, according to the gating strategy shown in Fig. 2a.

Table 2
List of the antibodies and working dilution for preparation of surface staining mixes to isolate T cell populations from PBMCs or CSF and tissue biopsy

Source	Antibody	Fluorophore	Clone
PBMCs	Anti-CD4	PE/Dazzle 594	RPA-T4
	Anti-CD45RA	BV650	HI100
	Anti-CD8	APC Fire750	RPA-T8
	Anti-CCR7	BV421	G043H7
	Anti-CD14	PE–Cy5	RMO52
	Anti-CD25	PE–Cy5	B1.49.9
	Anti-CD56	PE–Cy5	N901
	Anti-CD19	FITC	HIB19
CSF or tissue biopsy	Anti-CD3	BV785	UCHT1
	Anti-CD4	PE/Dazzle 594	RPA-T4
	Anti-CD8	APC Fire750	RPA-T8
	Anti-CD56	PE–Cy5	N901

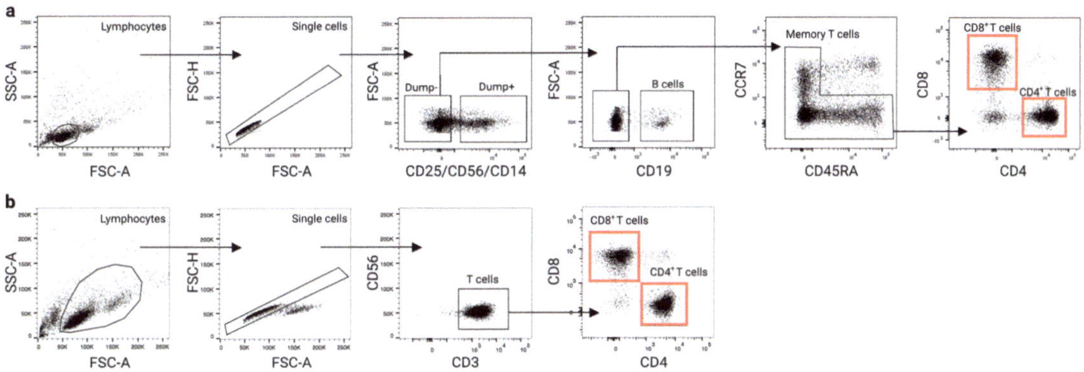

Fig. 2 Representative gating strategy used to isolate $CD4^+$ and $CD8^+$ memory T cells from $CD14^-$ cell fractions ex vivo from blood (**a**) and total $CD4^+$ and $CD8^+$ T cell populations from CSF or tissue biopsy after in vitro polyclonal expansion (**b**)

3.3 Cell Purification and Sorting of T Cell Populations from CSF or Tissue Biopsy

1. Collect tubes with cerebrospinal fluid (CSF) or tissue biopsy and process within a few hours of sampling.
2. For tissue biopsies only, place the sample in a sterile Petri dish with 1 mL of wash buffer, gently mince, and filter through a 40 μm cell strainer into a new sterile tube to obtain a single-cell suspension.
3. Wash cells by adding excess volume of RPMI with HEPES and centrifuge at 350 g for 7 min at room temperature.
4. Discard the supernatant and resuspend the cell pellet in CM + IL-2 medium containing 1 μg/mL PHA and irradiated allogeneic PMBCs (1×10^5 per well).
5. Seed cells in 200 μL/well in 96 U-bottom microplates and incubate at 37 °C with 5% CO_2.
6. After 1 week, harvest the expanded T cells and transfer them into 24-well plates (3×10^6 per well). Continue expanding them until around 2 weeks.
7. Count cells and determine viability by trypan blue exclusion or another suitable method.
8. Wash cells by adding excess volume of FACS buffer and centrifuge at 350 g for 5 min at 4 °C.
9. Discard the supernatant and resuspend the cells by gently tapping the tube.
10. Add 300 μL of surface staining mix (Table 2) and incubate for 15–20 min at 4 °C, protected from light (*see* **Note 6**).
11. Wash the cells by adding excess volume of FACS buffer and centrifuge at 350 g for 5 min at 4 °C.
12. Discard the supernatant, resuspend the stained cells in 300 μL of FACS buffer, and filter them through a 40 μm strainer to remove clumps.
13. Sort total $CD4^+$ and $CD8^+$ T cell populations by flow cytometry according to the gating strategy shown in Fig. 2b.
14. Seed the sorted cells in CM + IL-2 medium in 24-well plates (3×10^6 per well) and incubate at 37 °C with 5% CO_2 for further expansion until they achieve a resting state or freeze them as described in Subheading 3.1, **steps 10–14**.

3.4 Peptide Pools Preparation

1. Resuspend lyophilized peptides in 100% DMSO at a concentration of ≥20 mg/mL and prepare peptide pools as detailed in previously described procedures [9] and outlined below:
   (a) *Total Pool*: Combine each individual overlapping 15-mers peptide to create a total peptide pool spanning the full sequence of the selected self-antigen.
   (b) *Matrix Pools*: Create a peptide library of overlapping 15-mers vertical (V) and horizontal (H) peptide pools

where each peptide is contained in two different peptide pools (Table 3). This setup has the advantages of (i) reducing the number of total T cell screenings and (ii) each peptide is contained in exactly 2 peptide pools, which allows the identification of specific epitopes recognized by a T cell clone.

2. Store the total and matrix pools at −80 °C (*see* **Note 7**).

Table 3
Representative matrix epitope pools consisting of vertical (V) and horizontal (H) peptide pools of P0 (UniProtKB: P25189-1, A1-D12, $n = 48$), P2 (UniProtKB: P02689, E1-G1, $n = 25$) and PMP22 (UniProtKB: Q6FH25, G2-I7, $n = 30$) myelin proteins

Matrix pool P0 myelin protein								
	V1	V2	V3	V4	V5	V6	V7	V8
H1	A1	A2	A3	A4	A5	A6	A7	A8
H2	A9	A10	A11	A12	B1	B2	B3	B4
H3	B5	B6	B7	B8	B9	B10	B11	B12
H4	C1	C2	C3	C4	C5	C6	C7	C8
H5	C9	C10	C11	C12	D1	D2	D3	D4
H6	D5	D6	D7	D8	D9	D10	D11	D12

Matrix pool P2 myelin protein					
	V1	V2	V3	V4	V5
H1	E1	E2	E3	E4	E5
H2	E6	E7	E8	E9	E10
H3	E11	E12	F1	F2	F3
H4	F4	F5	F6	F7	F8
H5	F9	F10	F11	F12	G1

Matrix pool PMP22 myelin protein						
	V1	V2	V3	V4	V5	V6
H1	G2	G3	G4	G5	G6	G7
H2	G8	G9	G10	G11	G12	H1
H3	H2	H3	H4	H5	H6	H7
H4	H8	H9	H10	H11	H12	I1
H5	I2	I3	I4	I5	I6	I7

3.5 In Vitro Screening for Detection of Autoreactive T Cells

3.5.1 Antigen Pulsing of Antigen Presenting Cells (APCs)

1. Irradiate autologous $CD14^+$ cells on an X-ray source with 45 Gy.
2. Centrifuge the irradiated cells at 350 g for 5 min at room temperature, discard the supernatant, and resuspend the cell pellet in CM.
3. Seed 2.5×10^4 $CD14^+$ cells per well in 50 μL CM in 96-well U-bottom microplates and incubate for 1–3 h at 37 °C with 5% CO_2 for antigen pulsing. To assess reactivity against self-antigens, add peptide pools at a final concentration of 3 μg/mL. For the negative control, use CM alone or CM with DMSO (vehicle control). As a positive control for $CD4^+$ memory T cell response, add seasonal Flu vaccine at a final concentration of 5 μg/mL, and for $CD8^+$ memory T cell response, add EBV or CMV MHC class I peptide mega pools [9] at a final concentration of 1 μg/mL, as previously described [4, 7].

3.5.2 CFSE Labeling and Stimulation

1. Prepare a staining solution of 5 μM CFSE in pre-warmed DPBS.
2. Wash T cells twice with DPBS at 350 g for 5 min at room temperature to remove protein containing buffer.
3. Resuspend the cells at a concentration of 10^6/mL in CSFE staining solution in a 15 mL conical propylene tube and incubate them for 10 min in the cell incubator at 37 °C with 5% CO_2 (*see* **Notes 6** and **8**).
4. Stop the reaction by filling up the tube with CM.
5. Centrifuge at 350 g for 5 min at room temperature, discard the supernatant, and resuspend the cells in 10 mL of CM.
6. Repeat the centrifugation and resuspension two more times to ensure all excess CFSE is removed.
7. After the final wash, resuspend the T cells in CM and seed 5×10^4 cells per well in 150 μL per well at a ratio of 2:1, placing on top of the antigen pulsed APCs in a 96-well U-bottom plate (*see* **Note 9**).
8. Incubate the cell co-cultures at 37 °C with 5% CO_2 for 7–10 days.

3.5.3 Analysis and Sorting of Autoreactive T Cells

1. Harvest the cells from the cell co-culture plates.
2. Wash the cells by adding FACS buffer and centrifuge at 350 g for 5 min at room temperature.
3. Meanwhile, prepare live/dead exclusion dye (LIVE/DEAD™ Fixable Aqua) staining mix according to the manufacturer's instruction.

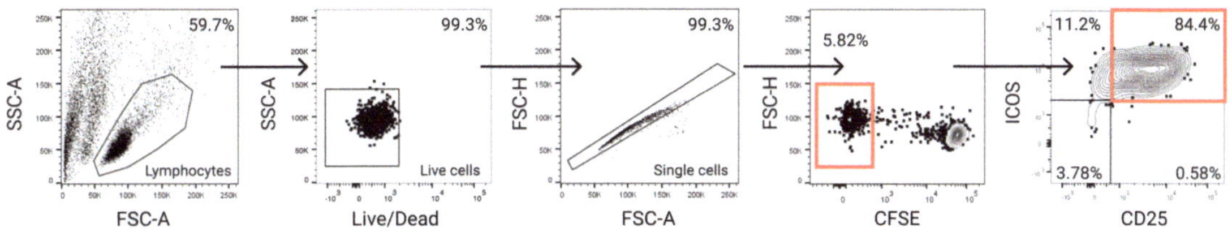

Fig. 3 Representative gating strategy used to identify and isolate autoreactive T cell clones based on CFSE dilution and expression of the activation markers CD25 and ICOS in response to autoantigens in in vitro stimulation assay

4. Discard supernatant, resuspend in 50 μL live/dead staining, and incubate for 30 min at room temperature, protected from light.
5. Wash the cells by adding an excess of cold FACS buffer and centrifuge at 350 g for 5 min at 4 °C.
6. Meanwhile, prepare surface staining mix containing CD25–PE (M-A251) and ICOS–Pacific Blue (H4A3) in a total volume of 50 μL FACS buffer for each condition.
7. Discard supernatant and stain the cells with the antibody cocktail. Incubate for 20 min at 4 °C, protected from light (*see* **Note 6**).
8. Wash the cells by filling up the tube with FACS buffer and centrifuge at 350 g for 5 min at 4 °C.
9. Discard supernatant, resuspend the cells in an appropriate volume of FACS buffer, and filter them through a 40 μm strainer to remove clumps.
10. Analyze or sort cells by flow cytometry (Fig. 3).

3.6 Isolation and Characterization of Autoreactive Single T Cell Clones

3.6.1 Generation of Single T Cell Clones by Limiting Dilution

1. Autoreactive $CD4^+$ or $CD8^+$ T cells are sorted as $CFSE^{low}CD25^+ICOS^+$ cells and seeded at a density of 0.5 cells per well in 50 μL in CM + IL-2 medium, containing 1 μg/mL PHA and irradiated allogeneic PBMCs (5×10^4 cells per well), in 384-well microplates.
2. Incubate in the cell incubator at 37 °C with 5% CO_2.
3. After around 12–15 days, the expanded T cell clones are collected and transferred to 96-well microplates, placing one clone per well.
4. Add 150 μL CM + IL-2 to each well and incubate the cell cultures at 37 °C with 5% CO_2 to allow further T cell clone expansion (*see* **Note 10**).
5. Split the growing T cell clones every 2–3 days as needed.
6. Continue this process until the cells become resting, typically on day 25–30.

3.6.2 T Cell Clone Specificity, Epitope Mapping, and HLA Restriction Screening

1. Seed 2.5×10^4 irradiated autologous APCs (*see* **Note 11**) per well in 50 μL CM alone or in the presence of the selected antigen in 96-well U-bottom microplates and incubate for 1–3 h at 37 °C with 5% CO_2 for antigen pulsing. To assess reactivity against self-antigens, add peptide pools at a final concentration of 3 μg/mL. For the negative control, use CM alone or CM with DMSO (vehicle control).
2. Meanwhile, count the T cell clones and collect the amount needed, considering 5×10^4 cells per condition.
3. Wash the T cell clones by adding pre-warmed wash buffer and centrifuge at 350 g for 5 min at room temperature.
4. Repeat the centrifugation and resuspension two more times to ensure the removal of residual IL-2.
5. Discard supernatant, resuspend the T cells in CM, and seed 5×10^4 cells per well in 150 μL per well at a ratio of 2:1, placing on top of the antigen pulsed irradiated autologous APCs in a 96-well U-bottom plate.
6. Incubate the cell co-cultures at 37 °C in 5% CO_2 incubator for 2 days.
7. On day 2, add CM containing 1 μCi/mL [methyl-^{3}H]-thymidine to each well and incubate the cell co-cultures at 37 °C in 5% CO_2 incubator for 16–18 h.
8. Harvest the cells using a FilterMate Harvester onto filtermats, place them into transparent foils, and add scintillation liquid according to the manufacturer's instruction.
9. Measure the radioactivity of each well on a Microbeta Microplate Counter.
10. Analyze the proliferation data based on the obtained counts per minute (c.p.m.) values.
11. Antigen-specific T clones are selected based on defined cut-off values [4, 7].
12. Define the self-epitope targeted and HLA-restriction of antigen specific T cell clones following the protocol described above in **steps 1–11**, with a few modifications:
    (a) On day 2 of co-culture, harvest 50 μL of culture supernatants for the quantification of cytokine release by autoreactive T cell clones using preferred method (i.e., multiplex bead-based immunoassay or ELISA assays), according to the manufacturer's instructions.
    (b) For epitope mapping analysis, irradiated autologous APCs (*see* **Note 11**) are pulsed with matrix peptides pools (Table 3), total pool or negative control, according to **step 1**.

(c) For HLA restriction analysis, co-cultures are performed in the absence or presence of 10 μg/mL blocking anti-MHC class II or class I monoclonal antibodies.

3.6.3 TCR Vβ Sequencing of Autoreactive T Cell Clones

Reverse Transcription

The reverse transcription of the mRNA of single-cell clones is carried out directly on lysed cells in 96-well plate, without any RNA extraction step.

1. Resuspend 10^3–10^4 cells in 8 μL lysis buffer containing 0.2% Triton X-100 and RNase Inhibitor at final concentration 2.5 U/μL.
2. Incubate 5 min on ice.
3. Meanwhile, prepare the master mix for reverse transcription containing 5 μM HPLC-purified oligo dT(25) primers, 25 U/sample Maxima H Minus reverse transcriptase, and 0.5 mM dNTPs.
4. Dispense 12 μL per sample in standard PCR tubes or 96-well PCR microplates and add 8 μL of cell lysate.
5. Run cDNA synthesis program on a thermocycler as follows: 60 min at 50 °C, 5 min at 85 °C; hold at 4 °C.
6. Proceed directly with PCR amplification or store the cDNA at −20 °C.

PCR Amplification

1. Prepare the PCR mix containing 0.25 mM dNTPs, 0.5 U/sample Q5 Hot Start High-Fidelity DNA Polymerase, 0.2 μM TCR Vβ-specific forward primer pools, and 0.2 μM Rev84 reverse primer pairing to C1–C2 β-chain constant region [4, 7, 10] (Table 4).
2. Dispense 20 μL of PCR mix per sample in standard PCR tubes or 96-well PCR microplates and add 5 μL of cDNA from single-cell clones.
3. Run PCR amplification program on a thermocycler as follows: initial denaturation, 1 min at 98 °C, denaturation, 10 s at 98 °C, primer annealing, 20 s at 55 °C, polymerase extension, 40 s at 72 °C; repeat denaturation to extension step 35 times; final extension, 2 min at 72 °C, and hold at 4 °C. The plates can be stored at −20 °C until the next step.

Agarose Gel Electrophoresis

Sequence amplification is assessed through agarose gel electrophoresis followed by sequencing and clonotype identification.

1. Cast a 1–1.5% agarose gel by dissolving ultrapure agarose in 1× TAE buffer.

Table 4
List of primers for TCR Vβ genes amplification and sequencing of individual T cell clones

Primers	
TBV-A	TCAGGTGTGATYCAATTTC
TBV-B	AGGTGTGATCCAATTTCG
TBV-C	TGTGTCC TGGTACCAACAG
TBV-D	GTATCGACAAGACCCAGG
TBV-E	GTATCG ACAAGACCYGGG
TBV-F	CTCACCTGAATGCCCCAA
TBV-G	ATGTTYT GGTAYCGTCAG
TBV-H	CCTTTACTGGTACCDGCAGA
TBV-I	ACAGAG ATGGGACAAGAAG
TBV-J	GCCATGTACTGGTAYMGA
TBV-K	CCCCAT CTCTAATCACTTATAC
TBV-L	ACATCAAACCCCAACCTATAC
TBV-M	ACCAGCAGAAGTCAAGTCA
TBV-N	TGTSTACTGGTACCARCAG
TBV-O	GGGAAGGACAGAAAGCAAAA
TBV-P	TTACTCAGTTCCCCAGCC
TBV-Q	AGATGCAGCCCAATGAAA
TBV-R	ACAGATGGGAAACGACAA
TBV-S	GTATCRACAAGAYCCAGGA
TRBC-Rev84	TGTGGCCTTTTGGGTGTGG
TRBC-Rev64	AGATCTCTGCTTCTGATGGC

2. Dilute PAGE GelRed to a 1:10,000 concentration in the melted agarose gel solution to stain DNA.
3. Pour the gel into a gel casting tray with a well comb and leave at room temperature to allow it to solidify.
4. Load the gel wells with 5 μL of PCR products premixed with TriTrack DNA gel-loading dye, allowing direct loading of the PCR reactions into the wells of the agarose gel.
5. Load the gel wells with 5 μL of molecular weight markers, as the Quick-Load 1 kb DNA Ladder and the GeneRuler 100 bp ladder.
6. Perform electrophoresis at 90 V for 30–45 min to separate the PCR products.

7. At the end of the electrophoresis, record the image of the agarose gel using gel documentation system to visualize bands and confirm the presence of PCR products (*see* **Note 12**).
8. Following a successful PCR reaction, the PCR products can be purified and sequenced by Sanger sequencing using the Rev64 reverse primer (Table 4).

Annotate the TCR Vβ sequences using the international ImMunoGeneTics information system (IMGT) database [11]. The output returns the CDR3β nucleotide and amino acid sequence information and paired TRBV, TRBJ, and TRBD usage data, allowing further downstream analyses.

4 Notes

1. All culture media should be prepared and handled under sterile conditions to prevent contamination. Store all culture media at 4 °C unless otherwise specified.
2. 0.01% (w/v) sodium azide can be added in the case of long-term storage.
3. Keep blood samples at room temperature and process within 24 h from collection to maintain cell integrity and viability. Samples processing should be handled under sterile conditions to prevent contamination.
4. Carefully layer the blood on top of the Ficoll solution to avoid mixing the layers.
5. While thawing cells, do not allow them to warm to 37 °C to avoid damaging the cells.
6. Ensure cells are well resuspended in staining buffers to achieve uniform staining.
7. Avoid freeze-thaw cycles to maintain integrity.
8. Maintain precise incubation timing to prevent over-staining, which can impact cell viability and downstream analysis.
9. Add 200 μL of PBS to any empty wells surrounding cell cultures to reduce evaporation.
10. Prepare fresh CM + IL-2 for T cell culture to support optimal growth, and pre-warm the media to 37 °C before use.
11. Use autologous $CD14^+$ cells as APCs, as described in Subheading 3.5.1. If $CD14^+$ cell quantities are limited, autologous EBV immortalized B cells can be used, prepared according to the protocol described in [12].
12. Successful amplification will show bands corresponding to the expected size range of TCR Vβ sequences.

Acknowledgments

DL is supported by the Young investigator Grant from Helmut Horten Foundation, the Novartis Foundation for Medical-Biological Research Grant and the 2024 Discovery Research Grant from the GBS|CIDP Foundation International (ETH Zurich), and the Career Development Award from the Giovanni Armenise Harvard Foundation and a Spark grant from the Swiss National Science Foundation (IRCCS Ospedale San Raffaele, Milan).

References

1. Cooper GS, Bynum ML, Somers EC (2009) Recent insights in the epidemiology of autoimmune diseases: improved prevalence estimates and understanding of clustering of diseases. J Autoimmun 33(3–4):197–207. https://doi.org/10.1016/j.jaut.2009.09.008
2. Pisetsky DS (2023) Pathogenesis of autoimmune disease. Nat Rev Nephrol 19(8): 509–524. https://doi.org/10.1038/s41581-023-00720-1
3. Sun L, Su Y, Jiao A, Wang X, Zhang B (2023) T cells in health and disease. Signal Transduct Target Ther 8(1):235. https://doi.org/10.1038/s41392-023-01471-y
4. Latorre D, Kallweit U, Armentani E, Foglierini M, Mele F, Cassotta A, Jovic S, Jarrossay D, Mathis J, Zellini F, Becher B, Lanzavecchia A, Khatami R, Manconi M, Tafti M, Bassetti CL, Sallusto F (2018) T cells in patients with narcolepsy target self-antigens of hypocretin neurons. Nature 562(7725): 63–68. https://doi.org/10.1038/s41586-018-0540-1
5. Latorre D, Sallusto F, Bassetti CLA, Kallweit U (2022) Narcolepsy: a model interaction between immune system, nervous system, and sleep-wake regulation. Semin Immunopathol 44(5):611–623. https://doi.org/10.1007/s00281-022-00933-9
6. Liblau RS, Latorre D, Kornum BR, Dauvilliers Y, Mignot EJ (2023) The immunopathogenesis of narcolepsy type 1. Nat Rev Immunol. https://doi.org/10.1038/s41577-023-00902-9
7. Súkeníková L, Mallone A, Schreiner B, Ripellino P, Nilsson J, Stoffel M, Ulbrich SE, Sallusto F, Latorre D (2024) Autoreactive T cells target peripheral nerves in Guillain–Barré syndrome. Nature. https://doi.org/10.1038/s41586-023-06916-6
8. Ripellino P, Schreiner B, Latorre D (2024) Expanding our understanding of Guillain-Barre syndrome: recent advances and clinical implications. Eur J Immunol:e2250336. https://doi.org/10.1002/eji.202250336
9. da Silva Antunes R, Weiskopf D, Sidney J, Rubiro P, Peters B, Lindestam Arlehamn CS, Grifoni A, Sette A (2023) The MegaPool approach to characterize adaptive CD4+ and CD8+ T cell responses. Curr Protoc 3(11): e934. https://doi.org/10.1002/cpz1.934
10. Becattini S, Latorre D, Mele F, Foglierini M, De Gregorio C, Cassotta A, Fernandez B, Kelderman S, Schumacher TN, Corti D, Lanzavecchia A, Sallusto F (2015) T cell immunity. Functional heterogeneity of human memory CD4(+) T cell clones primed by pathogens or vaccines. Science 347(6220):400–406. https://doi.org/10.1126/science.1260668
11. Lefranc MP, Giudicelli V, Ginestoux C, Jabado-Michaloud J, Folch G, Bellahcene F, Wu Y, Gemrot E, Brochet X, Lane J, Regnier L, Ehrenmann F, Lefranc G, Duroux P (2009) IMGT, the international ImMunoGeneTics information system. Nucleic Acids Res 37(Database issue):D1006–D1012. https://doi.org/10.1093/nar/gkn838
12. Traggiai E, Becker S, Subbarao K, Kolesnikova L, Uematsu Y, Gismondo MR, Murphy BR, Rappuoli R, Lanzavecchia A (2004) An efficient method to make human monoclonal antibodies from memory B cells: potent neutralization of SARS coronavirus. Nat Med 10(8):871–875. https://doi.org/10.1038/nm1080

Chapter 2

T Cell Antigen Discovery Using Cell-Based Epitope Libraries

Paul M. Zdinak, Sanya Arshad, Venkata Krishna Kanth Makani, and Alok V. Joglekar

Abstract

T cells carry out their functions by specifically recognizing antigens through their T cell receptors (TCRs), making the identification of TCR specificities crucial for understanding health and disease. Over the past decade, several antigen discovery methods have been developed to deconvolute TCRs. Our laboratory has established a cell-based antigen discovery platform based on chimeric receptors termed "SABRs" (Signaling and Antigen-presenting Bifunctional Receptors). SABRs are extracellular peptide-MHC complexes fused to intracellular signaling domains. SABRs present a covalently linked epitope to T cells and induce a readable signal upon their recognition. Moreover, large epitope libraries in SABRs can be constructed to screen the cognate epitopes of a given TCR. The modularity and continued refinement of SABRs allow antigen discovery for CD8+ and CD4+ T cells. In this chapter, we provide an overview of the SABR platform, followed by a detailed description of the steps involved in the generation of SABR libraries and performing SABR screens to deconvolute TCR specificities.

Key words Antigen discovery, Epitope libraries, SABR libraries, T cell receptor deconvolution

1 Introduction

T cells play a pivotal role in the initiation and progression of autoimmune diseases [1]. T cells recognize peptides bound to Major Histocompatibility Complex (MHC) molecules surface TCR. T cell repertoire generates a vast amount of TCR diversity through V(D)J recombination allowing for T cells to respond to vast array of antigens. Although the antigen specificity of a TCR is encoded within its amino acid sequence, it is currently impossible to predict TCR specificity based on sequence alone. The promiscuity of TCRs allows them to recognize a large sequence space of epitopes and presents a challenge in antigen discovery. T cell antigen discovery has been an active field of development for over

Paul M. Zdinak, Sanya Arshad, and Venkata Krishna Kanth Makani contributed equally and are free to change the order of appearance of their names on their CV.

Rami Bechara (ed.), *Autoimmunity: Methods and Protocols*, Methods in Molecular Biology, vol. 2990,
https://doi.org/10.1007/978-1-0716-4997-8_2,

3 decades [2]. There are two broad categories of experimental T cell antigen discovery methods: (1) antigen-directed methods that rely on reading out binding or functional response to a small set of antigens. These include pMHC multimers, Activation Induced Marker (AIM) assays, and ELISA/ELISPOT assays, and pMHC-driven targeted lentiviral entry methods such as RAPTR. These are often limited in scale at <1000 epitopes to be interrogated at a time. (2) TCR-directed methods that screen given TCRs against many epitopes. These include Yeast display, SABRs, Trogocytosis, T-scan, TCR-MAP, T-Syn, and MCR-TCR [3–10]. These methods often interrogate a small number of TCRs but can present >10,000 epitopes to them. Out of these methods, yeast display offers a truly unbiased way of screening the entire possible sequence space for an epitope but often leads to identifying mimotopes rather than physiological epitopes. While not fully degenerate, T-scan can present epitopes from genome-wide libraries, allowing interrogation of the entire proteome subject to antigen processing and presentation. MCR-TCR and T-Syn can present transcriptome wide, cDNA-derived, albeit not precisely designed libraries. Finally, SABRs, MCR-TCR, and Trogocytosis, among others, can present precisely defined large epitope libraries. Our lab developed and pioneered the SABR platform that allows the identification of MHC class I and class II-restricted epitopes [4, 11]. We have previously deployed it to identify the antigens of tumor- and self-reactive TCRs. In this chapter, we describe how SABRs can be used for antigen discovery, along with several considerations for deploying the platform for autoimmunity.

The SABR antigen discovery platform presents defined epitope libraries on reporter cells and uses SABR-driven signal to discriminate cognate epitopes from the libraries. First, custom designed and defined libraries are cloned en masse into SABRs to generate SABR libraries. Second, SABR libraries are expressed in an NFAT-GFP-Jurkat reporter cell line and coincubated with Jurkat cells expressing a TCR of interest. Third, SABR+GFP+ cells are sorted, and genomic DNA is extracted from sorted cells. Fourth, SABR constructs are sequenced, and enrichment of epitopes from the library is determined to identify putative cognate epitopes. Finally, epitopes are validated using single epitope SABRs. In this chapter, we describe in detail the steps involved in designing and performing SABR screens.

2 Materials

2.1 SABR Cloning Reagents

1. SABR Backbone with Stuffer (available through Addgene or through our lab upon request).

Table 1
Table of primers to be used in the study

ClassI-Oligo-Fwd	CAGGAGGGCTCGGCA
ClassI-Oligo-Rev	GGACCCTCCGCATCC
ClassII-Oligo-Fwd	CAGGAGGGCTCGGCA
ClassII-Oligo-Rev	GCTGCCACCACCTCC
MND_Input_Verify_F	GCTCCCCGAGCTCAATAAAAG
MND_Input_Verify_R	TGGCTAAGATCTACAGCTGCCTT
IDT-UD-SABR-C1-F (CLASS I) -1	ACACTCTTTCCCTACACGACGCTCTTCCGATCTGGCCTGC TTTGTTTGCC
IDT-UD-SABR-C1-R (CLASS I) -1	GTGACTGGAGTTCAGACGTGTGCTCTTCCGATCTCC TCCACCACCGCTACCTC
IDT-UD-SABR-C2-F (CLASS II)	ACACTCTTTCCCTACACGACGCTCTTCCGATCTGGCCTGC TTTGTTTGCC
IDT-UD-SABR-C2-R (CLASS II)	GTGACTGGAGTTCAGACGTGTGCTCTTCCGATCCCTCCACC TGAACCTCGC
UDI0001-R (representative of Index #1)	CAAGCAGAAGACGGCATACGAGAT**<u>AACCGCGG</u>**GTGACTGGAG TTCAGACGTGTGCTCTTCCGATC
UDI0001-F (representative of Index #1)	AATGATACGGCGACCACCGAGATCTACAC**<u>AGCGCTAG</u>**ACAC TCTTTCCCTACACGACGCTCTTCCGATCT

The UDI indexes (index 1) are highlighted in bold and underlined

2. Backtranslate.py script from Joglekar laboratory Github (https://github.com/joglekar-lab/SABR-II).
3. Single Stranded oligo library encoding the epitopes designed as described below.
4. BsmBI-v2.
5. NEB 3.1 rBuffer.
6. Nuclease-free water.
7. SABR oligo amplification primers (*see* Table 1).
8. KOD Hot Start Master Mix.
9. Thermocycler.
10. PCR/Gel Purification kit.
11. Agarose, Molecular Biology grade.
12. DNA binding dye such as SYBR Safe.
13. Gel electrophoresis equipment (for up to 48 samples).
14. Razor blades.
15. InFusion HD Mix.
16. Stable *E. coli*.

17. LB agar.
18. Petri dishes.
19. Carbenicillin.
20. Plasmid DNA Maxiprep kit.
21. OneTaq 2× Master Mix with dye included.
22. SABR oligo verification and sequencing primers (*see* Table 1).
23. DNA ladder such as Quick Load 1 kb + ladder.
24. Erlenmeyer flasks (1 L capacity).
25. Bacterial shaker set at 37 °C.

2.2 Library Virus Transfection and Transduction

1. HEK 293T/17 cells.
2. TC incubator, set at 37 °C, 5% CO_2, and humidity chamber.
3. 0.25% Trypsin.
4. Complete D-10: DMEM Base medium with L-Glutamine, 1× penicillin-streptomycin (100× stock),10% Fetal Bovine Serum.
5. 15 cm^2 TC treated Plates.
6. Mirus Bio™ TransIT™-293 Transfection Reagent.
7. pCMV-Δ8.9 (or psPAX2 or similar packaging plasmid).
8. pMDG-VSVG (or pCAGGS-VSVG or similar envelop plasmid).
9. Opti-MEM™.
10. NFAT-GFP Jurkat Cells (a kind gift from Dr. Arthur Weiss).
11. Spinning rotor centrifuge with buckets with biohazard lids.
12. Protamine Sulfate.
13. Complete R-10: RPMI1640 base medium with L-Glutamine, 1× penicillin-streptomycin (100× stock), 10% Fetal Bovine Serum.
14. 0.45 micron filters (Steriflip or equivalent).
15. 50 mg/mL G-418 sulfate.
16. T-75 flasks with vented cap.

2.3 SABR Screening

1. Jurkat cells expressing TCRs of interest.
2. Cell trace labeling dye such as CellTrace Violet.
3. 6-well TC treated plates.
4. FACS tubes.
5. Anti-human CD69-APC-Cy7 antibody (Clone FN50).
6. DPBS without calcium or magnesium.
7. FACS Buffer: DPBS, 2% Fetal Bovine Serum.

8. FACS Sorter such as BD Aria with 405, 488, and 640 nm lasers.
9. Genomic DNA extraction kit.
10. SABR library amplification and indexing primers (*see* Table 1).
11. KOD Hot Start Master Mix.
12. Thermocycler.
13. Agarose, Molecular Biology grade.
14. DNA binding dye such as SYBR Safe.
15. Gel Electrophoresis equipment (for up to 48 samples).
16. PCR Cleanup Kit.
17. DNA Lo-Bind tubes, 1.5 mL.
18. Commercial sequencing service using Illumina Sequencer such as Novaseq X Plus.
19. demultiplex_dual.py, extract_epitope.py, merge_counts_split.py scripts from the Joglekar lab Github (https://github.com/joglekar-lab/SABR-II).

3 Methods

3.1 SABR Library Design

SABR Libraries allow encoding for custom-designed lists of epitopes. In our hands, we have successfully made libraries up to 300,000 epitopes. Given the scalability and flexibility of the platform, there are different ways to design SABR libraries:

3.1.1 Literature and IEDB Searches

The Immune epitope database (IEDB) is an online resource that has curated information on epitopes from published literature. It has epitope-specific experimental assays, and every assay reflects the binding of an epitope-specific T cell receptor (TCR), antibody or MHC molecule to an experimentally tested antigen or epitope [12]. IEDB has advanced search tools where queries can be performed by organism, disease, antigen, MHC restriction, or immune response. It also allows for MHC binding prediction and has tools for identifying potential B cell or T cell epitopes in new proteins. These tools use algorithms to predict which regions of a protein are likely to be immunogenic based on factors like MHC binding affinity. While IEDB is useful to build libraries, it is important to consider that IEDB-derived libraries will only interrogate published epitopes.

3.1.2 RNA-Seq and Proteomics Data

Genes that may be potential antigens by virtue of being expressed in target tissue/cells can be identified using transcriptomics or proteomics. Once a set of genes is identified, MHC binding prediction algorithms like NetMHCpan, NetMHCIIpan, and MHCflurry, can be used to identify potential epitopes that can be presented on MHC alleles of interest [13–15]. Predicted epitopes can be then cloned into SABR libraries. It is important to consider the limitations of these algorithms, particularly with understudied MHC alleles/epitopes.

3.1.3 Immunopeptidomics

Immunopeptidomics is a method that isolates pMHC complexes from cells, elutes peptides presented on MHC, and identifies them using mass spectrometry. It is a reliable source to design a SABR library because it identifies peptides which are derived from actual proteome, processed under physiological conditions reflecting a cell's real time antigen presentation machinery. We have previously used immunopeptidomics successfully to make SABR libraries to interrogate autoimmune T cells (*see* **Note 1**). While it does involve additional labor and skill, it does yield the most likely epitopes presented on the target tissue.

3.1.4 In Silico Predictions

In cases where direct identification of epitopes by immunopeptidomics may not be readily feasible, e.g., for identifying altered neo- or neo-self-epitopes, in silico predictions can be used to build "theoretical" libraries of epitopes. For instance, in tumors, whole exome sequencing can be used to identify non-synonymous mutations that lead to neoepitopes, which can be incorporated into personalized SABR libraries. In addition, hard to detect epitopes such as Hybrid Insulin Peptides (HIPs) can also be predicted in silico and used to build SABR libraries. Furthermore, an in silico designed library where degenerate amino acids are incorporated into SABRs can be built to interrogate TCR cross-reactivity.

Ultimately, library size as a consideration will be at the discretion of the user. As size increases, each step of the process increases in its cost, labor, and resources. We find that libraries up to 100,000 epitopes are manageable, with the whole protocol being executed by 1–2 individuals. To date, our published libraries have ranged between a few thousands and tens of thousands of epitopes. However, facilities and logistics permitting, libraries exceeding one million epitopes are possible.

3.2 SABR Library Cloning

3.2.1 SABR Library Synthesis

1. Backtranslate protein sequences into nucleotide sequences using "*backtranslate.py*" script. This script takes a list of epitope amino acid sequences, backtranslates it into nucleotide sequences, removes duplicates, and returns the output.
2. Open the output file in Excel or a similar spreadsheet program.
3. Add overhangs to nucleotide sequences using the following commands:
   1. For class I oligos:
      = LEFT("CAGGAGGGCTCGGCA"&LOWER(**C1**)&"GGATGCGGAGGGTCC"&"G"&"GCGGCCGC"&"

TCATCTCGTGCGACATCAAGCTATACTCTAATA-TAGCATTCCGTTCGAGTATAGCAG",100), where cell "C1" has the backtranslated oligonucleotide sequence.

2. For class II oligos:
= LEFT("CAGGAGGGCTCGGCA"&LOWER(**C1**)&"GGAGGTGGTGGCAGC"&"G"&"GCGGCCGC"&"TCATCTCGTGCGACATCAAGCTATACTCTAATA-TAGCATTCCGTTCGAGTATAGCAG", 120), where cell "C1" has the backtranslated oligonucleotide sequence.

4. Synthesize the library using commercial services such as *oligo pools* from Twist Biosciences Inc.

3.2.2 First BsmBI Digest of SABR Backbone

1. In PCR strips, set up multiple digestions of the SABR backbone (*see* **Note 2**)

10× NEB r3.1 buffer	10 μL
BsmBI-v2 (10 units/μL)*	3 μL (30 units)
2 kb Backbone (10 μg per reaction)	× μL
Nuclease-free water	Up to 100 μL
	100 μL Rxn total

2. Use a thermocycler with the following program:

Step	Temp	Time
Digest	55 °C	6 h
Heat inactivation	80 °C	20 min
Hold/store	4 °C	

3. Purify with commercially available PCR purification kit.
   1. Reactions can be pooled if the DNA amount does not exceed kit/column limitations.
   2. Elution volume can be minimized for next step(s).
4. *Stoppage point*: DNA can be stored at 4 °C or long term at −20 °C.

3.2.3 Second BsmBI Digest of SABR Backbone

1. In PCR strips, set up digestion reactions of all the PCR purified product from the first digestion.

	1×
10× NEB r3.1 buffer	10 μL
BsmBI-v2 (10 units/μL)*	3 μL (30 units)

(continued)

PCR purified, BsmBI digested backbone (10 μg per reaction)	– μL
Nuclease-free water	Up to 100 μL
	100 μL Rxn total

2. Use a thermocycler with the following program:

Step	Temp	Time
Digest	55 °C	6 h
Heat inactivation	80 °C	20 min
Hold/store	4 °C	

3. Make a 1% agarose gel DNA binding dye with wells to accommodate the total volume of the second BsmBI digest reactions. Reactions can be pooled in wells.
4. Run the second digested product for 1 h at 100 V.
5. Excise the top band using a razor blade which is the SABR backbone ready for epitope insertion.
6. Use a commercially available gel purification kit to gel purify the excised backbone (top band) and discard the 2 kb stuffer (remaining gel). A representative BsmBI digestion image is in Fig. 1a.
7. Elute in smallest volume kit recommends. *Stoppage point*: DNA can be stored at 4 °C or long term at −20 °C.

3.2.4 Oligonucleotide Library Amplification

1. Aliquot the single-stranded oligo library to 10 ng/μL for use in the amplification step.
2. Set up multiple reactions +1 no template control reaction to double strand the oligo library.

2× KOD MM	12.5 μL
100 μM Class I OR II Oligo Fwd	0.33 μL
100 μM Class I OR II Oligo Rev	0.33 μL
Library ssOligo (10 ng/μL)	0.5 μL (5 ng)
Nuclease-free water	11.33 μL
	25 μL total

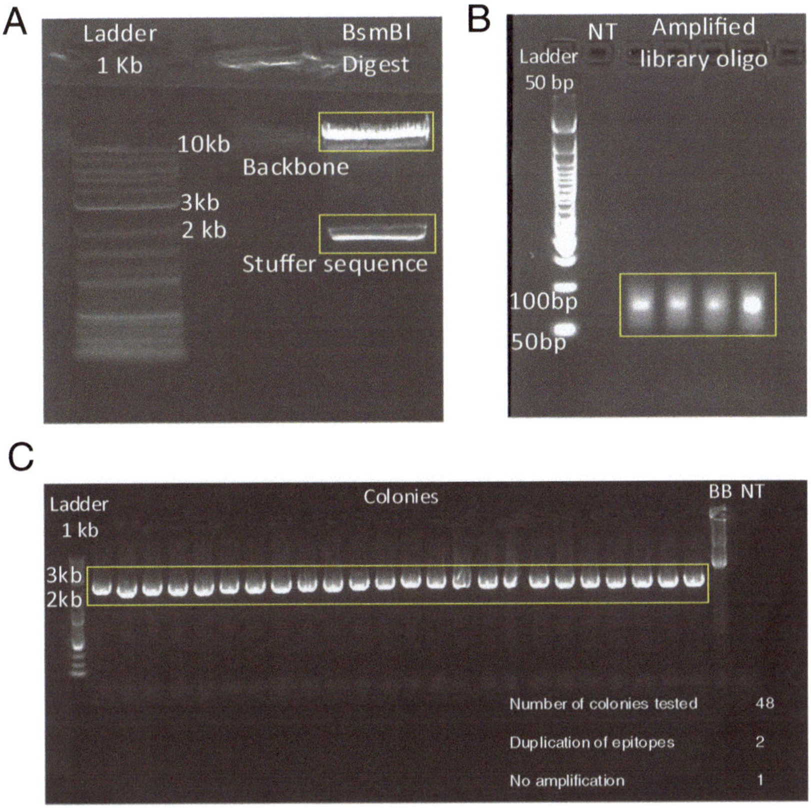

Fig. 1 Example of SABR library cloning and expected outcome. (**a**) Representative SABR backbone digestion with BSMBI (second digestion). The top band to be purified (backbone) is highlighted. The stuffer fragment to be discarded is also highlighted. (**b**) Representative amplification of pooled library oligos. NT = no template reaction. (**c**) Representative gel showing PCR amplification of colonies, and a summary of colonies tested. BB = reaction with the backbone as a positive control. NT = no template reaction. The inset shows a representative of a typical library quality check by Sanger sequencing. In all the figures, the appropriate ladders are indicated with specific sized bands

3. Use a thermocycler with the following program:

Step	Temp	Time
Initial denaturation	95 °C	2 min
10 Cycles: denature	95 °C	20 s
Anneal	59 °C	10 s
Extension	70 °C	15 s
Final extension	70 °C	1 min
Hold	4 °C	

4. Run 2–3 μL of PCR product on a 2% agarose gel to ensure concatenation did not occur. A representative image is in Fig. 1b.
5. Purify with commercially available PCR purification kit and elute using smallest volume. This purified product will be used as the insert for InFusion of the library into the SABR backbone. *Stoppage point*: DNA can be stored at 4 °C or long term at −20 °C.

3.2.5 Infusion Cloning and Transformation of Oligonucleotide Library into the SABR Backbone

1. Set up the appropriate number of InFusion reactions for the double-stranded oligo library and 2× digested SABR backbone (*see* **Note 3**)

5× In-Fusion mix	2 μL
Gel purified SABR backbone (*200 ng total*)	Unique
Double-stranded oligo library (*20 ng total*)	Unique
Nuclease-free water	Up to 10 μL
	10 μL total

2. Incubate (using thermocycler if possible) using the following program:

Step	Temp	Time
Incubation	50 °C	15 min
Hold	4 °C	

3. Transform 2 μL (5 transformations per Infusion reaction) of ligation mix per tube of NEB High Efficiency Stable Component *E. coli* (e.g., NEB C3040H) according to the manufacturer's protocol.
4. Combine the volume of bacteria from all transformation tubes into one tube.
5. Set up three- to tenfold serial dilutions for plating onto Luria Broth plates containing 1× carbenicillin. This will be used for the library coverage calculation.
6. Distribute the remaining bacterial volume by adding 5–250 mL of Luria Broth with 1× carbenicillin in an Erlenmeyer flask, incubating overnight at 37 °C on a shaker (200–250 rpm).

Next Day

7. Prior to plasmid prepping the liquid cultures, count the number of colonies on each of the countable (density) dilution plates and determine library coverage using the following equation:

$$\text{Coverage} = (\#\text{of Colonies counted} \times \text{Dilution factor} \times \text{Total mLs of bacteria prior to plating})/\text{Library Epitope Number}$$

Once it has been determined that the coverage is appropriate (>30×), the liquid cultures can be prepped via maxi-prep according to the manufacturer's protocol.

8. A colony PCR should be performed to ensure there is not overrepresentation within the library.
9. Pick 48 or more colonies into 15 μL of Luria Broth.
10. In PCR strips, set up the following reaction:

OneTaq 2× MM	12.5 μL
100 μM MND_Input_Verify_F	0.05 μL
100 μM MND_Input_Verify_R	0.05 μL
Colony in Luria Broth	2 μL colony
Nuclease-free water	10.5 μL
	25 μL total

*One reaction for no template (no colony), one reaction for the SABR backbone with stuffer as a positive control and reference for background colonies

11. Use a thermocycler with the following program:

Step	Temp	Time
Initial denaturation	94 °C	3 min
30 Cycles: denature	94 °C	20 s
Anneal	57 °C	30 s
Extension	68 °C	4 min
Final extension	68 °C	5 min
Hold	4 °C	

12. Run 2–3 μL of each reaction on a 1% agarose gel with DNA binding dye to ensure that the amplification was appropriate. A representative gel image is shown in Fig. 1c.
13. Send colony PCR products for Sanger sequencing to ensure epitope bias did not occur during cloning. Once quality control steps have been completed to satisfaction, the pDNA SABR library can be used as genomic material for lentiviral

packaging and introduction into NFAT-GFP Jurkat reporter cells. Stoppage point: The pDNA library can be stored at 4 °C or long term at −20 °C.

3.2.6 Lentiviral Packaging and Transduction of the SABR Library

1. *Plate cells by* harvesting 293T cells using trypsinization and plate at 10e6 cells per 15 cm plate in D-10 in 20 mL D10 media a day before transfection (*see* **Note 4**).
2. Allow Complete D-10 and Opti-MEM medias to warm to room temperature.
3. For *lentiviral transfection*, prepare a mixture of TransIT-293 and Opti-MEM and pulse vortex 2–3 times.

TransIT-293 + Opti-MEM	*1× 15 cm plate*
TransIT-293	100 μL
Opti-MEM	1 mL

4. Incubate this mixture at room temperature for 15 min, during this incubation, prepare the pDNA mixture by combining the packaging plasmids and SABR-library pDNA.

pDNA mixture	*1× 15 cm plate*
SABR-library pDNA	15 μg
pCMV-Δ8.9	15 μg
pMDG-VSVG	3 μg

5. Add TransIT-293/Opti-MEM mixture to the pDNA mixture, mix by pipetting up and down 4–5 times.
6. Incubate this mixture for 20 min at room temperature. During this incubation, carefully replace the media of the 293T cell plate with 20 mL of complete D-10 without disturbing the cell monolayer.
7. Transfer the entire transfection mixture by equally distributing dropwise to plated cells.
8. Slightly tilt the plate back and forth to completely mix.
9. Incubate the plate at 37 °C and 5% CO_2 for 72 h.
10. To transduce the NFAT-GFP Jurkat cells, count cells to transduce 2e6 cells for every 1 mL of virus.
11. Spin down cells at 350 g for 5 min at room temperature.
12. Resuspend NFAT-GFP Jurkat cells at 2e6/mL in complete R-10 containing 50–100 μg/mL protamine sulfate.

13. Distribute 5 mL of cells per T75 flask (can be scaled up to larger flask if desired). Return cells to incubator during viral harvest.
14. Collect viral supernatant from transfected 15 cm plates into 50 mL conical tubes.
15. Attach Steriflip to the top of the tube, flip, and connect to vacuum (or filter by desired method).
16. Turn on vacuum and allow supernatant to pass through.
17. 5 mL of filtered virus can be added to each 5 mL flask of cells.
18. Incubate at 37 °C and 5% CO_2 for 24 h.
19. After 24 h, add up to 25 mL of R-10 and 2 mg/mL of G418 to maintain the selection of the NFAT-GFP reporter.
20. Allow cells to continue to be transduced for 2 days before assessing their transduction by assaying the surface expression of the MHC allele of interest by flow cytometry.
21. Once the SABR-library has been transduced into the NFAT-GFP Jurkat cells, this new library cell line can be used for the TCR screening process.

3.2.7 SABR Screening via Coincubation and Sorting

Coincubation Setup

1. Count all cells and determine the proper volumes of cells from your flasks (*see* **Note 5**).
2. Label the SABR-Library NFAT-GFP Jurkat cells (library cells) with cell trace dye (or equivalent) according to manufacturer's protocol.
3. Resuspend library cells at 3e6/mL.
4. Collect and resuspend TCR Jurkat cells at 1e6/mL.
5. Plate 3 mL of the TCR Jurkat cells per well in 3 separate wells of a 6-well plate.
6. Add 1 mL of the labeled and resuspended library cells to the TCR Jurkat cells, mix thoroughly by pipetting.
7. Plate 3 wells of library cells alone, add complete RPMI-10 up to 4 mL (these wells will serve as the library alone control to determine the baseline distribution of epitopes within the experiment).
8. Plate at least 1 well with just library NFAT-GFP Jurkat cells alone.
9. Incubate at 37 °C and 5% CO_2 for 18–24 h (closer to 18 is desired, but up to 24 h is ok).

Cell Sorting

1. After 18 h, harvest each well into its own labeled FACS tube and harvest one well of the library alone to set the baseline during sort gating.
2. Spin down the cells at 350 g for 5 min 4 °C.
3. *During spin* generate anti-human CD69 antibody diluted in ice cold FACS buffer (dilute antibody 1:200, 200 μL per tube).
4. Resuspend cells in 200 μL of the antibody diluted in FACS buffer.
5. Incubate on ice for 20 min protected from light.
6. *During incubation*, label 1.5 mL snap cap tubes for each sort well and fill with at least 200 μL FACS bugger for collection of sorted cells.
7. After incubation, wash cells by adding 1 mL of FACS buffer to each tube.
8. Spin down the cells at 350 g for 5 min 4 °C.
9. Resuspend cells in 400 μL FACS buffer (or desired cellular concentration for sorting keeping in mind cells may have doubled during the incubation).
10. Filtered cells are ready for sorting.
11. Begin cytometer acquisition with the library alone sample, you should be able to gate on single cells, and observe the cell trace labeling.
12. Draw your sort gate-utilize a dot plot or contour plot to observe the GFP and anti-human CD69 fluorescence (draw the sort gate that encompasses the top 1–2% (of parent not total) of the upper right quadrant of double positive cells.
13. Sort sample for at least 15 min, attempting to get through as much of the sample as possible (20 min has been the standard so that 1-TCR per hour (3 replicates) can be screened).
14. Representative sort gates are shown in Fig. 2a.
15. After cells have been sorted, immediately proceed to genomic DNA isolation using commercially available column kits. *Stop-page point*: isolated genomic DNA can be stored at −20 °C until amplification.

3.2.8 Unique Dual Indexing and Sequencing

1. **Step 1** *PCR*—Amplify the epitopes (*see* **Note 6**).

   The following reaction will be required for each genomic DNA sample acquired during cell sorting. PCR strips or a 96 well plate.

   Set up the following reactions (x = number of genomic DNA samples + a no template control):

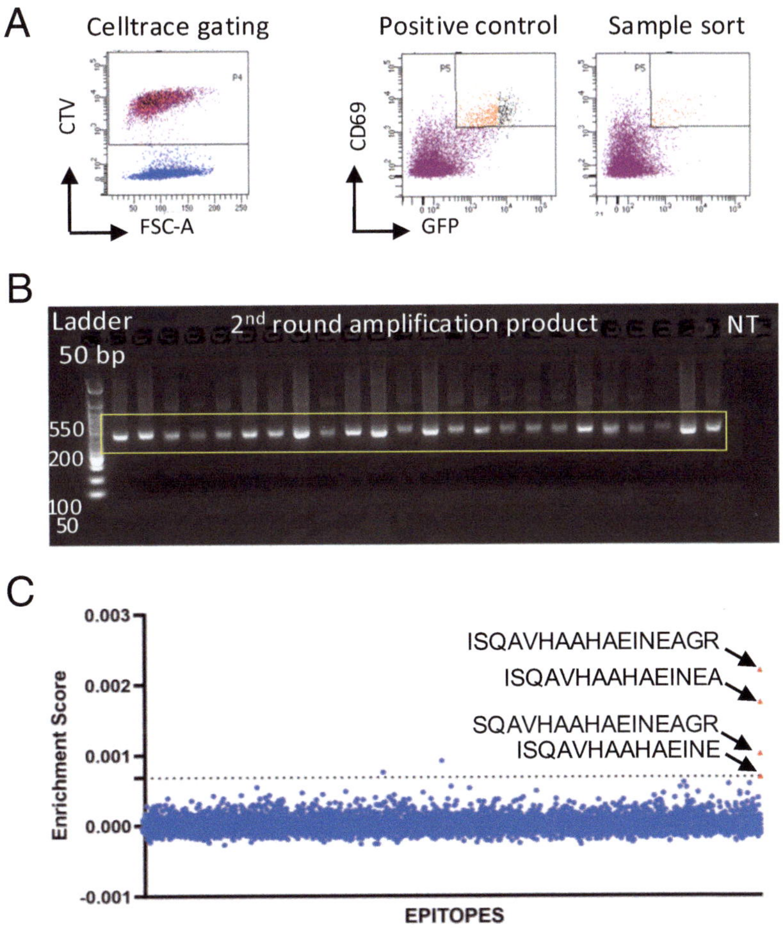

Fig. 2 Example of SABR screens for identifying cognate TCR-pMHC interactions. (**a**) Representative plots for library cells labeled with CTV (left panel), and further gated for proportion of cells in the sort gate (1–2% CD69+GFP+). The right panel shows a positive control TCR+epitope, and a sample sort with TCRs of unknown specificities. (**b**) Representative gel image of PCR product from second round of amplification from sorted cells. The PCR products are highlighted with the respective ladder. NT = no template reaction. (**c**) Representative Enrichment Score (ES) plot, with epitopes on X-axis and mean ES on Y-axis. The four red triangles indicate putative hits for the TCR screened. This screen was performed using the OT-II TCR, and the individual epitopes identified are indicated

	1×
2× KOD MM	25 μL
UDI-Fwd	0.15 μL
UDI-Rev	0.15
Template DNA (genomic DNA)	6 μL
Nuclease-free water	18.7 μL
	50 μL total

One reaction per sample and one no template.
Use a thermocycler with the following program:

Step	Temp	Time
Initial denaturation	95 °C	2 min
30 Cycles: denature	95 °C	20 s
Anneal	66 °C	10 s
Extension	70 °C	15 s
Final extension	70 °C	2 min
Hold	4 °C	

2. Optionally, up to 5 μL of the product can be run on a 2% agarose gel with a 100 bp ladder to ensure concatenation did not occur. However, this step is often skipped due to low product intensity.
3. **Step 2** *PCR*—Add indexes

   Set up the following reactions (x = number of genomic DNA samples + a no template control):

   (a) For this reaction, pick a unique dual index (e.g. UDI-1) for the no template control reaction*.

	1×
2× KOD MM	25 μL
10 μM UDI index pool (*use a unique index/sample*)	3 μL
Step 1 PCR (unpurified PCR product from **step 1** PCR)	2 μL
Nuclease-free water	20 μL
	50 μL total

*Keep a no template using UDI1
Cycle using the following program:

Step	Temp	Time
Initial denaturation	95 °C	2 min
25 Cycles: denature	95 °C	20 s
Anneal	66 °C	10 s
Extension	70 °C	15 s
Final extension	70 °C	2 min
Hold	4 °C	

4. Required: Run up to 5 μL of PCR product from each reaction on a 2% gel with a 100 bp ladder to ensure concatenation did not occur. A representative gel image is shown in Fig. 2b.
5. PCR purify each reaction individually using commercially available PCR cleanup kit (e.g., spin-based or vacuum-based kits).
6. Quantify the amount of DNA present in each reaction using nanodrop or preferred method of DNA quantification.
7. Pool purified DNA reactions in equal concentrations (molarity not required) to desired amount based on sequencing depth available.
8. Sequence using preferred Illumina platform, sequenced DNA are ready for analysis and putative hit calling (*see* **Note 7**).

3.2.9 Sequencing Analysis and Calling Hits

1. Download the demultiplexed sequencing data as *.fastq files to a cluster/laboratory server.
2. For each experiment, create a folder and move all the relevant fastq files to that folder.
3. Concatenate all the R1 and R2 fastq files for an experiment using:
   (a) zcat *R1.fq > read_1.fqZcat *R2.fq > read_2.fq
   (b) This will create two large fastq files called "read_1.fq" and "read_2.fq"
4. Demultiplex the concatenated files using the script "demultiplex_dual.py". This will generate a demultiplexed fastq file corresponding to each index.
5. Extract the epitope sequence and count epitopes using the "extract_epitope.py" script. This script searches for the sequences that flank the epitope, extracts the epitope sequence, and translates it into protein sequence. It takes a list of the epitopes in the library as an input and counts the number of reads corresponding to each epitope.
6. Combine the read counts for the entire experiment using the "merge_counts_split.py" script, which will output a table of epitopes and their corresponding read counts for each index.
7. Calculate enrichment scores using the template spreadsheet provided. The enrichment scores are calculated as follows:
   1. First, a slope (m) and intercept (c) are calculated based on the counts of a given TCR and the library.
   2. Second, the expected values for each epitope are calculated as follows:
   3. Expected value = $m \times$ (library count) + c

4. Third, the enrichment score for each epitope is calculated as:
5. Enrichment Score = Observed value – Expected value

8. Enrichment scores for all epitopes for a given TCR are plotted.
9. Using known positive and negative control epitopes, thresholds for calling putative hits are set (*see* **Note 8**).
10. Each putative hit is then cloned as a single epitope using steps described above and tested in a coincubation assay. Validation coincubations are typically performed in 96-well plates with 25,000–50,000 (1:1 ratio) of both TCR Jurkats and NFAT-GFP Jurkats with the single SABR containing the putative hit.

4 Notes

1. In our experience, Immunopeptidomics datasets offer the best chance of generating specific and high confidence SABR libraries. Depending on the depth of a proposed study, we would suggest starting with a literature search for immunopeptidomics datasets, and/or performing immunopeptidomics. In addition, having a known TCR-pMHC pair is a must, as it allows to benchmark the library and serves as a positive control, which is instrumental in troubleshooting. Finally, the inclusion of epitopes with differential affinity to the positive control TCR will allow for better defined putative hit calling.
2. SABR Backbone digestions can be done in parallel to library oligo amplification setup. For Subheading 2.2, it is beneficial to set up multiple digests of the SABR backbone to ensure enough material for pooled oligo Infusion. For Subheading 2.3, the number of reactions will be dictated by the total amount of DNA recovered after the first digest. All the backbone should be subject to a second digest to minimize background during cloning.
3. For optimal efficiency, transformation should immediately proceed after InFusion; however, the InFusion reaction can be stored at −20 °C. The number of InFusion reactions will be dictated by the oligo library size. The rule of thumb is 1 InFusion reaction per 10,000 epitopes, always rounded up to the next whole number of reactions (e.g., A library of 4000 epitopes would equate to 0.4 reactions, so 1 reaction would be performed). The molar ratio (100 ng of vector + 10 ng of insert) serves as an example and thus dictate the concentrations of the SABR backbone and oligos for InFusion. The relative amount of DNA can be altered between the two to accommodate different oligo/backbone molarities; however, it is not recommended to increase the total DNA beyond 200 ng for each 10 μL InFusion reaction.

4. The number of 293T plates for packaging can be increased to the cell needs. For reference, the virus generated from 1 to 15 cm plate will be used to transduce 40e6 NFAT-GFP Jurkat cells. This is our lab's strategy for generating high titer lentivirus. However, it should be noted that many commercially available transfection reagents and protocols could be used interchangeably for the generation of SABR-library containing lentiviral particles. *All lentiviral work should be done with the appropriate precautions for BSL2+.*
5. Cell numbers planning is as follows. We typically utilize a 1:1 ratio of cells, 3e6 each, and run 6 TCR Jurkat cell lines per sort (5 experimental TCR, 1 known TCR).
   - 12e6 total of each TCR (1× more than needed, important for known/+control TCR).
   - 60e6 of the library (Typically 5-T75 flasks nearing 7e5–8e5 cells/mL).
6. This step is performed as two separate PCR reactions. The first reaction amplifies the epitopes and adds sequence handles for the second reaction, which adds the Illumina unique dual indexes to identify each genomic DNA reaction. The primer pair for the first PCR reaction will be dependent on the MHC class which is being screened. SABR screens can be multiplexed by class I and II alleles as long as post alignment analysis is performed for each allele separately. Historically, amplification of genomic DNA from class I and class II screens has occurred in separate reactions. Reactions can be performed as sorts are performed or can be done once a desired number of samples have been sorted, up to 95 (1 reserved for a no template control reaction) as per the unique dual indices available.
7. While sequencing, ensure that there is at least 1000× coverage of your library. For example, for 1000 epitopes, sequence with sufficient depth to get at least one million reads. As the sequencing libraries are inherently low diversity, it is recommended to spike in up to 30% PhiX. Set the UDI indexes according to the sequencing platform to demultiplex the data. It is typically convenient to perform 3–4 experiments with ~20–24 samples each, and multiplex all of them into one sequencing lane.
8. The thresholds for calling hits are flexible, and over time as more screens are performed, researchers should get better at choosing them. While selecting putative hits, it is preferable to draw generous thresholds in order to not miss out on possible epitopes. Each epitope is cloned using the same protocol described above by synthesizing and cloning single oligonucleotides.

References

1. James EA et al (2023) The beta cell-immune cell interface in type 1 diabetes (T1D). Mol Metab 78:101809. https://doi.org/10.1016/j.molmet.2023.101809
2. Joglekar AV, Li G (2021) T cell antigen discovery. Nat Methods 18(8):873–880. https://doi.org/10.1038/s41592-020-0867-z
3. Birnbaum ME et al (2014) Deconstructing the peptide-MHC specificity of T cell recognition. Cell 157(5):1073–1087. https://doi.org/10.1016/j.cell.2014.03.047
4. Joglekar AV et al (2019) T cell antigen discovery via signaling and antigen-presenting bifunctional receptors. Nat Methods 16(2): 191–198. https://doi.org/10.1038/s41592-018-0304-8
5. Li G et al (2019) T cell antigen discovery via trogocytosis. Nat Methods 16(2):183–190. https://doi.org/10.1038/s41592-018-0305-7
6. Kula T et al (2019) T-scan: a genome-wide method for the systematic discovery of T cell epitopes. Cell 178(4):1016–1028.e1013. https://doi.org/10.1016/j.cell.2019.07.009
7. Dezfulian MH et al (2023) TScan-II: a genome-scale platform for the de novo identification of CD4(+) T cell epitopes. Cell 186(25):5569–5586.e5521. https://doi.org/10.1016/j.cell.2023.10.024
8. Kohlgruber AC et al (2024) High-throughput discovery of MHC class I- and II-restricted T cell epitopes using synthetic cellular circuits. Nat Biotechnol. https://doi.org/10.1038/s41587-024-02248-6
9. Jin Y et al (2024) Tsyn-Seq: a T-cell synapse-based antigen identification platform. Cancer Immunol Res 12(5):530–543. https://doi.org/10.1158/2326-6066.Cir-23-0467
10. Kisielow J, Obermair FJ, Kopf M (2019) Deciphering CD4(+) T cell specificity using novel MHC-TCR chimeric receptors. Nat Immunol 20(5):652–662. https://doi.org/10.1038/s41590-019-0335-z
11. Zdinak PM et al (2024) De novo identification of CD4(+) T cell epitopes. Nat Methods 21(5): 846–856. https://doi.org/10.1038/s41592-024-02255-0
12. Vita R et al (2015) The immune epitope database (IEDB) 3.0. Nucleic Acids Res 43(Database issue):D405–D412. https://doi.org/10.1093/nar/gku938
13. Hoof I et al (2009) NetMHCpan, a method for MHC class I binding prediction beyond humans. Immunogenetics 61(1):1–13. https://doi.org/10.1007/s00251-008-0341-z
14. Karosiene E et al (2013) NetMHCIIpan-3.0, a common pan-specific MHC class II prediction method including all three human MHC class II isotypes, HLA-DR, HLA-DP and HLA-DQ. Immunogenetics 65(10):711–724. https://doi.org/10.1007/s00251-013-0720-y
15. O'Donnell TJ, Rubinsteyn A, Laserson U (2020) MHCflurry 2.0: improved pan-allele prediction of MHC class I-presented peptides by incorporating antigen processing. Cell Syst 11(1):42–48.e47. https://doi.org/10.1016/j.cels.2020.06.010

Chapter 3

Purification, Phenotypic and Functional Analyses of Regulatory T Cells in Autoimmunity

Nainisha Shah, Annemarie Fortuin, Georgia Afonso, and Pascale Louis-Plence

Abstract

Regulatory T cells (Tregs) are essential in maintaining immune tolerance and preventing autoimmune responses. In this chapter, we describe methods for isolating highly purified human CD4$^+$ Tregs from peripheral blood mononuclear cells and synovial fluid mononuclear cells of patients with rheumatoid arthritis, as an example of autoimmune disease. We outline the methods for flow cytometry-based cell sorting to separate two distinct Treg populations (naïve and effector), their phenotypic analysis through surface and intracellular staining, and their functional characterization using an in vitro suppression assay. Additionally, we discuss strategies for expanding and genetically engineering Tregs for potential therapeutic applications in autoimmune diseases.

Key words Regulatory T cells, Cell isolation, Cell sorting, In vitro suppression assay, Synovial fluid, Rheumatoid arthritis

1 Introduction

Naturally occurring CD4$^+$ regulatory T cells (Tregs) play a crucial role in immune regulation by modulating immune responses and maintaining homeostasis. Tregs express the X-linked transcription factor Foxp3, and mutations in *Foxp3* gene can cause fatal multi-organ autoimmune and inflammatory disease featuring lymphadenopathy and splenomegaly, eosinophilia, hyper-IgE syndrome, and markedly increased systemic levels of a wide range of pro-inflammatory cytokine [1]. The maintenance of an appropriate balance between Tregs and conventional T cells (Tconvs) is critical for immune homeostasis. An increased presence of FoxP3$^+$ Tregs within the tumor microenvironment contributes to immune evasion by suppressing anti-tumor responses, whereas their depletion has been associated with enhanced cancer immunity [2].

Rami Bechara (ed.), *Autoimmunity: Methods and Protocols*, Methods in Molecular Biology, vol. 2990,
https://doi.org/10.1007/978-1-0716-4997-8_3,

The loss or dysfunction of Tregs—or the resistance of Tconvs to Treg-mediated suppression—has been implicated in allergic disorders and systemic autoimmune diseases [3]. These Treg-dependent immune dysregulations include conditions such as type 1 diabetes, multiple sclerosis, systemic lupus erythematosus, myasthenia gravis, and rheumatoid arthritis (RA) [4–7].

Human Tregs can be studied from various biological sources, including peripheral blood (PB), biofluids, and tissues. In this chapter, we focus on the analysis of Tregs in PB and synovial fluid (SF) as representative examples, particularly in the context of RA as an autoimmune disease model [8].

We describe protocols for Treg isolation, purification, and stability assessment using flow cytometry immunophenotyping. Furthermore, we outline in vitro suppressive assays, in coculture with Tconvs, to evaluate Treg suppressive function. We detail protocols for Treg expansion and genetic engineering, as a strategy to modulate immune responses in autoimmune diseases. However, while these functional assays provide valuable insights, they do not offer a comprehensive characterization of Tregs. Therefore, their interpretation should be integrated with molecular techniques to obtain a more holistic understanding of Treg functional potential.

2 Materials

2.1 Isolation of Peripheral Blood Mononuclear Cells (PBMC)/Synovial Fluid Mononuclear Cells (SFMCs)

1. Ficoll®-Paque Plus.
2. 40 μm Nylon Cell Strainer.
3. Phosphate-Buffered Saline (PBS) or Physiological Saline.
4. Trypan Blue Solution (0.4%).

2.2 Purification of Tregs

1. EasySep™ Human $CD4^{+}CD127^{low}CD25^{+}$ Regulatory T Cell Isolation Kit.
2. EasySep™ Magnet (*see* **Note 1**).
3. PBE buffer: PBS Ca^{2+}/Mg^{2+}-free, 2% Fetal Calf Serum (FCS), 1 mM EDTA (*see* **Note 2**).

2.3 Cytometer-Based Cell Sorting of Tregs

1. Fluorescent-Conjugated Antibodies (BD Biosciences) (*see* **Note 3**):
   - BD Horizon™ BV786 Anti-Human CD3.
   - BD Horizon™ BV711Anti-Human CD25.
   - BD Horizon™ BV650 Anti-Human CD4.
   - PE-CF594 Anti-Human CD127.

- BD Horizon™ BV605Anti-Human CD45RA.
- DAPI—for dead cell exclusion.

2. PBE Buffer: PBS, 2% FCS, 1 mM EDTA.
3. FACS Sorting Tubes (Sterile 5 mL Polystyrene Round-Bottom Tubes, pre-coated with FCS overnight on rotation).
4. Cell Sorter equipped with at least 3 lasers for the staining (405 nm, 488 nm, or 561 nm for PE-CF594 and 355 nm for DAPI).

2.4 Intracellular Staining for Treg Markers

1. Blocking and Permeabilization Reagents:
   - PBA Buffer: PBS, 1% BSA, 0.1% NaN3.
   - Rat Serum.
   - Human Fc Block™ (BD Biosciences).
   - Live/Dead™ Fixable Blue Dead Cell Stain kit Dye, for UV excitation (Invitrogen).
2. Preparation of Working Solutions:
   - Fixation/Permeabilization Solution (1×): 1:4 ratio of Fixation/Permeabilization Concentrate (4×) (eBioscience) to Fixation/Permeabilization Diluent (eBioscience).
   - Permeabilization Buffer (1×): 1:10 ratio of Permeabilization Buffer (10×) (eBioscience) to distilled water.
3. Fluorescent-Conjugated Antibodies (BD Biosciences) (*see* **Note 3**):
   - BD Horizon™ BV786 Anti-Human CD3.
   - BD Horizon™ BV711Anti-Human CD25.
   - BD Horizon™ BV650 Anti-Human CD4.
   - PE-CF594 Anti-Human CD127.
   - BD Horizon™ BV605Anti-Human CD45RA.
   - FITC Anti-Human FoxP3 (BioLegend).
   - BD Horizon™ BV421 Anti-Human CTLA-4/CD152.
   - Alexa Fluor® 647Anti-Human Helios.
4. Flow Cytometer (for flow cytometric analysis of stained Treg markers) equipped with 5 lasers (488, 561, 640, 405, and 355 nm).

2.5 In Vitro Suppressive Tests of Tregs

1. Treg Suppression Inspector, Human (Miltenyi Biotec).
2. Sorted Tregs and purified Tconvs using EasySep™ Human $CD4^{+}CD127^{low}CD25^{+}$ Regulatory T Cell Isolation Kit.
3. Complete Iscove's Modified Dulbecco's Medium (IMDM) medium: IMDM supplemented with 1% Penicillin/Streptomycin, 1% Glutamine, 1% AANE, and 10% FCS. 50–100 μM β-Mercaptoethanol added during use (*see* **Note 4**).

4. 96-Well Round bottom Culture Plates.
5. CellTrace Violet™ (CTV) stock solution 5 mM in DMSO (20 μL in 1 vial), final solution 2.5 μM in PBS solution.
6. Flow Cytometer (*see* previous section **step 4**).

2.6 Expansion of Naïve and Effector Tregs

1. Complete IMDM medium same as Subheading 2.5, **step 3**.
2. PROLEUKIN® 18 millions U.I., IL-2: 80 U.I./mL for Tconvs, 300 U.I./mL for Tregs.
3. Dynabeads™ Human T-Activator CD3/CD28 beads for T cell expansion.

2.7 Therapeutic Approaches by Genetically Engineering Tregs

1. Retronectin coating:
   - RetroNectin® (TAKARA Bio Inc) stock at 1 mg/mL.
   - 96 well flat bottom plate.
   - PBS.
2. Treg transduction:
   - Sorted Tregs.
   - Complete IMDM medium same as Subheading 2.5, **step 3**.
   - Trypan Blue Solution (0.4%).
   - Lentiviral vector for transgene expression.
3. Characterization of transgene expression:
   - DAPI for dead cell exclusion.
   - Antibodies for CAR detection (anti-CAR G4S linker, Cell Signaling Technology).
   - Antibodies for Treg phenotyping (*see* Subheading 2.4).
   - Flow cytometer (*see* previous section).

3 Methods

3.1 Isolation of PBMC/ Synovial Fluid Mononuclear Cells (SFMCs)

1. Dilute blood and/or synovial fluid at a ratio of 1:3 with physiological saline or PBS at room temperature (RT) (*see* **Note 5**).
2. In 50 mL Falcon tubes, add 15 mL Ficoll®-Paque Plus (*see* **Note 6**) and carefully layer 30 mL of diluted blood or SF on top. Ensure that the two phases do not mix.
3. Centrifuge at 400 g for 30 min at 20 °C (acceleration: 2, brake: 2).
4. Carefully remove the supernatant and transfer the mononuclear cell layer (buffy coat) into a fresh 50 mL Falcon tube.
5. Adjust the volume to 50 mL with physiological saline and centrifuge at 900 g for 12 min at 20 °C (acceleration: 7, brake: 4) to pellet the cells.

6. Discard the supernatant carefully.
7. Resuspend the cell pellet in 1 mL saline, then add 44 mL of saline/PBS solution and centrifuge at 170 g for 15 min at 20 °C to remove platelets (acceleration: 9, brake: 1) (*see* **Note 7**).
8. Discard the supernatant, resuspend the pellet in 1 mL saline, and adjust the final volume to 15 mL using either saline or PBE solution if planning to perform Treg/Tconv isolation.
9. Perform a total cell count before proceeding with further experiments.
10. Cells can be used immediately or frozen in 90% FCS 10% DMSO (*see* **Note 8**).

3.2 Purification of Tregs

Phenotypic identification of Tregs is challenging due to the expression of CD25 (IL-2 receptor α) and FoxP3 on both Tregs and activated Tconvs. In human Tregs, CD127 (IL-7 receptor α) expression is either absent or significantly reduced, making $CD25^{+}CD127^{-/low}$ a reliable marker for Treg purification [9].

This protocol isolates both human Tconvs and Tregs using the EasySep™ Human $CD4^{+}CD127^{low}CD25^{+}$ Regulatory T Cell Isolation Kit, which employs a column-free immunomagnetic separation (*see* **Note 9**).

1. Resuspend PBMCs/SFMCs at a concentration of 5×10^7 cells/mL in appropriate volume of PBE buffer.
2. Add 50 μL/mL of CD25 Positive Selection Cocktail to the sample, mix gently, and incubate at RT for 5 min (*see* **Note 10**).
3. Vortex Releasable RapidSpheres™ for 30 s to ensure even dispersion, then add 30 μL/mL to the sample and mix.
4. Add 50 μL/mL of $CD4^{+}$ T Cell Enrichment Cocktail and incubate at RT for 5 min.
5. Adjust the sample volume with PBE and mix gently by pipetting.
6. Place the tube (without lid) into the EasySep™ magnet and incubate for 10 min at RT.
7. For volumes between 6 and 40 mL, carefully pipette the supernatant into a new tube. For volumes between 0.25 and 6 mL, invert the magnet and tube in one continuous motion, allowing the supernatant to collect in a new tube (*see* **Note 11**).
8. Vortex Releasable RapidSpheres™ and add 15 μL/mL of the original sample volume to the supernatant. Mix thoroughly and incubate for 4 min at RT.
9. Transfer the supernatant back into the tube in the magnet and incubate for another 10 min at RT.

10. Carefully pipette the supernatant into a new tube. This fraction can be used for the isolation of $CD4^{+}CD25^{-}$ Tconvs used in the in vitro suppression assay.
11. Remove the tube from the magnet and resuspend the cells in fresh medium.
12. Repeat **steps 10–12** for a total of four magnetic separations.
13. Remove the tube from the magnet and resuspend the cells in medium to match the original starting volume.
14. Add 100 μL/mL of Release Buffer and mix vigorously by pipetting.
15. Add 50 μL/mL of CD127high Depletion Cocktail and incubate for 5 min at RT.
16. Vortex Dextran RapidSpheres™ for 30 s, then add the appropriate volume and mix thoroughly. Incubate for 5 min at RT.
17. Adjust the sample volume with medium and incubate in the magnet for 10 min at RT.
18. Carefully pipette the enriched cell suspension into a new tube. The isolated Tregs are now ready for downstream applications (*see* **Notes 12** and **13**).
19. Transfer the supernatant from **step 10** into a new tube.
20. Vortex Dextran RapidSpheres™ for 30 s to ensure even dispersion (*see* **Note 14**).
21. Add 180 μL/mL of the original sample volume of Dextran RapidSpheres™ to the sample and mix. Incubate at RT for 5 min.
22. Place the tube (without lid) into the magnet and incubate for 10 min at RT.
23. Carefully pipette the enriched cell suspension into a new tube.
24. Remove the tube from the magnet and repeat the separation step for another 10-min incubation.
25. Carefully pipette the final enriched cell suspension into a new tube. The isolated Tconvs are now ready for use (*see* **Note 12**).

3.3 Cytometer-Based Cell Sorting of Tregs

Following pre-enrichment, FACS sorting enables the isolation of Treg subsets based on additional markers. One key marker, CD45RA, distinguishes between naïve ($CD45RA^{+}$) and effector ($CD45RA^{-}$) Tregs, each with distinct functional properties [10]. This method ensures a high-purity population by eliminating contaminating Tconvs.

1. Wash the cells with PBE and pellet them by centrifugation at 480 g for 5 min.
2. Resuspend 2 million cells in staining buffer and label with the following antibodies (*see* **Note 15**):

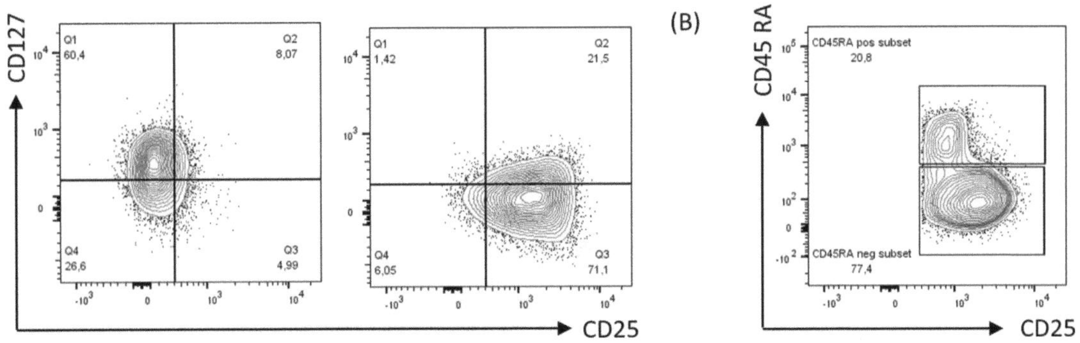

Fig. 1 Cell sorting strategy for Treg and Tconv populations. (**a**) Representative density plot showing CD25 and CD127 expression following isolation of $CD4^+$ Tconv (left panel) and Tregs (right panel) from PBMCs using EasySep™ Human $CD4^+CD127^{low}CD25^+$ Regulatory T Cell Isolation Kit. (**b**) Density plot of CD45RA and CD25 labeling on isolated Tregs. Strategy of gating for the purification of $CD45RA^+$ naïve Treg and $CD45RA^-$ effector Tregs

- 5 μL of BD Horizon™ BV786 Anti-Human CD3.
- 5 μL of BD Horizon™ BV711Anti-Human CD25.
- 5 μL of BD Horizon™ BV650 Anti-Human CD4.
- 5 μL of PE-CF594 Anti-Human CD127.
- 2 μL of BD Horizon™ BV605Anti-Human CD45RA.

3. Incubate for 20 min at 4 °C in the dark.
4. Wash out unbound antibodies by centrifugation (480 g, 5 min) and resuspend in PBE at 5 million cells/mL.
5. Use overnight FCS-coated tubes for cell collection.(see Note 16).
6. Remove excess of FCS.
7. Add 1 mL of complete IMDM to each collection tube.
8. Just before sorting, add DAPI to stain dead cells (concentration should be adapted for the cytometer used).
9. Primary Gating: SSC-A/FSC-A cell size (*see* Fig. 1).
10. Exclusion of doublet and Dead Cells.
11. Gating on lymphocytes $CD3^+$ $CD4^+$.
12. Regulatory T Cell Selection: $CD25^+$ $CD127^{-/low}$.
13. Subset Sorting:
    - Naïve Tregs: $CD45RA^+$.
    - Effector Tregs: $CD45RA^-$.
14. Sort cells using Cell Sorter based on the defined gating strategy (*see* **Note 17**).
15. Collect sorted cells into prepared FCS-coated tubes containing complete IMDM.

3.4 Intracellular Staining for Phenotypic Analyses

To better understand and characterize Tregs, intracellular staining is used to assess key markers like FoxP3, Helios, and CTLA-4, alongside surface markers. FoxP3 regulates Treg function, Helios is linked to Treg stability, and CTLA-4 is crucial for immune suppression [11, 12].

1. Take purified 2×10^5 Tregs/Tconv cells per cytometry tube.
2. Take 2×10^4 unmarked cells per cytometry tube.
3. Centrifuge at 480 g for 5 min at 4 °C, discard the supernatant, and wash the cells in PBS (*see* **Note 18**).
4. Centrifuge at 480 g for 5 min at 4 °C, discard the supernatant.
5. Prepare LIVE/DEAD Fixable Blue Dye by resuspending the entire stock vial in 50 μL DMSO (*see* **Note 19**).
6. Dilute 2 μL of resuspended dye in 198 μL serum-free PBS (1: 100 dilution).
7. Resuspend cells in 100 μL of diluted LIVE/DEAD dye.
8. Incubate for 15 min at RT in the dark.
9. Wash cells with 2 mL PBE and centrifuge at 480 g for 5 min at 4 °C.
10. Prepare Human Fc Block™ by diluting 1 μL of stock solution in 49 μL PBS.
11. Resuspend pellet in 50 μL of diluted Fc Block and incubate for 10 min at RT.
12. Add the following surface antibodies (*see* **Note 15**):
    - 2 μL of BD Horizon™ BV786 Anti-Human CD3.
    - 2 μL of BD Horizon™ BV711Anti-Human CD25.
    - 2 μL of BD Horizon™ BV650 Anti-Human CD4.
    - 2 μL of PE-CF594 Anti-Human CD127.
    - 2 μL of BD Horizon™ BV605Anti-Human CD45RA.
13. Incubate for 20 min at RT in the dark.
14. Wash cells with 2–3 mL PBE, centrifuge at 480 g for 5 min at 4 °C, and discard the supernatant.
15. Add 800 μL of Fixation/Permeabilization Solution (1:3 concentrate to diluent ratio).
16. Vortex and incubate for 45 min at RT in the dark (*see* **Note 20**).
17. Wash with 2 mL Permeabilization Buffer (1×).
18. Centrifuge at 600 g for 5 min at 4 °C, discard the supernatant.
19. Repeat wash with 2 mL Permeabilization Buffer (1×).
20. Centrifuge at 600 g for 5 min at 4 °C, discard the supernatant.

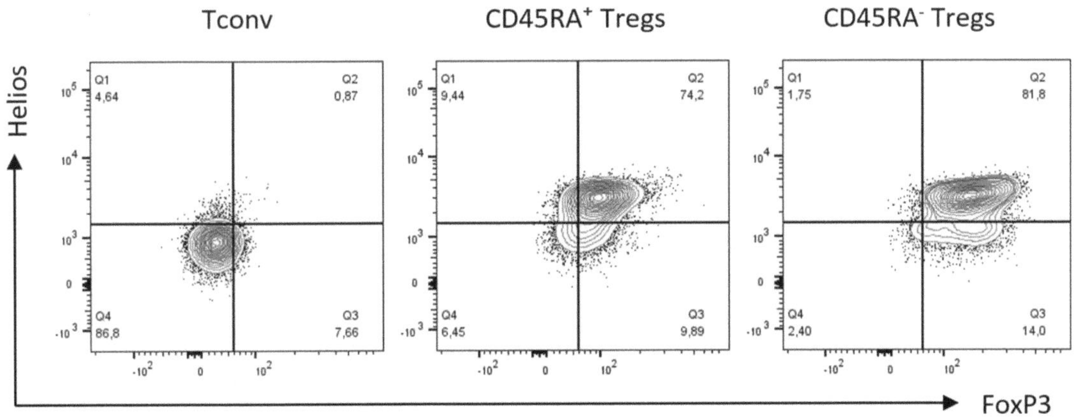

Fig. 2 Intracellular staining of FoxP3 and Helios. Flow cytometry analysis showing FoxP3 and Helios expression across different T cell subsets. Plots display staining in conventional T cells (Tconv, FoxP3$^-$), cell-sorted naïve Tregs (CD45RA$^+$), and cell-sorted effector Tregs (CD45RA$^-$), highlighting the double-positive FoxP3$^+$Helios$^+$ phenotype in both Treg populations

21. Add 40 μL PermWash containing 2% Rat Serum and incubate 5 min at RT in the dark.
22. Add the following intracellular antibodies (*see* Fig. 2):
    - 2 μL of FITC Anti-Human FoxP3 (BioLegend).
    - 2 μL of BD Horizon™ BV421 Anti-Human CTLA-4/CD152.
    - 2 μL of Alexa Fluor® 647Anti-Human Helios.
23. Vortex and incubate for 30 min at RT in the dark.
24. Wash with 2 mL Permeabilization Buffer (1×).
25. Centrifuge at 600 g for 5 min at 4 °C, discard the supernatant.
26. Resuspend in 200 μL PBE.

3.5 In Vitro Suppressive Tests of Tregs

The in vitro suppression assay is used to assess the functional capacity of isolated Tregs by measuring their ability to inhibit Tconv proliferation upon activation [13]. Tconvs, labeled with CTV, are cocultured with Tregs in the presence of Treg Suppression Inspector (MACSiBeads), an optimized stimulation reagent consisting of Anti-Biotin MACSiBead™ particles loaded with biotinylated CD2, CD3, and CD28 antibodies. Treg-mediated suppression is then quantified using flow cytometry, providing a clear measure of Treg suppressive function.

1. Determine cell concentration of Tregs and Tconv and prepare:
    - Tconv cells: 9×10^5 cells in 1800 μL complete culture medium (final concentration: 5×10^5 cells/mL).
    - Treg cells: 6×10^5 cells in 1200 μL complete culture medium (final concentration: 5×10^5 cells/mL).

Table 1
Experimental setup for the Treg suppression assay

Tconv:Treg ratio	Tconv cells (μL, 5×10^5 cells/mL)	Treg cells (μL, 5×10^5 cells/mL)	MACSiBead particles (μL, 1×10^7/mL)	Culture medium (μL)	Total volume (μL)
1:0	100	–	5	105	210
0:1	–	100	5	105	210
1:1	100	100	10	–	210
2:1	100	50	7.5	52.5	210
4:1	100	25	6.5	78.5	210
8:1	100	12.5	6	91.5	210
Control (Tconv only)	100	–	–	110	210
Control (Treg only)	–	100	–	110	210

2. Aliquot appropriate volumes into a 96-well U-bottom plate according to Table 1.
3. CTV Solution: 2.5 μM CTV dye in PBS (*see* **Notes 21** and **22**).
4. Resuspend Tconv cells in CTV solution (*see* **Note 23**) and incubate at 37 °C in a water bath for 20 min.
5. Immediately quench staining by adding complete medium and incubating at 4 °C for 15 min.
6. Wash cells twice with complete medium, resuspend, and proceed with coculture.
7. Resuspend 60 μL of MACSiBead Particles (Treg Suppression Inspector).
8. Wash by adding 0.3–0.6 mL culture medium and centrifuging at 300 g for 5 min.
9. Remove the supernatant and resuspend in 120 μL culture medium (final concentration: 1×10^7 MACSiBeads/mL).
10. Add Treg Suppression Inspector (MACSiBeads) at a 1:1 bead-to-cell ratio (see Table 1).
11. Adjust the final volume to 210 μL with complete culture medium.
12. Incubate at 37 °C, 5–7% CO_2 for 4–5 days (*see* **Note 24**).
13. Assess Tconv cell proliferation via CTV dilution using flow cytometry (*see* Fig. 3).

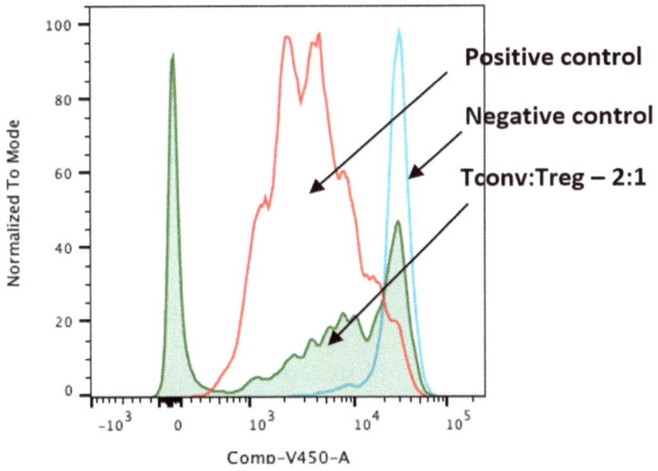

Fig. 3 Treg-mediated inhibition of T cell proliferation. Histograms displaying the proliferation of CTV-labeled Tconv in the presence of Tregs at a 2:1 ratio (Tconv: Treg). Tconv cells were either activated (positive control) or left unstimulated (negative control). This demonstrates the suppressive effect of Tregs on Tconv proliferation, as evidenced by reduced proliferation in the coculture condition

14. For each Tconv/Treg ratio, calculate the percentage of suppression using the following formula [14]:

$$\%\text{Suppression} = \frac{[\text{Log2}(y)(\text{Tconvs alone}) - \text{Log2}(y)(\text{Tconv} + \text{Tregs})]}{\text{Log2}(y)\ \text{of (Tconv alone)}} \times 100$$

where the *y* value corresponds to the MFI of the CTV of the whole Tconv cell population when Tconv cells are cocultured with Tregs, divided by the MFI of the CTV of undivided Tconv cells cultured alone.

3.6 Expansion of Naïve and Effector Tregs

Due to the limited number of purified Tregs, expansion of $CD45RA^+$ and $CD45RA^-$ Tregs can be used for Treg-based therapies. Both subsets of Tregs are expanded using activation beads in the presence of exogenous IL-2. This expansion can generate antigen-specific T cells to assess immune activation, evaluate their functional responses and suppressive capacity.

- Day 0: Seed at 0.4–1 × 10^6 cells/mL, stimulate with CD3/CD28 beads (ratio 1:1), and supplement with IL-2 (300 UI/mL). Add fresh media, with fresh IL-2 every 2–3 days (*see* **Note 25**).
- Day 9: Count cells, restimulate 2.5 × 10^5 cells/mL using the same conditions.
- Day 18: Continue expansion with media replenishment as needed.

3.7 Therapeutic Approaches by Genetically Engineering Tregs

Treg therapies for the treatment of autoimmune diseases aim to enhance cell specificity, function, and proliferation. This can be done by directly administering polyclonal Tregs, antigen-specific Tregs or transduced Tregs, engineered to express high-affinity TCRs or chimeric antigen receptors (CAR) specific for autoantigens [15]. Here, we present a detailed protocol for transducing Tregs, utilizing ZsGreen encoding lentiviral vectors to monitor the efficiency of transduction.

1. Prepare the coating solution by diluting RetroNectin® stock to 20 μg/mL in PBS the day after Treg sorting and activation (*see* **Note 26**).
2. Distribute 200 μL of the coating solution to each well of a flat-bottom 96-well plate.
3. Seal the covered-plate with parafilm and incubate at 4 °C for 24 h.
4. Two days after sorting and activation, count the activated Tregs and resuspend them at 0.5×10^6 cells/mL in complete IMDM supplemented with IL-2.
5. Remove PBS from the RetroNectin®-coated plate and add 1×10^5 cells/well (200 μL).
6. Calculate the appropriate amount of lentiviral vector to use based on the desired MOI (*see* **Note 27**).
7. Thaw the lentiviral vector at RT (*see* **Note 28**).
8. Add the viral supernatant to the wells and gently mix by pipetting.
9. Incubate the plate for 4–16 h at 37 °C/5% CO_2 in a humidified incubator.
10. Centrifuge the plate at 480 g for 5 min and carefully discard the supernatant containing the viral suspension.
11. Resuspend the cells in 200 μL of complete IMDM medium supplemented with IL-2.
12. Incubate the cells at 37 °C/5% CO_2 for at least 48 h before assessing transgene expression.
13. The Zsgreen expression can be checked using flow cytometry day 4 post transductions, DAPI can be added as a viability marker (*see* Fig. 4).
14. In the case of CAR transduction, use the CAR-specific transduction marker or an anti-CAR linker antibody like an anti-G4S Linker antibody.
15. The transduced Tregs can be stained as described in Subheading 3.4 to check phenotype using appropriate antibodies.

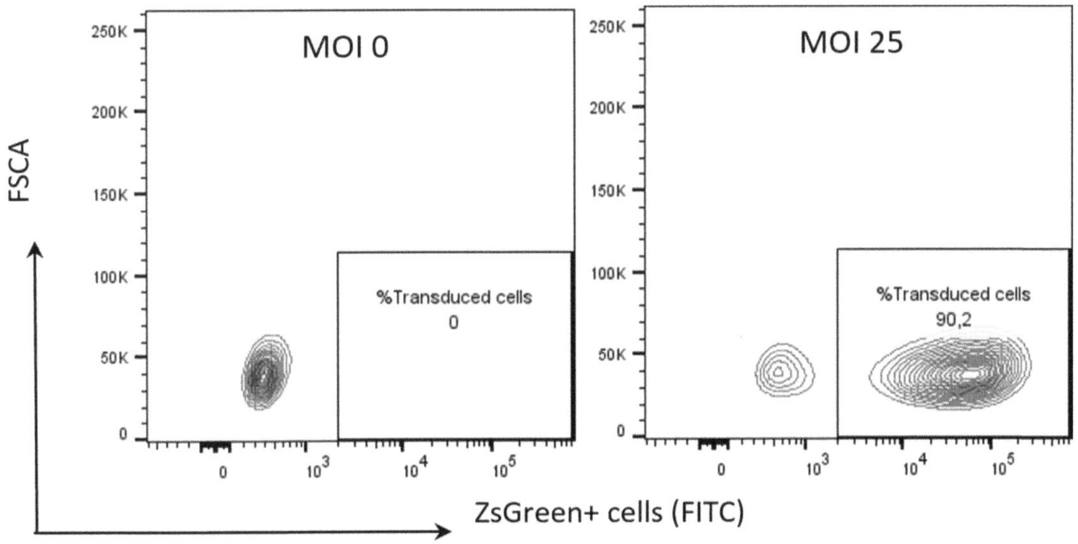

Fig. 4 Characterization of ZsGreen expression by Treg transduced cells. Tregs were activated at day 1 with Dynabeads™ and lentivirally transduced at day 3 with a lentiviral vector encoding the ZsGreen protein. Six days after transduction, Tregs were stained with DAPI for dead cell exclusion and analyzed by flow cytometry to evaluate the percentage of transduced cells

4 Notes

1. Use an appropriately sized magnet based on the volume of the starting sample.
2. Filter the PBE buffer to ensure sterility and store it at 4 °C.
3. Centrifuge the antibodies and pipette them to prepare a staining master mix. Store the mix at 4 °C in the dark.
4. β-Mercaptoethanol can be used to preserve cell viability in cases of rapid cell growth.
5. The dilution of synovial fluid may vary depending on its viscosity. If necessary, pass the synovial fluid through a 40 μm cell strainer to remove clumps.
6. Ficoll®-Paque Plus must be at RT (20–25 °C) before being added to the tube.
7. This is an optional step for PBMCs and can be skipped for SFMCs. It is recommended depending on the clarity of the supernatant of PBMC and number of platelets.
8. When thawing, place the vial in a 37 °C water bath until only a small piece of ice remains. Pool the cells with complete medium, centrifuge at 480× *g* for 5 min, discard the supernatant, and resuspend the pellet.

9. This protocol includes a pre-enrichment step, recommended for samples with a high initial cell count. For samples with a low cell count, this step may be bypassed, proceeding directly to cytometric-based cell sorting.
10. Avoid vortexing the selection and depletion cocktails.
11. Whether to pipette or decant by tilting the magnet depends on the starting sample volume and the magnet used. Refer to the kit protocol for further instructions.
12. Isolated cells can be stored at 4 °C in PBE for a maximum of 4–5 h before further use.
13. The expected purity of isolated Tregs is typically >85%, though this may vary depending on donor variability and sample quality.
14. When using the magnet for a 50 mL tube, request two additional vials of EasySep™ Dextran RapidSpheres™ 50100.
15. When preparing a staining master mix, prepare extra volume (+0.5–1 sample) to account for pipetting inaccuracies.
16. Cell sorting can be performed in 5 mL or 15 mL tubes, coated overnight, depending on the adapter available for Cytek.
17. The pressure, flow rate, and efficiency must be adjusted based on the experimental requirements.
18. PBS buffer is used to quench the cells and ensure proper live/dead staining.
19. After resuspension, the LIVE/DEAD™ Fixable Blue Dye can be stored in 2 μL aliquots at −20 °C until further use.
20. Cells can be fixed overnight at 4 °C if necessary, though this is not recommended unless time constraints require it.
21. The concentration of the CTV can be adjusted from 2 to 5 μM.
22. PBS must be cold before use.
23. When adding CTV to PBS-pelleted cells, it is preferable to use CTV at 37 °C.
24. Cells can be acquired on the day 3, 4, or 5, depending on fluorescence intensity, to prevent overlap with unstained suppressive Tregs.
25. The density of the cells is also important and should not exceed 1×10^5–1.5×10^5 cells/cm^2.
26. Freezing and thawing of the RetroNectin® can be repeated up to 10 times. Do not mix the solution vigorously and do not vortex.
27. Optimal MOI (>90% positive cells) may vary depending on the vector. We recommend testing several MOI to find the most efficient one. We use MOI 25 for the transduction of Tregs.

28. For a volume of viral supernatant between 100 μL and 1 mL, thaw at RT, for lower volumes thaw on ice, and for higher volumes thaw in a water bath. After thawing, the viral supernatant has to be homogenized by pipetting and not by inverting the tube.

Acknowledgments

This work was supported by Immun4Cure IHU "Institute for innovative immunotherapies in autoimmune diseases" (France 2030/ANR-23-IHUA-0009) and the INBS EcellFrance (ANR-11-INBS-0005) for the technical support and equipment. We thank FlashTherapeutics (Christine Dutoit & Alexandra Iché) for the ZsGreen lentiviral production.

References

1. Bennett CL, Christie J, Ramsdell F et al (2001) The immune dysregulation, polyendocrinopathy, enteropathy, X-linked syndrome (IPEX) is caused by mutations of FOXP3. Nat Genet 27: 20–21
2. Sakaguchi S, Yamaguchi T, Nomura T et al (2008) Regulatory T cells and immune tolerance. Cell 133:775–787
3. Basdeo SA, Moran B, Cluxton D et al (2015) Polyfunctional, pathogenic CD161+ Th17 lineage cells are resistant to regulatory T cell–mediated suppression in the context of autoimmunity. J Immunol 195:528–540
4. Bonelli M, Savitskaya A, Von Dalwigk K et al (2008) Quantitative and qualitative deficiencies of regulatory T cells in patients with systemic lupus erythematosus (SLE). Int Immunol 20:861–868
5. Viglietta V, Baecher-Allan C, Weiner HL et al (2004) Loss of functional suppression by CD4+CD25+ regulatory T cells in patients with multiple sclerosis. J Exp Med 199:971–979
6. Brusko TM, Wasserfall CH, Clare-Salzler MJ et al (2005) Functional defects and the influence of age on the frequency of CD4+CD25+ T-cells in type 1 diabetes. Diabetes 54:1407–1414
7. Lindley S, Dayan CM, Bishop A et al (2005) Defective suppressor function in CD4+CD25+ T-cells from patients with type 1 diabetes. Diabetes 54:92–99
8. Jiang Q, Yang G, Liu Q et al (2021) Function and role of regulatory T cells in rheumatoid arthritis. Front Immunol 12:626193
9. Liu W, Putnam AL, Xu-yu Z et al (2006) CD127 expression inversely correlates with FoxP3 and suppressive function of human CD4+ T reg cells. J Exp Med 203:1701–1711
10. Miyara M, Yoshioka Y, Kitoh A et al (2009) Functional delineation and differentiation dynamics of human CD4+ T cells expressing the FoxP3 transcription factor. Immunity 30: 899–911
11. Sakaguchi S, Mikami N, Wing JB et al (2020) Regulatory T cells and human disease. Annu Rev Immunol 38:541–566
12. Fontenot JD, Gavin MA, Rudensky AY (2003) Foxp3 programs the development and function of CD4+CD25+ regulatory T cells. Nat Immunol 4:330–336
13. Takahashi T, Kuniyasu Y, Toda M et al (1998) Immunologic self-tolerance maintained by CD25+CD4+ naturally anergic and suppressive T cells: induction of autoimmune disease by breaking their anergic/suppressive state. Int Immunol 10:1969–1980
14. Zaragoza B, Chen X, Oppenheim JJ et al (2016) Suppressive activity of human regulatory T cells is maintained in the presence of TNF. Nat Med 22:16–17
15. Eggenhuizen PJ, Ng BH, Ooi JD (2020) Treg enhancing therapies to treat autoimmune diseases. Int J Mol Sci 21:7015

Chapter 4

Culturing Murine Bone Marrow-Derived Macrophages

Becky Lin and Giraldina Trevejo-Nuñez

Abstract

Bone marrow-derived macrophages (BMDMs) mimic several properties of macrophages, such as phagocytosis and antigen processing and presentation, and may be used to study and characterize macrophage biology in response to specific stimuli. BMDMs are generated through isolation of cells from murine femur and tibia and cultured with L929-conditioned media, which contain factors essential for differentiation into macrophages. This protocol details the steps of retrieving bone marrow from mice, collecting L929-condition media, and culturing bone marrow cells to differentiate into macrophages.

Key words Bone marrow-derived macrophages (BMDMs), Murine bone marrow, Monocyte differentiation, Macrophage colony-stimulating factor (M-CSF), In vitro cell culture

1 Introduction

Macrophages are a vital component of the innate immune system, playing key roles in pathogen phagocytosis, releasing inflammatory mediators, and activating T cells through antigen presentation [1]. They can be differentiated from bone marrow cells via circulating monocytes upon exposure to growth factors such as macrophage colony-stimulating factor (M-CSF) and granulocyte-macrophage colony-stimulating factor (GM-CSF) [2, 3]. A method to culture bone marrow-derived macrophages (BMDMs) is to harvest bone marrow from mice and culture the cells with cultured media from L929 cells. L929 cells are immortalized mouse fibroblast cells that secrete M-CSF, along with other proteins [4].

Deriving macrophages from murine bone marrow may be advantageous over monocyte and macrophage cell lines due to absence of immortalization and continuous cell growth that may affect the cells. For example, immortalization can alter cellular phenotypic behavior or can cause epigenetic modifications [5]. BMDMs are also an attractive option because they can be obtained from genetically modified mice strains. Although

Rami Bechara (ed.), *Autoimmunity: Methods and Protocols*, Methods in Molecular Biology, vol. 2990,
https://doi.org/10.1007/978-1-0716-4997-8_4,

macrophages differentiated from bone marrow may differ in behavior from tissue-resident macrophages, such as Kupffer cells and alveolar macrophages [6], BMDMs remain a valuable alternative when large numbers of homogeneous cells with prototypical macrophage features are required.

Here we highlight how to collect bone marrow from the femur and tibia of mice to culture with a mixture of RPMI 1640 and L929-conditioned media to differentiate hematopoietic precursors into macrophages. L929-conditioned media is collected from L929 cells after 10 days of culture. This method yields a substantial number of macrophage cells that may be studied with downstream manipulation. These in vitro–differentiated macrophages can serve as a powerful tool to investigate macrophage-mediated mechanisms in autoimmune diseases, particularly their roles in antigen presentation, cytokine production, and interaction with autoreactive lymphocytes.

2 Materials

Prepare all solutions using proper aseptic technique and autoclave all tools to be used for harvesting bone marrow. Follow local and institutional waste disposal regulations.

2.1 Culture Media

1. 10% DMEM medium for L929-conditioned media: DMEM (with 4.5 g/L glucose, L-Glutamine, sodium pyruvate), 10% Fetal Bovine Serum (FBS). Remove 50 mL of media from a pre-prepared bottle of 500 mL DMEM. Add 50 mL of FBS to the bottle and filter the entire bottle through a 0.2 μm PES vacuum filter unit (*see* **Note 1**). Store media at 4 °C until use.
2. 10% RPMI medium: RPMI 1640 (with L-Glutamine), 10% FBS, 0.01 M HEPES, with antibiotics and antimycotic (*see* **Note 2**). Remove 60 mL of media from a pre-prepared bottle of 500 mL RPMI 1640. Add 50 mL of FBS, 5 mL of 1 M HEPES, and 5 mL of 100× antibiotic/antimycotic solution (penicillin, streptomycin, amphotericin B) to the bottle. Filter the entire bottle through a 0.2 μm PES vacuum filter unit. Store media at 4 °C until use.

2.2 Cell Culture Materials

1. Serological pipets.
2. Pipet filler.
3. Pipette and sterile pipette tips.
4. 0.2 μm PES vacuum filter unit
5. Cell culture treated flasks with filter cap (T-25 cm^2, T-75 cm^2, T-175 cm^2).
6. Cell culture petri dishes (58.1 cm^2).

7. Conical tubes.
8. Petri dishes.
9. Cell scraper.
10. 1 mL microcentrifuge tube.
11. 1.5 mL microcentrifuge tube.
12. 0.25% Trypsin-EDTA.
13. Dulbecco's Phosphate-Buffered Salt Solution (DPBS) 1×.
14. ACK lysing buffer.
15. 70 μm cell strainers.

2.3 Tools for Collecting Mouse Bone Marrow

1. Razor blades.
2. 45° angled forceps.
3. Dissecting scissors.
4. 70% ethanol in spray bottle.
5. 22 gauge needles.
6. 10 mL syringe.

3 Methods

Ensure to follow all institutional animal care protocols when using animals for research.

3.1 Preparation of L929 Media to Differentiate Cells into Macrophages

1. Warm cryovial with frozen L929 cells in a 37 °C water bath until 70% thaw, ensuring the cap does not submerge. In a biosafety cabinet, resuspend the cells in 500 μL DPBS by gently pipetting up and down.
2. Add 10 mL DPBS to a 50 mL conical tube and transfer the cells from cryovial to the tube using a pipette. Pipette up and down to mix. Centrifuge at 300× *g*, 4 °C for 5 min.
3. The cells will form a pellet at the bottom of the tube. Remove the liquid without disturbing the pellet (*see* **Note 3**). Resuspend the cell pellet in 5 mL DMEM10%. Transfer the cell suspension to a T-25 cell culture flask (*see* **Note 4**) and gently rock the flask gently horizontally to evenly distribute the cells. Then, incubate the flask at 37 °C, 5% CO_2.
4. Monitor the cells daily under a light microscope. Once the flask reaches 70% confluency, prepare to split the cells. Remove the media from the flask without disturbing the attached cells. Gently wash the cells with 5 mL DPBS, then remove the DBPS. Add 2.5 mL of 0.25% Trypsin-EDTA, gently rock the flask to distribute the solution, and incubate at 37 °C for 5 min. Tap the flask gently to detach the cells. Under the microscope,

the cells should be round and floating. Add 5 mL of 10% DMEM to the flask (*see* **Note 5**) to stop Trypsin-EDTA activity and resuspend the cells. Transfer cell suspension to a 50 mL conical tube and centrifuge at 300× *g*, 4 °C for 5 min. Remove the supernatant, resuspend the cell pellet in 10 mL 10% DMEM, and transfer the suspension to a T-75 flask. Incubate the cells at 37 °C, 5% CO_2.

5. Repeat **step 4** to transfer the cells to a T-175 flask: remove media, add 5 mL Trypsin-EDTA, incubate for 5 min, add 10 mL DMEM10%, centrifuge, remove the supernatant, resuspend in 20 mL DMEM10% and transfer to a T-175 flask. Incubate the cells at 37 °C, 5% CO_2.
6. Repeat **step 5** to transfer the cells to five T-175 flasks (*see* **Note 6**): remove the media, add 10 mL Trypsin, incubate for 5 min, add 10 mL DMEM10%, centrifuge, remove the supernatant, resuspend in 25 mL DMEM10%, and transfer 5 mL of the cell suspension to each T-175 flask. Add 40 mL DMEM10% to each flask for a total volume of 45 mL. Incubate the cells at 37 ° C, 5% CO_2.
7. After 10 days in the T-175 flask, carefully remove the conditioned media from each flask without disturbing the cells using a serological pipet. Transfer the media to a 0.2 μm PES vacuum filter unit and filter the entire volume of L929-condited media. Mix the pooled media to ensure homogeneity within the batch. Transfer 40 mL to a 50 mL conical tube and label accordingly (*see* **Note 7**). Store at −20 °C until use.

3.2 Harvest Femurs and Tibia from Mice

1. Sacrifice mice following IACUC protocol (i.e., CO_2 inhalation), plus a secondary method (i.e., cervical dislocation).
2. Douse mice in 70% ethanol. Remove fur and skin from the legs by lifting the skin at the torso and making a small incision. Pull the skin away using tweezers down each leg and over each foot and firmly tug until skin is removed (*see* **Note 8**).
3. Remove the leg by exposing the hip joint of each leg. Cut across the joint using dissection scissors. Cut away the foot and repeat the same procedure with the other leg. Place the legs in a tube containing 10 mL of RPMI media on ice. Discard mouse and all excess tissues according to IACUC protocol.

3.3 Isolate Bone Marrow Cells

1. Prior to culturing bone marrow from the harvested legs, ensure all materials are sterile and tools have been autoclaved.
2. In a biosafety cabinet, prepare three petri dishes: two filled halfway with 70% ethanol and one halfway with DPBS.
3. Pour out the medium containing the legs into an empty petri dish. Remove the excess muscle from the bone using forceps and a razor blade. Once all the tissue has been removed,

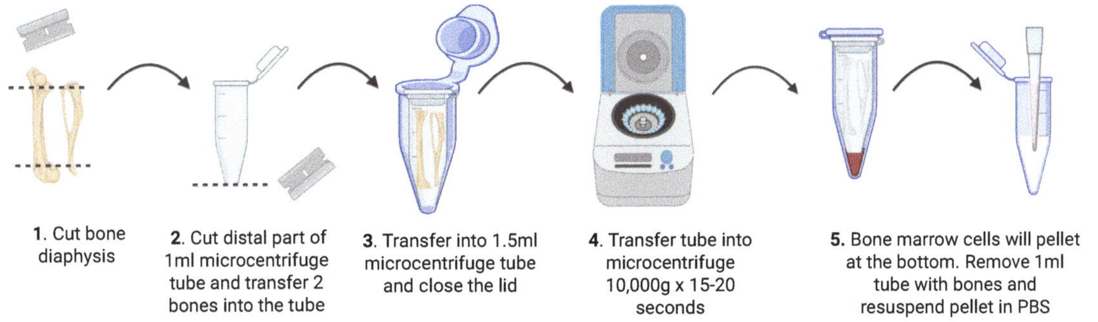

Fig. 1 Rapid method of extracting bone marrow cells by microcentrifugation. After bones are retrieved, they are cut at the distal part with blades and placed in a 1 mL microcentrifuge tube, which also has been cut down distally (no more than two bones at a time). This 1 mL tube with bones is snugged into a 1.5 mL tube, placed in a microcentrifuge, and spun at 10,000 g for 15–20 s. Bone marrow cells will be pelleted, and bones will turn white. Pellet is ready for resuspension. (Created in BioRender. Trevejo-Nuñez, G. (2025) https:/BioRender.com/v15v180)

separate the femur from the tibia (*see* **Note 9**). Use a Kimwipe to clean any remaining muscle from the bone if necessary.

4. Using forceps, rinse each bone in the petri dish with 70% ethanol for 5 s. Repeat the rinse again in the second petri dish with 70% ethanol, then place the bone in DPBS. Repeat this step for all collected bones.
5. Remove a bone from the DPBS with forceps and cut both distal cartilages with a razor blade and follow **steps 6–9** of this protocol (Fig. 1) [7, 8].
6. Cut the distal part of a 1 mL microcentrifuge tube (approximately 5 mm) with a razor blade and place two pieces of bone inside the tube and close the lid.
7. Place the 1 mL microcentrifuge tube (**step 6**) inside a 1.5 mL microcentrifuge tube and centrifuge at 10,000× *g* for 10–20 s. The bone marrow cells will pellet at the bottom of the tube.
8. Remove the 1 mL microcentrifuge tube with the bones which will turn from red to white after the centrifugation.
9. Resuspend the bone marrow cells pellet with 1 mL DPBS and transfer it in a labeled 50 mL conical tube. Add 10 mL of DPBS and resuspend again and proceed to **step 13**.
10. Alternatively, **steps 11** and **12** may be performed after **step 5**.
11. Thread the bone using a 22 gauge needle and 10 mL syringe filled with DPBS. Flush out the bone marrow into a 50 mL conical tube or a clean petri dish until it turns from red to white (*see* **Note 10**). Repeat this step with the remaining bones.
12. Resuspend the cells to remove clumps. If the cells are in a petri dish, transfer them to a labeled conical tube. Keep tubes on ice

until bone marrow has been isolated from all samples (repeat **steps 3–6**).

13. Centrifuge the tubes at 300× *g*, 4 °C for 5 min. Carefully remove the supernatant and discard it without disturbing the pellet. Add 5 mL ACK lysing buffer to each tube to lyse red blood cells. Resuspend the cells using a serological pipet. Incubate for 5 min on ice. Add 5 mL RPMI10% to each tube to stop the reaction. Centrifuge at 300× *g*, 4 °C for 5 min.
14. Discard the supernatant and resuspend cells in 10 mL RPMI10%. Strain the cells through a 70 μm cell strainer (*see* **Note 11**) into a new 50 mL conical tube. Centrifuge at 300× *g*, 4 °C for 5 min.
15. Discard the supernatant and resuspend cells in 35 mL RPMI10% supplemented with 30% L929-conditioned media: 24.5 mL RPMI10% + 10.5 mL L929-conditioned media (*see* **Note 12**).
16. For each sample (2 legs from 1 mouse), prepare 7 cell culture petri dishes. Divide the cell suspension between each dish by adding 5 mL/plate. Add 5 mL of RPMI10% supplemented with 30% L929 to each plate for a total volume of 10 mL/plate. Swirl the plates carefully to evenly distribute the cells.
17. Incubate at 37 °C and 5% CO_2. Visualize the cells each day under a light microscope.
18. On day 3, add 3 mL of RPMI10% with 30% L929 to each plate.
19. Repeat **step 11** on day 5 (*see* **Note 13**).
20. Split the cells on day 6: remove the media from each plate. Add 5 mL of DPBS to each plate. Use a cell scraper to detach the cells from each plate (*see* **Note 14**). Remember to use a new cell scraper between samples. Transfer cells to a 50 mL conical tube, pooling the 7 plates together for each sample. Rinse each plate with a small amount of DPBS and transfer to the tube. Centrifuge at 300× *g*, 4 °C for 5 min.
21. Remove the supernatant from the tube. Resuspend in appropriate amount of RPMI10% media and divide cells into preferred plates and configuration for your experiment (*see* **Note 15**).
22. Incubate at 37 °C and 5% CO_2. The cells should be 70% confluent the next day for your experiment.

4 Notes

1. An alternative to using the PES vacuum filtration unit is using a syringe filter to filter the FBS. However, we have found that

multiple syringe filters are needed for the entire 50 mL volume, and this method may compromise the sterility of the media.

2. We added antibiotic antimycotic solution here but not to the D10% media because we also used the D10% to inoculate the cells with bacteria downstream. Antibiotics and antimycotics are added to media as a precautionary against possible contamination. We suggest considering your experiment goals and comfort level in aseptic technique when debating whether to add antibiotics and antimycotics to the media.
3. Liquid from the conical tubes may be removed by using a vacuum system or by using a serological pipet and discarding the liquid.
4. Depending on the number of viable cells that were in the cryovial, a T-75 flask may be used. If there were less than 10^5 cells, err on the side of caution and use a T-25.
5. Trypsin is toxic to cells with prolonged exposure. The FBS in the D10% will inactivate the trypsin.
6. This is the last splitting step before collecting the culture media. If you wish to prepare a larger volume of L929-conditioned media, you may continue splitting the cells until you reach your desired amount.
7. It is important to make note of the batch of conditioned media and the passage number of the L929 cells. One weakness of this method is that one cannot guarantee the batches of L929 are similar between experiments. However, we have found that using M-CSF alone for differentiation resulted in significantly less macrophages.
8. A tip for removing the skin is after making the first incision, using two forceps to pull the skin away from the entire circumference of the torso. After that, pull the skin downward toward the feet at the front and back of the body. Then, remove the skin from the legs one side at a time.
9. Try to remove as much muscle tissue from the bone as possible using the razor without breaking the bones. More muscle can be cut away after separating the femur and the tibia. To separate the femur and the tibia, find the joint by bending the knee. Using two forceps, one on each bone, snap the joint and separate the two bones. Some muscles may pull away from the bone. Remove the rest of the tissue using the razor and Kimwipe.
10. When flushing out the bone, we suggest using pulses to push the marrow out of the bone. You may use more DPBS, just be cognizant of the volume of your tube. Also, we prefer to flush into a petri dish because then we can resuspend the bone marrow using a syringe. We found that this breaks up the cell

clumps better than a serological pipet. Wash the petri dish with DPBS after transferring to a conical tube to collect as many cells as possible.

11. When straining cells, make sure to wet the cell strainer beforehand to prevent cells sticking to the strainer. After passing the cells through the strainer, rinse the strainer again with media.
12. We have experimented with adding 15% L929-conditioned media instead of 30% and have produced similar results. Also, if you are mixing a large volume of media to accommodate for all your samples, remember to account for some volume loss (i.e., 4 samples = 2 × 140 mL. Make 300 mL media).
13. More recently, we have been supplementing the cells with 10 mL media on day 3 and splitting on day 5. Rather than the timing, it is more important to maintain consistency between experiments.
14. When debating between using trypsin or a cell scraper to detach cells, we have found the cell scraper to be more time efficient, as the macrophages have a strong affinity to attach to the plates. However, using a cell scraper may affect flow cytometry experiments.
15. For our experiments, we use 6-well plates, with 1×10^6 cells/well (5×10^5 cells/mL).

References

1. Morrissette N, Gold E, Aderem A (1999) The macrophage–a cell for all seasons. Trends Cell Biol 9(5):199–201
2. Kennedy DW, Abkowitz JL (1998) Mature monocytic cells enter tissues and engraft. Proc Natl Acad Sci USA 95(25):14944–14949
3. Akagawa KS (2002) Functional heterogeneity of colony-stimulating factor-induced human monocyte-derived macrophages. Int J Hematol 76(1):27–34
4. Heap RE, Marín-Rubio JL, Peltier J et al (2021) Proteomics characterisation of the L929 cell supernatant and its role in BMDM differentiation. Life Sci Alliance 4(6):e202000957
5. Maqsood MI, Matin MM, Bahrami AR et al (2013) Immortality of cell lines: challenges and advantages of establishment. Cell Biol Int 37(10):1038–1045
6. Ginhoux F, Guilliams M (2016) Tissue-resident macrophage ontogeny and homeostasis. Immunity 44(3):439–449
7. Amend SR, Valkenburg KC, Pienta KJ (2016) Murine hind limb long bone dissection and bone marrow isolation. J Vis Exp 110:53936
8. Heib T, Gross C, Muller ML et al (2021) Isolation of murine bone marrow by centrifugation or flushing for the analysis of hematopoietic cells – a comparative study. Platelets 32(5):601–607

Chapter 5

Macrophage Cell Culture and Characterization by Serum Human AB in Health and Disease

Audrey Paoletti, Bineta Ly, Samuel Bitoun, Gaetane Nocturne, and Xavier Mariette

Abstract

This chapter focuses on generating macrophages from monocytes and highlights their crucial role in the immune system, particularly their involvement in the pathogenesis of rheumatoid arthritis (RA), the most frequent systemic autoimmune disease. It addresses the limitations of traditional in vitro models, which often rely on high-dose cytokines, and introduces human AB serum as a more physiological option for monocyte-to-macrophage differentiation. Monocyte-derived macrophages (MDMs) by SAB are presented as a relevant model for studying macrophages biology and function. A detailed protocol is provided for differentiating human blood monocytes into anti-inflammatory macrophages, covering steps for cell isolation, polarization, and phenotypic characterization.

Finally, this model is evaluated for its potential use in other pathologies, such as infections where MDMs heterogeneity closely reflects physiological conditions, offering a promising tool for research and therapeutic applications.

Key words Monocyte-derived macrophages, Serum human AB, Rheumatoid arthritis, Healthy donors, Polarization

1 Introduction

Monocytes and macrophages are essential components of the immune system, serving pivotal roles in both innate and adaptive immunity [1, 2]. Monocytes circulate in the bloodstream and differentiate into macrophages or dendritic cells upon migration into tissues, triggered by signals of infection, inflammation, or tissue damage [3–7].

Macrophages, derived from monocytes, are highly specialized phagocytic cells distributed throughout various tissues in the body [8]. They are fundamental for maintaining tissue homeostasis, as they not only eliminate pathogens but also clear dead cells and debris. Furthermore, macrophages exhibit significant plasticity,

Rami Bechara (ed.), *Autoimmunity: Methods and Protocols*, Methods in Molecular Biology, vol. 2990, https://doi.org/10.1007/978-1-0716-4997-8_5, adopting diverse functional states in response to environmental cues [9–12]. For instance, in inflammatory contexts, macrophages can assume a pro-inflammatory phenotype, producing cytokines such as TNF-α and IL-1β to orchestrate the immune response. Conversely, during tissue repair and resolution phases, macrophages can shift to an anti-inflammatory profile, promoting healing and tissue regeneration [13–16].

Monocytes and macrophages also play central roles in modulating immune responses by presenting antigens to T cells and producing a broad array of cytokines and chemokines that coordinate immune cell recruitment and activation [8, 12, 15, 16]. Given their diverse functionalities and dynamic responses to the tissue microenvironment, monocytes and macrophages are not only critical in infection control but also in various pathological conditions, including autoimmune diseases, cancer, and chronic inflammatory disorders [15, 17, 18].

However, studying these cells in vitro presents significant challenges, as it often fails to accurately mimic physiological conditions, particularly due to the use of high concentrations of cytokines such as M-CSF, IL-4, IL-10, GM-CSF, IFN-γ, and LPS for differentiation of monocytes in macrophages [13, 19–21]. Research involving HIV-infected macrophages has emphasized the utility of human AB serum as a physiological source of multiple cytokines, facilitating a more natural differentiation of blood monocytes into macrophages [22–24]. This method enables a more representative evaluation of their differentiation, which is particularly relevant in the context of autoimmune diseases [25, 26].

Rheumatoid arthritis (RA) is the most prevalent systemic autoimmune disease involving 0.5% of the general population, primarily affecting multiple joints and leading to pain, stiffness, and progressive destruction of bone and cartilage if left untreated. Without timely and adequate intervention, RA can cause severe disability, underscoring the need for increased awareness, improved access to care, and ongoing research aimed at developing more effective therapeutic strategies [27–30].

Monocytes and macrophages are central to the pathogenesis of RA, with significant accumulation in the synovial tissue [31–35]. Beyond the resident macrophages, blood monocytes can differentiate into macrophages—known as monocyte-derived macrophages (MDMs)—that may infiltrate the synovium and exhibit distinct phenotypes and functions [36–39].

Synovial biopsies, however, are invasive and challenging to obtain in certain countries. Consequently, the use of MDMs in a physiologically relevant manner is critical for accurately studying the phenotype and function of these diverse tissue-infiltrating macrophages, using their original source—blood monocytes.

Here, we present a protocol to drive the differentiation of human blood monocytes into anti-inflammatory macrophages

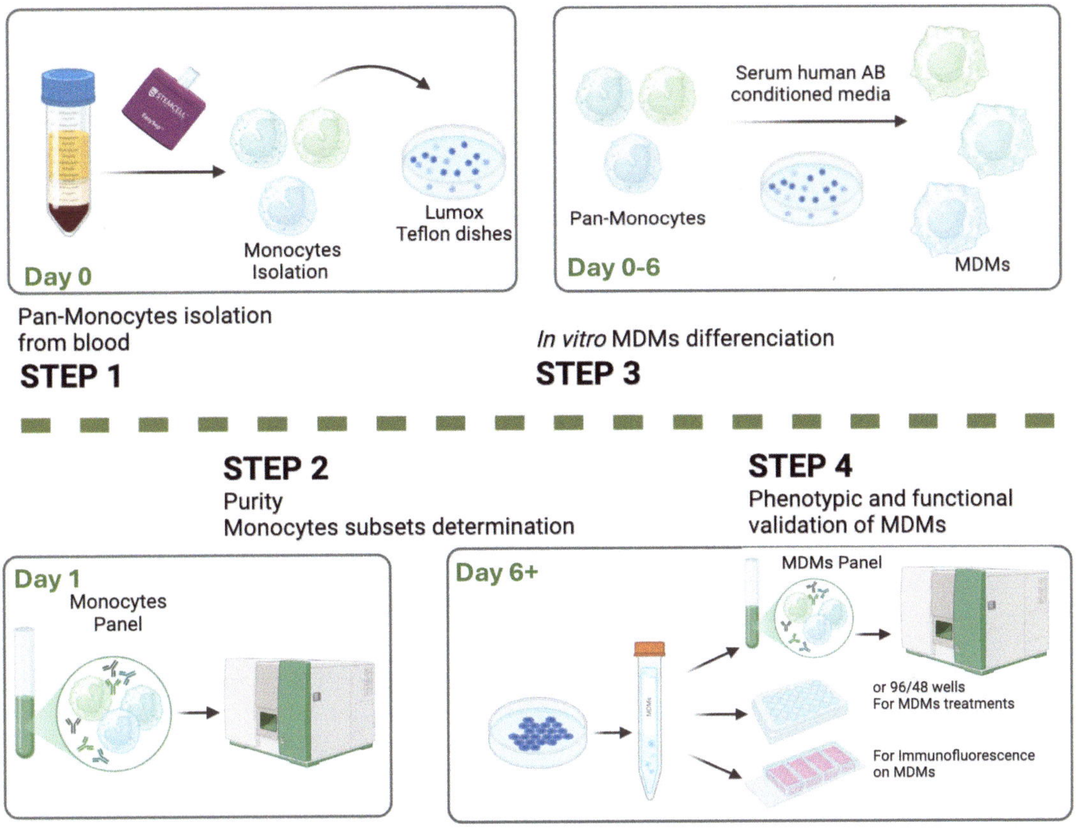

Fig. 1 Protocol to drive human monocytes-to-macrophages polarization in vitro using human serum AB media

using human AB serum-conditioned media, followed by phenotypic and functional characterization of these anti-inflammatory macrophages in vitro. We describe the isolation of monocytes from blood and detail the procedure to induce macrophage polarization. Finally, we outline the morphological assessment of macrophages and validate their phenotype in both healthy donors and RA patients (Fig. 1).

2 Materials

2.1 Institutional Permissions for Blood Sample

Study approval must be obtained for any work involving human biological samples prior to the commencement of the study. This study was approved for rheumatoid arthritis patients by the RIPH3 CPP LIMA under number 2020-A00509-30. Additionally, the collection of blood from healthy individuals is coordinated through the Etablissement Français du Sang (EFS) under an agreement with INSERM.

A minimum of 15 mL of blood is required to ensure an adequate number of monocytes for optimal differentiation.

Furthermore, patients undergoing corticosteroid therapy at a dose ≥10 mg per day are excluded, due to its potential impact on monocyte activity.

Appropriate health and safety procedures must be followed, including adherence to institutional Standard Operating Procedures (SOPs) for the handling and disposal of human blood samples using designated sterile laminar flow hoods. Additionally, users may need to have appropriate vaccinations to safely work with human biological samples, especially if the samples have not been screened for known human pathogens.

2.2 Plasma Storage

1. Cryovial tubes.

2.3 PBMCs Extraction

1. PBS 1× ($^{-/-}$CaCl/$MgCl_2$).
2. Pancoll Prep (density 1.077 g/mL; room temperature).
3. Buffer for PBMCs and monocytes sorting: 500 mL of PBS 1×, supplemented with 0.5% BSA (Bovine Serum Albumin, heat shock fraction) and 2 mM EDTA (0.5 M UltraPure).
4. C-CHIP hemocytometer or equivalent for PBMC counting, using Trypan Blue: diluted 1:2 in PBS 1× and filtered through a 0.2 μm filter.

2.4 Monocytes Magnetics Sorting

1. Monocytes without CD16 depletion are isolated from the PBMCs population using an immunomagnetic negative selection system to obtain untouched monocytes, Kits allow for the simultaneous enrichment of classical ($CD14^+CD16^-$), and intermediate ($CD14^+CD16^+$) monocytes. While the study of the classical and intermediate monocyte subpopulations is relevant in certain pathologies, the transformation of monocytes into macrophages via SAB can also be achieved using positive $CD14^+$ sorting.
2. Magnet.
3. Filters 40 μm.

2.5 Monocytes Polarization into MDMs

1. Macrophages medium: for 50 mL of macrophages medium composition: *RPMI 1640*, supplemented with *100 UI/mL Penicillin-Streptomycin* + *10 mM Hepes buffer solution 1 M* + *10 mM Sodium pyruvate 100 mM (100×)* + *1% MEM NEAA 100×* + *1% MEM vitamin solution 100×* + *15% serum human AB Male HIV-Tested* + *50 μM B-Mercepto 50 mM*.
2. For heat-inactivated human Serum AB or FBS, place the serum bottle in a water bath at 56 °C for 1 h. Then, pass it through a 10 mL syringe fitted with a 40 micron filter. Since the serum AB is very lipid-rich, multiple filters will be necessary. After obtaining pre-filtered serum at 40 microns, repeat the same procedure with a 20 micron filter.

3. Lumox Teflon 50 mm or 35 mm.

2.6 Purity Antibodies

1. Polypropylene tubes and compensation UltraComp eBeads.
2. Antibodies for monocytes purity on ZE5 24 colors cytometer: *CD45 PerCP* clone HI30 *-CD16 APC* clone 3G8 *-CD14 PE* clone 63D3 *-Live dead UV + Fc blocking.*

2.7 Harvest and Phenotyping of MDMs

1. To harvest macrophages and store the supernatant from a 6 day culture, the following are required: Yellow Cell Scraper TPP, Cryovial tubes, and Polypropylene tubes.
2. Compensation UltraComp eBeads.
3. Antibodies for MDMs phenotyping on ZE5 24 colors cytometer: *CD45 SBUV665– CD163 Vio Bright FITC– CD11b PerCP-Vio 700– CD71 PE-Vio615– CD206 PE-Vio770 -Live dead UV- Tandem Signal Enhancer human–Super Bright Complete Staining buffer and FcR Blocking.*

2.8 Immunofluorescence

1. 8-well tissue culture chambers.
2. 4% paraformaldehyde.
3. Permeabilization solution: 0.3% Triton X-100 in 1× PBS.
4. Blocking solution: PBS 1× + FBS 20% + BSA (1 mg/mL).
5. Antibodies for MDMs Immunofluorescence: anti-CD68 (1/100 dilution), anti-iNOS (1/200 dilution), or anti-Arginase-1 (1/50 dilution). Anti-rabbit and anti-mouse IgG conjugated to Alexa Fluor 488 or 546 fluorochromes (1/500 dilution), supplemented with DAPI (1/1000 dilution) is used as secondary antibodies.
6. Fluoromount G medium.

3 Methods

3.1 PBMCs Isolation from Whole Blood

1. Peripheral blood mononuclear cells (PBMCs) from healthy donors or RA patients are obtained from anonymized whole blood samples collected in heparin or EDTA tubes, provided by the EFS (Paris) or the Rheumatology Department (Le Kremlin-Bicêtre). An initial centrifugation at 650 g for 10 min at room temperature (21 °C), with acceleration set to 9 and brake set to 9, can be performed to collect the plasma for biobanking. The plasma can then be stored at −80 °C.
2. Dilute 15 mL of whole blood with 30 mL of pre-cooled PBS 1× (1:2 dilution) ($^{-/-}CaCl_2/MgCl_2$) in a standard 50 mL tube.

3. Gently transfer the entire volume of diluted blood into a standard 50 mL Falcon tube, carefully layering it on top of 15 mL of Pancoll Prep (density 1.077 g/mL; room temperature).
4. Centrifuge in a swinging-bucket centrifuge at 970 g for 20 min at room temperature (21 °C) with acceleration set to 2 and brake set to 2.
5. Using a 5 mL Pasteur pipette, carefully remove all liquid above the PBMCs layer to facilitate recovery.
6. Collect the PBMCs from the white interface layer using 1000 μL tips and transfer them into a 50 mL Falcon tube containing buffer.
7. Centrifuge at 300 g for 10 min at 4 °C with normal brake.
8. Add 50 mL of buffer and determine the number of viable cells using a C-CHIP hemocytometer with Trypan Blue. A 1:2 dilution of PBMCs from 15 mL of whole blood provides a suitable density for counting. Mix well before loading into the chamber. Count the cells and determine the total number of PBMCs.
9. Centrifuge at 120 g for 10 min at 4 °C with brake off to perform a more thorough platelet removal for Monocyte Isolation Kit.

3.2 Monocytes Human Isolation Kit

1. Resuspend PBMCs in buffer at a cell density of 5×10^7 PBMCs/mL. Add 50 μL of *Enrichment Cocktail* to the sample. Mix well and incubate for 10 min in the refrigerator (4 °C).
2. Add 50 μL of *Magnetic Particles.* Vortex well and incubate for 5 min in the refrigerator (4 °C).
3. Top up the sample with buffer to a total volume of 2.5 mL. Mix by gently pipetting. Place the tube (without lid) into the magnet and incubate at room temperature for 2.5 min.
4. Pick up the magnet, then, in one continuous motion, invert the magnet and tube, pouring the enriched cell suspension into a new tube. The isolated cells are now ready for use.
5. The isolated monocytes are washed in 1× PBS by centrifugation at 450 g for 5 min.
6. Determine the number of live cells using a C-CHIP hemocytometer with Trypan Blue. Resuspend the cells at a concentration of 1×10^6 cells/mL in Macrophage medium, plate 1–2 mL of the cell suspension in a 35 mm Lumox Teflon dish or 4–6 mL in a 50 mm Lumox Teflon dish and incubate 6 days at 37 °C with 5% CO_2.

3.3 Determination of Monocyte Purity by Flow Cytometry

1. Take a portion of isolated monocytes (approximately 50,000 cells or 25 μL) after the 10 mL PBS 1× wash step.

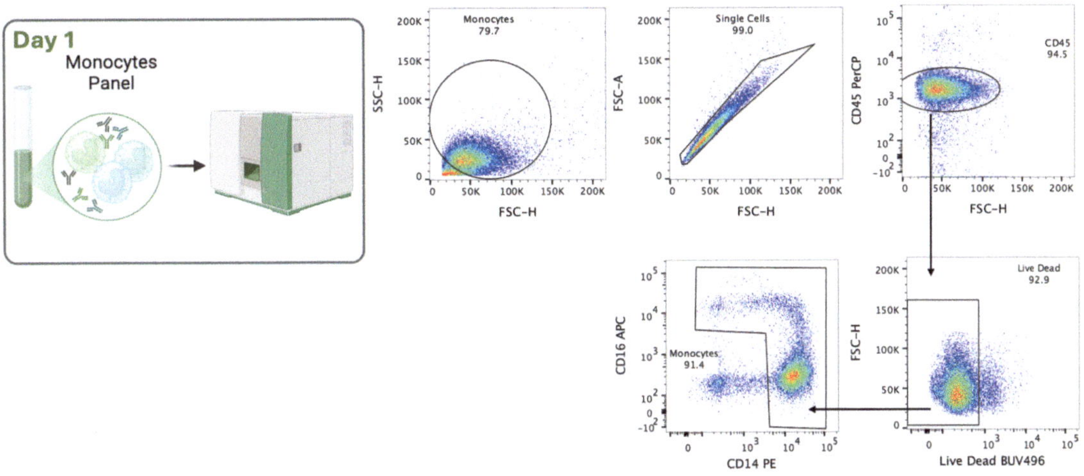

Fig. 2 Flow cytometry gating strategy for purified Pan monocytes populations from human PBMC populations

2. Resuspend in a solution containing anti-human CD14 and anti-human CD16 antibodies (dilution 1:20), a fixable viability dye to irreversibly stain dead cells (Live/Dead), and FcR blocking reagent in polypropylene tubes. Perform the staining in a total volume of 100 μL.
3. Gently vortex to mix. Incubate for 30 min at 4 °C in the dark.
4. Wash once with 1× PBS. Centrifuge at 450 g for 5 min at 4 °C. Fix cells in 1% formaldehyde (10 min room temperature incubation), centrifuge and resuspend in 200 μL of buffer kept at 4 °C if you wish to analyze later or the following day. Acquire data on a flow cytometer (Fig. 2).
5. In parallel, prepare compensation beads to correct for fluorescence spillover from the fluorochrome-conjugated antibodies. Add one drop each of "blank" and "anti-mouse" IgG beads and add the same dilution of the CD14 and CD16 antibodies. Gently vortex to mix. For the Live/Dead dye, ARC beads can be used for compensation. Bring the volume up to 200 μL with PBS 1×.

3.4 Macrophage Differentiation Using Human SAB-Conditioned Media

This step utilizes human AB serum-conditioned media to induce in vitro macrophage differentiation and polarization. Published studies have shown that human AB serum provides a physiological source of major cytokines for monocyte polarization into anti-inflammatory macrophages, leading to various macrophage morphologies. These macrophages are used in vitro for HIV-1 infections, with a good infection success rate due to the low level of restriction factors such as SAMHD1, TRIM22, or members of the APOBEC3 family [22–24]. Additionally, they are employed in studies of diseases such as rheumatoid arthritis (RA), where a specific defect in this polarization process can be observed [25, 26].

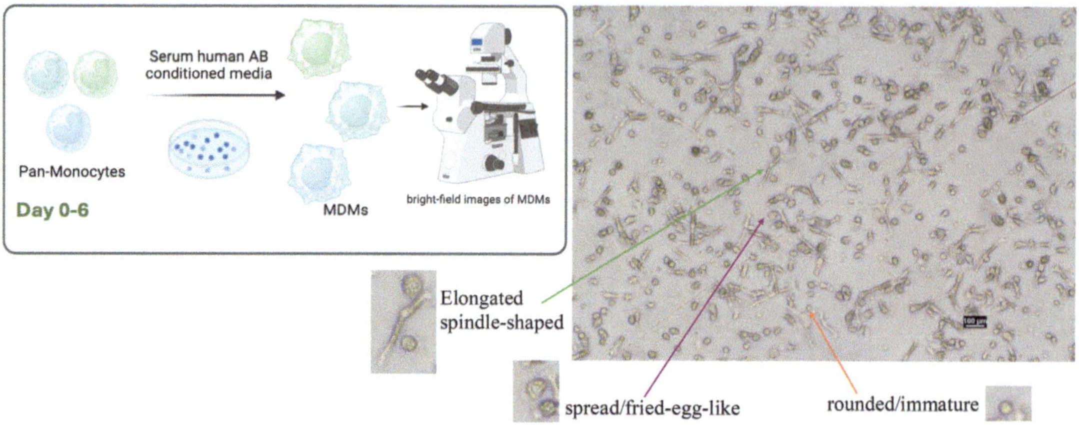

Fig. 3 Assessment of macrophage morphology after polarization with human SAB

1. *Day 0*

   Seed 1 × 10^6 Monocytes in 1 mL of macrophage medium in a 35 mm (maximum 2 mL) or 50 mm (maximum 6 mL) Lumox Teflon dish and incubate 6 days at 37 °C, 5% CO_2.

2. *Day 6*
   1. Acquire bright-field images of macrophages at 20× magnification to analyze MDMs morphological differences.
   2. Cells can be manually categorized into distinct subsets, such as elongated/spindle-shaped, spread/fried-egg-like, or rounded/immature (Fig. 3).
   3. Collect macrophage supernatants from all conditions and transfer them into separate PP tubes (including those from non-adherent macrophages). Add 1 mL of heat-inactivated FBS (*see* **Note 6**).
   4. Centrifuge at 450 g for 5 min.
   5. Collect the supernatant in cryovial and freeze at −80 °C.
   6. Scrape the Teflon dishes containing 1 mL of FBS and transfer the macrophages to the PP tubes using a TPP Cell Scrapper.
   7. Centrifuge tubes at 450 g for 5 min.
   8. Add 1 mL of macrophage medium and count the MDMs.

   The MDMs can be analyzed in different ways at this stage:

1. *Part A*: Phenotypic analysis.
2. *Part B*: Functional analysis/Immunofluorescence.

Part A: Phenotypic Analysis

1. Spin down samples at 450 g for 5 min 4 °C, and wash twice in buffer.

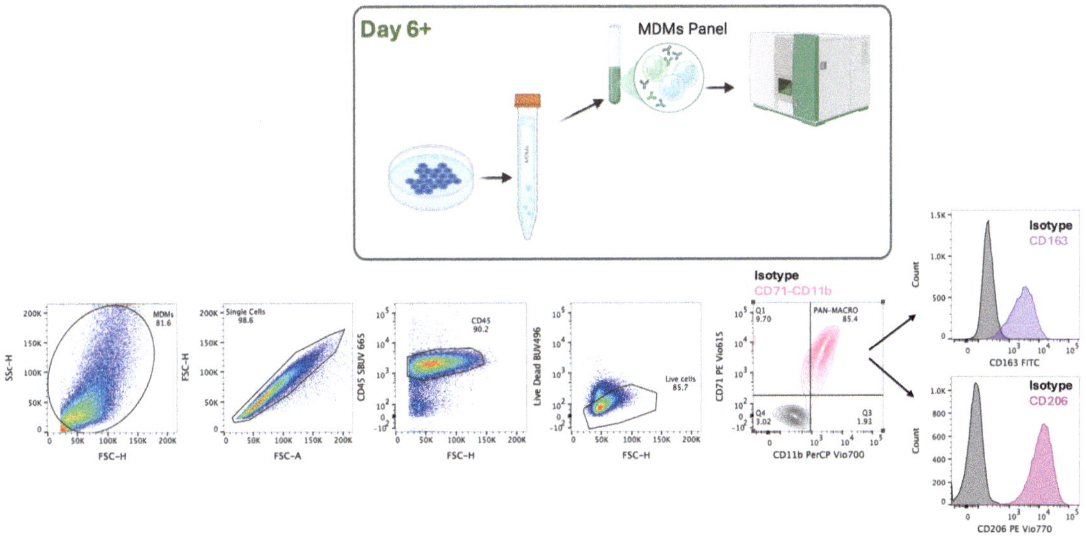

Fig. 4 Polarization of macrophages by human serum AB and gating strategy

2. Resuspend in buffer containing anti-human CD11b, anti-human CD71, anti-human CD206, anti-human CD163 for a basic anti-inflammatory phenotyping. Also add a fixable viability dye to irreversibly stain dead cells (Live/Dead), and FcR blocking reagent in polypropylene tubes. Perform the staining in a total volume of 200 μL.
3. For any flow cytometry analysis of monocytes or macrophages, it is essential to include isotype controls for each sample to assist in gating strategies. Despite the use of Fc blocking, monocytes and macrophages express high levels of Fc-gamma receptors. Therefore, the analysis requires five tubes:
   - *Unstained tube*: no staining to set the baseline for autofluorescence.
   - *Live/Dead tube*: only the viability dye to assess live/dead discrimination.
   - *Isotype 1 tube*: containing the isotype controls for CD11b and CD71.
   - *Isotype 2 tube*: containing CD11b and CD71 markers along with the isotype controls for CD206 and CD163.
   - *Full stain tube*: containing all the specific markers.
4. Gently vortex to mix. Incubate for 30 min at 4 °C in the dark.
5. Wash once with 1× PBS. Centrifuge at 450 g for 5 min at 4 °C. Fix cells in 1% formaldehyde (10 min room temperature incubation), centrifuge and resuspend in 200 μL of buffer kept at 4 °C if you wish to analyze later or the following day. Acquire data on a flow cytometer (Bio-Rad ZE5 Fig. 4).

6. In parallel, prepare compensation beads to correct for fluorescence spillover from the fluorochrome-conjugated antibodies. Add one drop each of "blank" and "anti-mouse" IgG beads and add the same dilution of the antibodies. Gently vortex to mix. For the Live/Dead dye, ARC beads can be used for compensation. Bring the volume up to 200 μL with PBS 1×.
7. Compensated profiles show that 91–96% of MDMs express both differentiation markers (CD11b and CD71) and anti-inflammatory macrophage markers (CD163 and CD206).

Part B: Functional Analysis/Immunofluorescence

1. At this stage, macrophages can be distributed with excellent cell survival. Seed 0.2×10^6 macrophages in 400 μL of macrophage medium in a 24-well plate and incubate at 37 °C, 5% CO_2. Alternatively, if a larger number of cells is required for functional assays following monocyte-to-macrophage differentiation, you can scale up the culture conditions as follows:
   - 0.4×10^6 in a T12 well (0.8 mL Macrophages medium).
   - 0.8×10^6 in a T6 well (1.6 mL Macrophages medium).
   - 10,000 in a 96 well plate (100 μL Macrophages medium).
2. Macrophages can also be plated in 8-well tissue culture chambers (0.1×10^6 in 200 μL of Macrophages medium) for immunofluorescence (IF).
3. After plating macrophages in plates or 8-well chambers, wait between 4 and 24 h to ensure the macrophages are in optimal condition for treatments or immunofluorescence.
4. *For IF*, the 8-well tissue culture chamber were fixed in 4% paraformaldehyde after three 1× PBS washes (5 min each), permeabilized for 5 min with 0.3% Triton X-100 in 1× PBS, washed 3 times with 1× PBS, and blocked for 1 h at room temperature in PBS-FBS 20%-BSA (1 mg/mL).
5. Macrophages were then incubated overnight at 4 °C with anti-CD68 (1/100 dilution), anti-iNOS (1/200 dilution), or anti-Arginase-1 (1/50 dilution). After three washes in 1× PBS, secondary antibodies, anti-rabbit and anti-mouse IgG conjugated to Alexa Fluor 488 or 546 fluorochromes (1/500 dilution), supplemented with DAPI (1/1000 dilution) for nuclei detection in PBS-FBS 20%-BSA (1 mg/mL), were added to the samples for 30 min at room temperature.

   8-well tissue culture chamber were then mounted with Fluoromount G medium. MDMs were imaged using an epifluorescence microscope with an 8-bit configuration, using a 40× dry objective or 100× oil objective (Fig. 5).

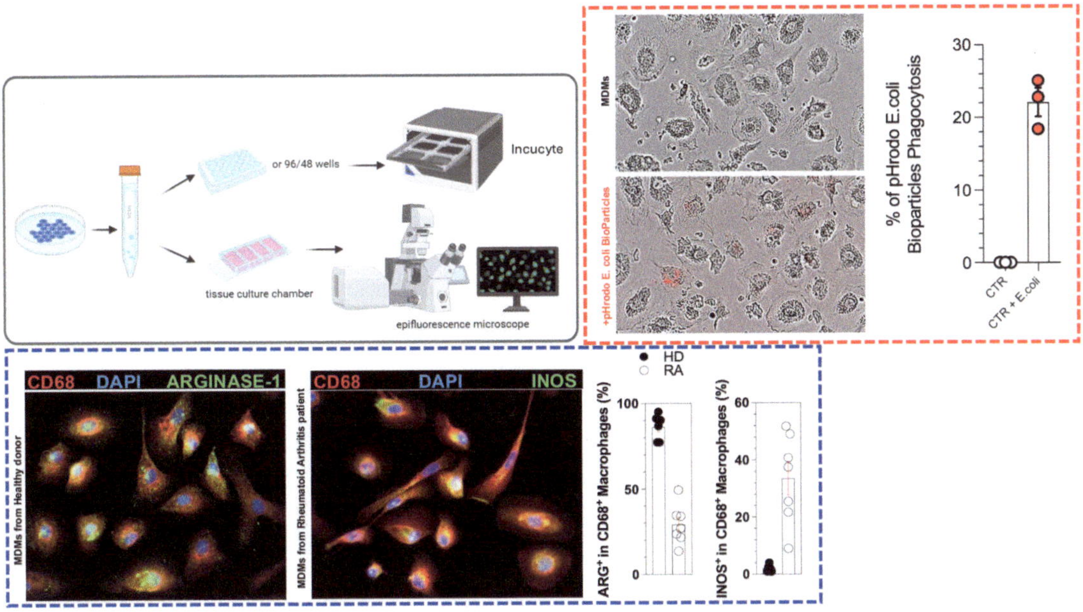

Fig. 5 Functional analysis/Immunofluorescence

6. To determine a *classical functionality of macrophages*, a phagocytosis assay using pHrodo *E. coli* BioParticles can be performed. MDMs are plated at 10,000 cells/well (100 μL) in a 96-well plate and incubated overnight to ensure proper adhesion. The wells are then washed three times with macrophage medium.

After adding 0.1 μg/mL of pHrodo *E. coli* BioParticles, phagocytosis by macrophages will be directly analyzed using the Incucyte live-cell imaging system. Image acquisition will be scheduled every 10 min for 2 h, followed by one image per hour for 24 h (Fig. 5).

The expected outcomes of this protocol, using human SAB blood monocytes for polarization, are to study classical and intermediate monocytes subsets of blood monocytes isolated from PBMCs. Flow cytometry is employed to verify population purity through analysis of $CD14^+/CD16^+$ staining (Fig. 2).

After 6 days in culture, monocyte-derived macrophages will begin to exhibit changes in phenotypic markers, along with alterations in morphology and function in RA patients. FACS analysis will determine the extent of cell-surface marker expression, such as the acquisition of alternative-activation markers $CD163^+CD206^+$ (Fig. 4).

MDMs morphology can be assessed using brightfield imaging. Several studies do not consider macrophage morphology, as M-CSF and GM-CSF are thought to drive monocyte-to-macrophage differentiation in a highly homogeneous manner. However, using the SAB technique, morphology emerges as a valuable source of information regarding the transformation of MDMs in both health and disease (Fig. 3).

Macrophage attachment and spreading increase during the differentiation process. Therefore, morphological assessment and immunofluorescence (IF) of intracellular markers of MDMs can complement phenotypic characterization.

MDMs obtained through polarization with human AB serum maintain their primary functionalities unaltered. However, these functionalities remain to be fully explored, as most studies preferentially use high-dose cytokines, since this approach is simpler to implement compared to the technique involving human AB serum with Lumox Teflon dish.

4 Notes

1. Buffy coat or LRSC cones can also be used.
2. For Ficoll, it is essential to work quickly to maintain good cell viability. The first few milliliters of diluted whole blood should be added very slowly onto the Ficoll-Paque layer, tilting the tube at a 45° angle to avoid disrupting the interface and ensuring optimal PBMCs separation.
3. Use a swinging bucket rotor and ensure the brake deceleration is set to MAX 2 to avoid disruption of the density gradient separation. The temperature should be at room temperature (RT) for effective Ficoll separation.
4. During washes, rather than aspirating the supernatant, which can be harsh, use a beaker containing a Virkon solution and gently pour the supernatant into this beaker.
5. For heat-inactivated human Serum AB or FBS, place the serum bottle in a water bath at 56 °C for 1 h. Then, pass it through a 10 mL syringe fitted with a 40 micron filter. Since the serum AB is very lipid-rich, multiple filters will be necessary. After obtaining pre-filtered serum at 40 microns, repeat the same procedure with a 20 micron filter.
6. It is vital to recognize that while the study of all three monocyte subpopulations holds relevance in certain pathologies, not all commercially available kits provide for the isolation of all three; some may only sort the classical and intermediate populations. Therefore, careful selection of kits is necessary. Additionally, the transformation of monocytes into macrophages via the SAB method can also be accomplished through positive $CD14^+$ sorting.

7. Monocytes sorted cells are washed 2 times in PBS 1× at 450 g for 5 min. It is crucial to remove EDTA as it will interfere with the proper transformation of monocytes into macrophages. This step is critical for successful differentiation.
8. Always maintain a proportion of 1×10^6 monocytes per mL. If the concentration is too low or too high, macrophage differentiation may be impaired.
9. Several protocols suggest performing the entire procedure on ice, but monocyte viability is lower under these conditions. The entire procedure can be performed at room temperature (18–21 °C), provided reagents such as the sorting kit and antibodies for cell staining are kept on ice as needed.
10. Keep a small proportion of monocytes for unstained and Live-dead single-stained controls. Use PP tubes to prevent monocytes from sticking to the tubes. In all FACS experiments, unstained controls should be used to identify forward/side scatter parameters of the cell population and to determine background fluorescence. Single-stain controls should be used to adjust laser voltages and correct for spectral overlap. Isotype controls can also be used to assist in gating strategies.
11. Any cytometer can be used for purity assessment or phenotyping of MDMs; however, certain instruments may require the inclusion of CD45 for optimal debris exclusion.
12. Successful monocyte isolation results in a population purity of approximately 80% of total monocytes.
13. Primary cells are sensitive, and monocyte polarization should be performed in an incubator with minimal door openings. The process should be as gentle as possible during the 6 day culture, with particular care given to the first 48 h.
14. For 35 mm Lumox Teflon dishes, as the medium is not replaced during these 6 days, a minimum volume of 1 mL is recommended, and 3 mL for 50 mm dishes. Additionally, do not exceed 2 mL in the 35 mm dishes to avoid the liquid touching the lid upon closure and potentially spilling over.
15. This is an in vitro model that provides informative results under controlled conditions. There is significant inter-donor variability for RA but also for HD, so experiments often need more than 5 donors' inclusion.
16. In terms of cell morphology, a rounded shape does not necessarily indicate dead cells, as flow cytometry analysis shows they are negative for the live/dead viability dye. However, these cells exhibit a relatively immature phenotype, characterized by low to intermediate expression of CD71 and CD11b.
17. For the phenotyping of successful monocyte polarization into macrophages, the overexpression of CD11b and CD71 is

sufficient. For anti-inflammatory macrophages, CD206 and CD163 are also sufficient, but the panel can be further expanded with Folate Receptor Beta for anti-inflammatory markers, and CD80, CD86, or HLA-DR for pro-inflammatory markers. This technique using human AB serum does not produce dendritic cells, as no cells express CD11c.

18. The Incucyte live-cell imaging system can be replaced by classical microscopy for imaging.

Acknowledgments

This work was supported by the Arthritis R&D and the Department of Rheumatology of CHU Kremlin Bicetre, AP-HP. The graphical abstract was created with BioRender.com.

References

1. Parihar A, Eubank TD, Doseff AI (2010) Monocytes and macrophages regulate immunity through dynamic networks of survival and cell death. J Innate Immun 2(3):204–215
2. Gonzalez-Mejia ME, Doseff AI (2009) Regulation of monocytes and macrophages cell fate. Front Biosci 14(7):2413–2431
3. Grage-Griebenow E, Flad HD, Ernst M (2001) Heterogeneity of human peripheral blood monocyte subsets. J Leukoc Biol 69(1): 11–20
4. Ziegler-Heitbrock L (2007) The CD14+ CD16+ blood monocytes: their role in infection and inflammation. J Leukoc Biol 81(3): 584–592
5. Geissmann F, Jung S, Littman DR (2003) Blood monocytes consist of two principal subsets with distinct migratory properties. Immunity 19(1):71–82
6. Geissmann F, Manz MG, Jung S, Sieweke MH, Merad M, Ley K (2010) Development of monocytes, macrophages, and dendritic cells. Science 327(5966):656–661
7. Gordon S, Taylor PR (2005) Monocyte and macrophage heterogeneity. Nat Rev Immunol 5(12):953–964
8. Lendeckel U, Venz S, Wolke C (2022) Macrophages: shapes and functions. ChemTexts 8(2): 12
9. Pidwill GR, Gibson JF, Cole J, Renshaw SA, Foster SJ (2021) The role of macrophages in *Staphylococcus aureus* infection. Front Immunol 19(11):620339
10. Taylor PR, Martinez-Pomares L, Stacey M, Lin HH, Brown GD, Gordon S (2005) Macrophage receptors and immune recognition. Annu Rev Immunol 23:901–944
11. Brown GC (2024) Cell death by phagocytosis. Nat Rev Immunol 24(2):91–102
12. Dash SP, Gupta S, Sarangi PP (2024) Monocytes and macrophages: origin, homing, differentiation, and functionality during inflammation. Heliyon 10(8):e29686
13. Yunna C, Mengru H, Lei W, Weidong C (2020) Macrophage M1/M2 polarization. Eur J Pharmacol 15(877):173090
14. Orecchioni M, Ghosheh Y, Pramod AB, Ley K (2019) Macrophage polarization: different gene signatures in M1(LPS+) vs. classically and M2(LPS-) vs. alternatively activated macrophages. Front Immunol 24(10):1084
15. Shapouri-Moghaddam A, Mohammadian S, Vazini H, Taghadosi M, Esmaeili SA, Mardani F, Seifi B, Mohammadi A, Afshari JT, Sahebkar A (2018) Macrophage plasticity, polarization, and function in health and disease. J Cell Physiol 233(9):6425–6440
16. Locati M, Curtale G, Mantovani A (2020) Diversity, mechanisms, and significance of macrophage plasticity. Annu Rev Pathol 24(15): 123–147
17. Funes SC, Rios M, Escobar-Vera J, Kalergis AM (2018) Implications of macrophage polarization in autoimmunity. Immunology 154(2): 186–195

18. Cassetta L, Pollard JW (2018) Targeting macrophages: therapeutic approaches in cancer. Nat Rev Drug Discov 17(12):887–904
19. Gordon S (2003) Alternative activation of macrophages. Nat Rev Immunol 3(1):23–35
20. Trus E, Basta S, Gee K (2020) Who's in charge here? Macrophage colony stimulating factor and granulocyte macrophage colony stimulating factor: competing factors in macrophage polarization. Cytokine 127:154939
21. Hamilton TA, Zhao C, Pavicic PG Jr, Datta S (2014) Myeloid colony-stimulating factors as regulators of macrophage polarization. Front Immunol 21(5):554
22. Allouch A, David A, Amie SM, Lahouassa H, Chartier L, Margottin-Goguet F, Barré-Sinoussi F, Kim B, Sáez-Cirión A, Pancino G (2013) p21-mediated RNR2 repression restricts HIV-1 replication in macrophages by inhibiting dNTP biosynthesis pathway. Proc Natl Acad Sci USA 110(42):E3997–E4006
23. Paoletti A, Allouch A, Caillet M, Saïdi H, Subra F, Nardacci R, Wu Q, Muradova Z, Voisin L, Raza SQ, Law F, Thoreau M, Dakhli H, Delelis O, Poirier-Beaudouin B, Dereuddre-Bosquet N, Le Grand R, Lambotte O, Saez-Cirion A, Pancino G, Ojcius DM, Solary E, Deutsch E, Piacentini M, Gougeon ML, Kroemer G, Perfettini JL (2019) HIV-1 envelope overcomes NLRP3-mediated inhibition of F-actin polymerization for viral entry. Cell Rep 28(13):3381–3394.e7
24. Allouch A, Di Primio C, Paoletti A, Lê-Bury G, Subra F, Quercioli V, Nardacci R, David A, Saïdi H, Cereseto A, Ojcius DM, Montagnac G, Niedergang F, Pancino G, Saez-Cirion A, Piacentini M, Gougeon ML, Kroemer G, Perfettini JL (2020) SUGT1 controls susceptibility to HIV-1 infection by stabilizing microtubule plus-ends. Cell Death Differ 27(12):3243–3257
25. Paoletti A, Ly B, Bitoun S, Nocturne G, Rivière E, Manson JJ, Matucci A, Pallardy M, De Vries N, Mariette X (2022) Restoration of default blood monocyte-derived macrophage polarization with adalimumab but not Etanercept in rheumatoid arthritis. Front Immunol 23(13):832117
26. Paoletti A, Rohmer J, Ly B, Pascaud J, Rivière E, Seror R, Le Goff B, Nocturne G, Mariette X (2019) Monocyte/macrophage abnormalities specific to rheumatoid arthritis are linked to miR-155 and are differentially modulated by different TNF inhibitors. J Immunol 203(7):1766–1775
27. McInnes IB, Schett G (2017) Pathogenetic insights from the treatment of rheumatoid arthritis. Lancet 389(10086):2328–2337
28. Demoruelle MK, Deane KD (2012) Treatment strategies in early rheumatoid arthritis and prevention of rheumatoid arthritis. Curr Rheumatol Rep 14(5):472–480
29. Aletaha D, Smolen JS (2018) Diagnosis and management of rheumatoid arthritis: a review. JAMA 320(13):1360–1372
30. Burmester GR, Pope JE (2017) Novel treatment strategies in rheumatoid arthritis. Lancet 389(10086):2338–2348
31. Smith MD (2011) The normal synovium. Open Rheumatol J 5:100–106
32. Tu J, Wang X, Gong X, Hong W, Han D, Fang Y, Guo Y, Wei W (2020) Synovial macrophages in rheumatoid arthritis: the past, present, and future. Mediat Inflamm 13(2020): 1583647
33. Alivernini S, MacDonald L, Elmesmari A, Finlay S, Tolusso B, Gigante MR, Petricca L, Di Mario C, Bui L, Perniola S, Attar M, Gessi M, Fedele AL, Chilaka S, Somma D, Sansom SN, Filer A, McSharry C, Millar NL, Kirschner K, Nerviani A, Lewis MJ, Pitzalis C, Clark AR, Ferraccioli G, Udalova I, Buckley CD, Gremese E, McInnes IB, Otto TD, Kurowska-Stolarska M (2020) Distinct synovial tissue macrophage subsets regulate inflammation and remission in rheumatoid arthritis. Nat Med 26(8):1295–1306
34. Kurowska-Stolarska M, Alivernini S (2017) Synovial tissue macrophages: friend or foe? RMD Open 3(2):e000527
35. Bresnihan B, Gerlag DM, Rooney T, Smeets TJ, Wijbrandts CA, Boyle D, Fitzgerald O, Kirkham BW, McInnes IB, Smith M, Ulfgren AK, Veale DJ, Tak PP (2007) Synovial macrophages as a biomarker of response to therapeutic intervention in rheumatoid arthritis: standardization and consistency across centers. J Rheumatol 34(3):620–622
36. Roszkowski L, Ciechomska M (2021) Tuning monocytes and macrophages for personalized therapy and diagnostic challenge in rheumatoid arthritis. Cells 10(8):1860
37. Kawanaka N, Yamamura M, Aita T, Morita Y, Okamoto A, Kawashima M, Iwahashi M, Ueno A, Ohmoto Y, Makino H (2002) CD14+,CD16+ blood monocytes and joint inflammation in rheumatoid arthritis. Arthritis Rheum 46(10):2578–2586

38. Rana AK, Li Y, Dang Q, Yang F (2018) Monocytes in rheumatoid arthritis: Circulating precursors of macrophages and osteoclasts and, their heterogeneity and plasticity role in RA pathogenesis. Int Immunopharmacol 65:348–359

39. Yano R, Yamamura M, Sunahori K, Takasugi K, Yamana J, Kawashima M, Makino H (2007) Recruitment of CD16+ monocytes into synovial tissues is mediated by fractalkine and CX3CR1 in rheumatoid arthritis patients. Acta Med Okayama 61(2):89–98

Chapter 6

Isolation and Culture of Rheumatoid Arthritis Synoviocytes

Mélissa Noack

Abstract

Cell culture is a fundamental and a routine method for research, providing the means to maintain and propagate cells for subsequent experimental studies. In rheumatoid arthritis, stromal cells, particularly synoviocytes, are now recognized as pivotal contributors to disease pathogenesis. Consequently, isolating and culturing these primary cells is of significant importance. This protocol outlines a method for isolating synoviocytes through mechanical adhesion.

Key words Cell culture, Primary cells, Mechanical isolation, Synovial tissue biopsy, Synoviocytes

1 Introduction

Stromal cells play a crucial role in the pathogenesis of various inflammatory diseases, including rheumatoid arthritis (RA). Synoviocytes are the primary stromal cells of the synovium, and their role in RA is now well established [1–3]. Consequently, there is a real interest in studying these cells, with the first step being the culture of synoviocytes. Currently, the synovium remains the primary and traditional source of synoviocytes, which can only be accessed during arthroplasty or arthroscopy surgery. However, advances in RA treatments have greatly benefited patients, leading to a decline in the frequency of surgical procedures. Patient-derived synoviocytes are therefore highly valuable, as they are primary cells—non-immortalized and limited in the number of passages they can undergo. It is therefore important to recover and process surgical biopsies with care, in order to obtain viable synoviocytes for research purposes.

The cell culture technique is a routine and widely used method in research laboratories, spanning fields from cell biology to medical research. Primary cell cultures are established derived directly from tissue fragments, and offer significant advantages, as they remain unmodified and closely resemble the in vivo state compared

Rami Bechara (ed.), *Autoimmunity: Methods and Protocols*, Methods in Molecular Biology, vol. 2990,
https://doi.org/10.1007/978-1-0716-4997-8_6,

to established cell lines. These cells reflect the physiological conditions of human cells, making them invaluable for studying disease mechanisms. Specifically, using synoviocytes directly derived from RA patients provides a unique opportunity to closely mimic in vivo conditions and investigate pathological processes in relevant cells.

Tissue can be homogenized either mechanically (explant culture) or with the aid of proteolytic enzymes. A commonly used method for isolating synoviocytes is the enzymatic approach. Tissue is cut into 1 mm fragments and incubated with collagenase to facilitate the release of synoviocytes. After centrifugation, the cell pellet is cultured, monitored for growth, and characterized before use in experiments. In our laboratory, we have chosen mechanical method (explant culture). Here, we describe the protocol followed, from biopsy receipt to use of cells in experiments.

2 Materials

Sterility is required while performing this protocol.

2.1 Reagents and Consumables

Trypsine 0.5 g/L.

EDTA 0.2 g/L.

Scalpel.

Petri dish.

6-well plates.

Culture flasks.

Staining antibodies and staining buffer.

2.2 Cell Media

Complete medium: Dulbecco's modified Eagle's medium (DMEM) supplemented with 10% fetal bovine serum (FBS), 1% L-glutamine, 2% penicillin/streptomycin, 1% amphotericin B and plasmocin prophylactic (2.5 mg/500 mL of medium).

Stop medium: DMEM supplemented with 10% FBS.

2.3 Instrumentation

1. Vortex.
2. Cell counter.
3. Refrigerated centrifuge.
4. Cell culture incubator, with 5% CO_2, at 37 °C.

3 Methods

3.1 Patients and Tissue Samples

Synovial tissues are obtained from RA patients undergoing joint surgery and who fulfilled the American College of Rheumatology criteria for RA [4]. Biopsies are received in sterile containers on the day of surgery and must be processed as soon as they are received.

3.2 Process

1. Cut the synovial tissue in small pieces of about 3–4 mm (white and hard parts) in PBS in Petri dish (at least 100 mm diameter).
2. Put the pieces in 6-well plates (Fig. 1a).
3. Let the pieces dry to get a good adhesion on the plastic.
4. When the piece is well adhered, add 1 drop of complete medium and incubate at least 1 h at 37 °C.
5. Add 2 mL of complete medium to cover the pieces (Fig. 1b).
6. Place the plate in the incubator and change medium twice per week.

   Observe under the microscope each time the medium is changed.
7. Leave the piece until the adherent cells are well visible (Fig. 1c).

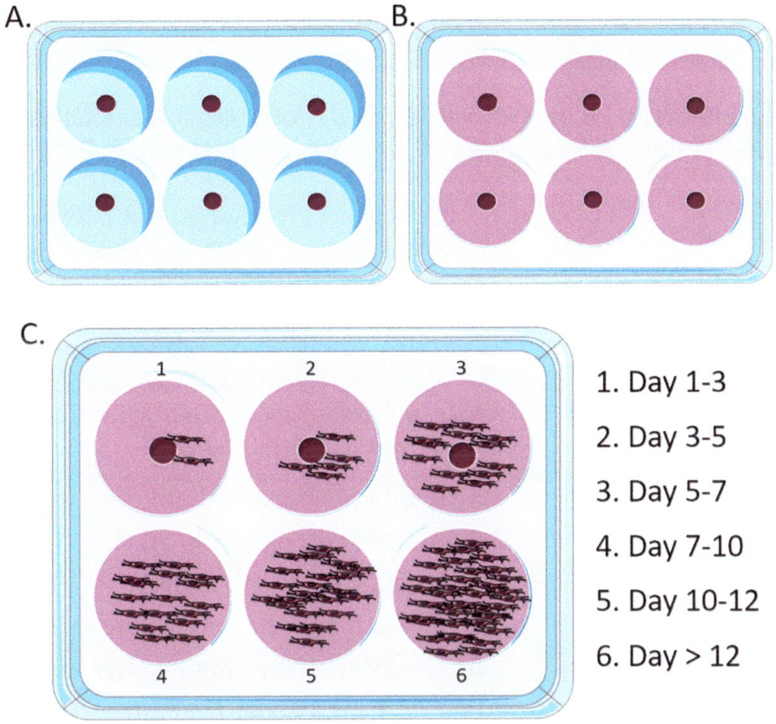

Fig. 1 Isolation of synoviocytes by adhesion. (**a**) Small pieces about 3 mm of synovial tissue are adhered into 6-well plates. (**b**) Complete medium is added to cover the pieces. (**c**) Cells begin to exit from biopsy (1–2) and grow (3). The piece of synovial tissue can be removed when cells are well visible (4). Cells continue to proliferate (5) and can be transferred to culture flasks when they have reached confluence (6)

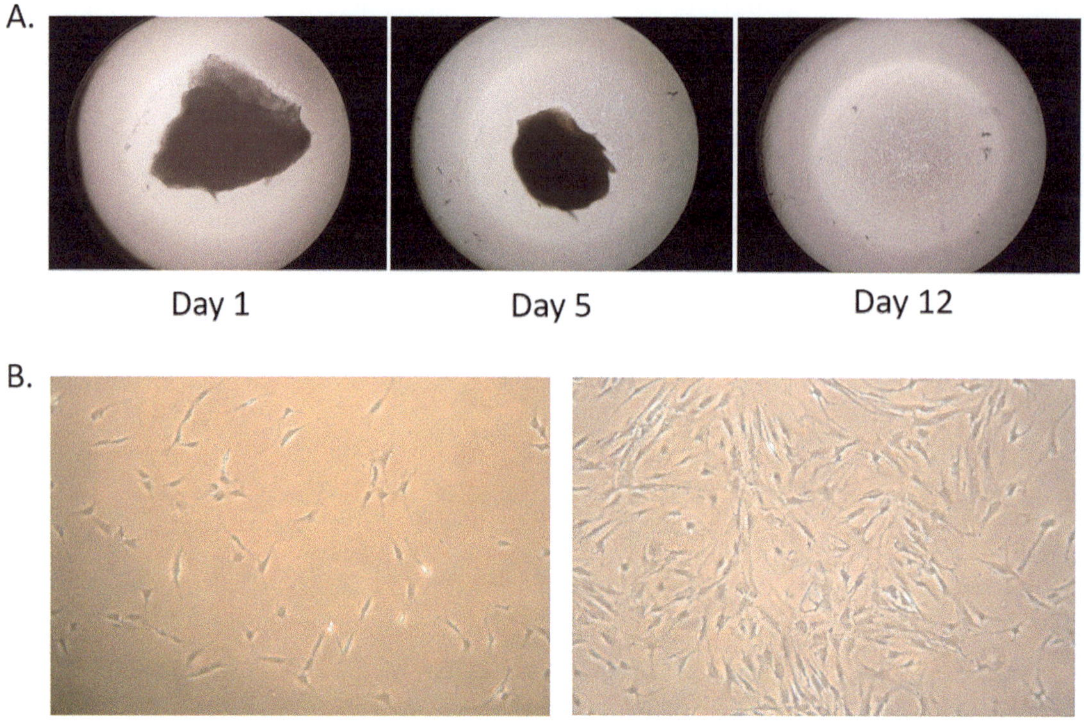

Fig. 2 Pictures of pieces of synovial tissue and synoviocytes in culture. (**a**) Pictures of a piece of synovial tissue biopsy in culture in 6-well plate at different days after treatment. (**b**) Pictures of synoviocytes in culture

Cells generally begin to exit between day 1 and day 3. Between the days 3 and 7, cells become clearly visible and begin to proliferate (*see* **Note 1**). The piece can then be removed (*see* **Note 2**). The cells then continue to proliferate until they reach confluence, at which point they can be recovered for culture (Fig. 2).

3.3 Culture of Cells

Once the cells have reached confluence in the wells, they can be transferred to culture flasks.

1. Remove culture medium from wells.
2. Add 500 μL of trypsin-EDTA.
3. Place plate 10 min at 37 °C.
4. Check under the microscope that cells are detached.
5. Gently tap the plate to ensure that the cells are well detached.
6. Add 1 mL of stop medium.
7. Recover cells in a 15 mL tube.
8. Centrifuge 10 min at 450 g.
9. Empty supernatant after checking that cell pellet is visible.
10. Resuspend cells in 1 mL of complete medium.

11. Count the cells and distribute approx. 200,000 cells per flask 75.
12. Place flasks in incubator, 5% CO_2, at 37 °C.
13. Change medium twice a week.
14. When cells are confluent, pass to 1/2 or 1/3 (same protocol as for recovering cells from 6-well plates, with 5 mL of trypsin-EDTA and 10 mL of stop medium).

3.4 Checking of Synoviocytes

3.4.1 Contaminations

The quality of research results depends on the health of cell cultures, and with this in mind, the manipulator must be vigilant in the face of various contaminants. Bacterial, fungal (mushrooms and molds), and yeast contamination is usually visible to the naked eye, as evidenced by the sudden appearance of cloudiness and color change in the culture medium. The presence of penicillin/streptomycin and amphotericin B in the complete medium is usually sufficient to prevent this kind of contamination. Mycoplasma contamination, on the other hand, is invisible to the eye. Mycoplasma is the most devastating and widespread of all contaminants. Infection reduces cell productivity, growth rate, and viability. Although they do not generally kill the mammalian cells they infect, mycoplasmas can seriously disrupt cultures by altering cell metabolism, reducing cell productivity and growth rate. In short, they can profoundly affect the results of most experiments carried out with contaminated cells. It is therefore crucial to regularly test cell cultures for mycoplasma. If the cells are tested positive, they are treated according to the manufacturer's recommendations.

3.4.2 Staining of Synoviocytes

In the third passage, the cells are stained to check synoviocyte purity. Synoviocytes are stromal cells and as defined, they must express CD105, CD7, and CD90, and lack the expression of CD45 and CD34 [5].

- Perform a surface staining for: CD45, CD34, CD73, CD90, and CD105. Synoviocytes have to be >99% negative for CD45 and CD34, and positive for CD73, CD90, and CD105.
- Suspend cells in 100 μL of buffer (50% complete medium/50% PBS).
- Add the antibodies (according to the manufacturer's recommendations) and incubate for at least 30 min in the dark at 4 °C.
- Wash the cells by adding buffer, 500 μL per tubes.
- Centrifuge at 450 g at 4 °C for 5 min.
- Discard the supernatant and repeat the wash.
- Resuspend cells in 300–500 μL of buffer.
- Acquire data on the flow cytometer.

3.5 Use of Synoviocytes

Once synoviocytes have tested negative for mycoplasma and are good at marker levels, they can be used in experiments. As they are not cell lines but primary cells, synoviocytes are not immortal and should be used between passages 4 and 9.

4 Notes

1. Days indicated are for guidance only and may vary from one donor to another. Cells from some biopsies may exit in 1 day, others in 10, and others not at all. There can also be variability between wells on the same plate, which is why it is important to examine all the wells each time medium is changed.
2. If the cells exit quickly and the piece is still in good condition, it can be placed into a new plate to give more cells.

References

1. Bartok B, Firestein GS (2010) Fibroblast-like synoviocytes: key effector cells in rheumatoid arthritis. Immunol Rev 233:233–255
2. Dakin SG, Coles M, Sherlock JP, Powrie F, Carr AJ, Buckley CD (2018) Pathogenic stromal cells as therapeutic targets in joint inflammation. Nat Rev Rheumatol 14:714–726
3. Nygaard G, Firestein GS (2020) Restoring synovial homeostasis in rheumatoid arthritis by targeting fibroblast-like synoviocytes. Nat Rev Rheumatol 16:316–333
4. Aletaha D, Neogi T, Silman AJ, Funovits J, Felson DT, Bingham CO 3rd, Birnbaum NS, Burmester GR, Bykerk VP, Cohen MD et al (2010) 2010 Rheumatoid arthritis classification criteria: an American College of Rheumatology/European League Against Rheumatism collaborative initiative. Arthritis Rheum 62:2569–2581
5. Dominici M, Le Blanc K, Mueller I, Slaper-Cortenbach I, Marini F, Krause D, Deans R, Keating A, Prockop D, Horwitz E (2006) Minimal criteria for defining multipotent mesenchymal stromal cells. The International Society for Cellular Therapy position statement. Cytotherapy 8:315–317

Chapter 7

Comprehensive Assessment of Neutrophil Function in Autoimmune Diseases

Quiterie Labaye, Daouda Abba Moussa, and Mar Naranjo-Gomez

Abstract

Neutrophils play a crucial role in autoimmune diseases beyond their well-characterized effector functions like phagocytosis and NETosis. Comprehensive assessment of neutrophil function is essential for understanding their diverse contributions to autoimmunity, including their roles in cytokine production and reactive oxygen species (ROS) generation. This chapter provides detailed methodologies for evaluating neutrophil immunomodulatory functions relevant to autoimmune pathology. We describe assays for quantifying cytokine and chemokine profiling by multiplex approaches and ROS production using flow cytometry. These methods enable researchers to dissect the nuanced contributions of neutrophils to autoimmune diseases, facilitating the identification of novel therapeutic targets.

Key words Neutrophils, Cell purification, Activation, Cytokine bead array, Multiparametric analyses, ROS production

1 Introduction

Neutrophils are essential innate immune cells that serve as first responders to infection [1–4] and inflammation [5, 6]. Their role in autoimmunity extends beyond their classical antimicrobial functions, with increasing evidence linking neutrophil dysfunction to disease pathogenesis [7]. While neutrophil extracellular trap (NET) formation (NETosis) has been extensively studied [8–12], a broader functional assessment is necessary to fully understand their contributions to autoimmune diseases.

Neutrophils exhibit a diverse range of effector functions, including phagocytosis, production of reactive oxygen species (ROS), degranulation, and cytokine secretion. Dysregulation in these processes has been implicated in diseases such as systemic lupus erythematosus (SLE) [13–15], rheumatoid arthritis (RA) [16–20], and vasculitis [21, 22], where neutrophils contribute to tissue damage, chronic inflammation, and immune

Rami Bechara (ed.), *Autoimmunity: Methods and Protocols*, Methods in Molecular Biology, vol. 2990,
https://doi.org/10.1007/978-1-0716-4997-8_7,

dysregulation [23]. As such, detailed functional assays are critical for identifying pathogenic neutrophil phenotypes and potential therapeutic targets.

Here, we outline flow cytometry-based methods for assessing cytokine and chemokine secretion and reactive oxygen species (ROS) production in neutrophils. These assays can provide a detailed functional profile of neutrophils in both health and disease. By leveraging these techniques, researchers can gain deeper insights into neutrophil-driven mechanisms in autoimmunity, facilitating the discovery of novel diagnostic biomarkers and therapeutic targets.

2 Materials

2.1 Density Gradient

The density gradient separation was performed using two solutions with defined densities to separate mononuclear cells based on buoyant density.

- Low-density medium ($\rho \approx 1.077$ g/mL) or Histopaque 1077: This medium consists of a polysucrose and sodium diatrizoate solution, typically used to isolate mononuclear cells such as lymphocytes and monocytes from whole blood. The osmolarity and pH are adjusted to physiological conditions (typically ~290–300 mOsm/kg and pH 7.0–7.2), allowing for optimal cell viability during centrifugation.
- High-density medium ($\rho \approx 1.119$ g/mL) or Histopaque 1119: This solution is similar in composition, containing polysucrose and sodium diatrizoate, but at a higher concentration, resulting in a greater density. It is used to allow better resolution between mononuclear cells and polymorphonuclear cells, such as granulocytes.
- Cell culture plates: Cells were cultured in sterile 96-well round-bottom (U-bottom) tissue culture plates, which facilitate cell aggregation and are commonly used for suspension cell cultures and stimulation assays.

2.2 Purification

Cell selection buffer (PBS/EDTA/FBS): For magnetic cell separation, a buffer composed of phosphate-buffered saline (1×), 0.2 mM EDTA, and 2% heat-inactivated fetal bovine serum (FBS) is used. The EDTA prevents cell clumping by chelating divalent cations that mediate cell adhesion, while the heat-inactivated FBS provides protein to reduce nonspecific binding and protect cell membranes during processing.

Magnetic bead-based selection system targeting CD15+ cells: Neutrophils are isolated using antibody-conjugated magnetic beads directed against the CD15 surface antigen. Depending on the protocol, either positive or negative selection strategies are employed: In positive selection, CD15+ cells are retained by the magnetic column, whereas in negative selection, all non-CD15+ cells are removed, enriching the flow-through for the desired population.

2.3 Cell Culture

Complete culture medium (R10): Cells were maintained in RPMI 1640 medium supplemented with 10% heat-inactivated fetal bovine serum (FBS) to provide essential nutrients and growth factors and 1% penicillin-streptomycin to prevent contamination.

Cell culture reagents. Granulocyte colony-stimulating factor (G-CSF): A recombinant cytokine used to promote the survival of neutrophils. G-CSF was diluted from a 50 μg/mL stock to a final concentration of 10 ng/mL in culture. Lipopolysaccharide (LPS): A bacterial endotoxin used to stimulate an innate immune response, particularly via Toll-like receptor 4 (TLR4). LPS was prepared from a 10 mg/mL stock concentration and added to cultures at a final concentration of 0.5 μg/mL. Tumor necrosis factor-alpha (TNF-α): A pro-inflammatory cytokine used to activate immune cells in vitro. TNF-α was used at a final concentration of 1 ng/mL, diluted from a 1 mg/mL stock solution.

Cell culture plates. Cells were cultured in sterile 96-well round-bottom (U-bottom) tissue culture plates, which facilitate cell aggregation and are commonly used for suspension cell cultures and stimulation assays.

2.4 Mitochondrial and Reactive Oxygen Species (ROS) Probes

Mitochondrial membrane potential dye: A far-red fluorescent probe selective for actively respiring mitochondria was used to assess mitochondrial mass and potential. The dye was diluted from a 1 mM stock to a working concentration of 50 nM and incubated with cells according to the manufacturer's protocol.

Mitochondrial superoxide indicator: A mitochondrial-targeted fluorogenic compound, sensitive to superoxide anions, was used to measure mitochondrial ROS production. The dye was prepared from a 5 mM stock and applied to cells at a final concentration of 5 μM.

General ROS detection assay: A cell-permeable fluorogenic substrate reactive to a broad range of reactive oxygen species was used to measure total intracellular oxidative stress. Cells were stained using a commercial kit following the manufacturer's recommended protocol.

3 Methods

This protocol is described for one blood tube of 5 mL EDTA tube but can be adapted for larger volumes by maintaining the appropriate proportions. All procedures must be carried out at room temperature unless otherwise specified.

3.1 Plasma Collection

1. Draw whole blood (WB) in a 15 mL Falcon tube (F15).
2. Transfer a 50 μL aliquot of WB into a FACS tube and set aside for later analysis.
3. Centrifuge the F15 tube at 1500 rpm for 8 min at room temperature (RT).
4. Carefully collect the supernatant fraction (corresponding to the plasma), without disturbing the remaining blood.
5. Transfer the plasma to a new sterile tube.
6. Use fresh plasma immediately or store it appropriately at −20 ° C (short term) or −80 °C (long term) for future analysis.
7. If freezing, aliquot plasma into small volumes to avoid repeated freeze-thaw cycles.
8. Add PBS (1×) to restore the original volume, then gently mix to homogenize.

3.2 Sequential Double-Density Gradient Separation (See Note 1)

1. Dilute WB with PBS (1×) to achieve a final volume of 12 mL (corresponding to at least 1:3 dilution).
2. In a 50 mL Falcon tube (F50), carefully layer 18 mL of Histopaque-1077 (H1077).
3. Gently overlay the 12 mL diluted blood onto the H1077 layer, ensuring minimal mixing.
4. Centrifuge at 800× *g* for 20 min at RT (22–24 °C) with acceleration 6 and brake 2.
5. Carefully remove the upper plasma layer without disturbing the interface.
6. Collect the Peripheral Blood Mononuclear Cells (PBMC) fraction into F15
7. Add PES to a final volume of 14 mL and centrifuge at 800× *g* for 5 min at RT.
8. Discard the supernatant and resuspend the pellet in PES for further analysis (e.g., characterization of low-density granulocytes (LDG)).
9. Discard the remaining H1077 fraction.
10. Resuspend the erythrocytes and granulocyte (E + G) fraction in PBS to a total volume of 12 mL.
11. In a new 50 mL Falcon tube (F50), carefully layer 18 mL of Histopaque-1119 (H1119).
12. Gently overlay the 12 mL of E + G fraction onto the H119 layer, ensuring minimal mixing.

13. Centrifugate at 800× *g* for 20 mn at RT (24 °C) with acceleration 6 and brake 2.
14. Carefully remove and discard the supernatant.
15. Collect the granulocyte fraction in two F15 tubes.
16. Add PES to a final volume of 14 mL and centrifugate at 800× *g* for 8 mn at RT with acceleration 9 and brake 9.
17. Discard the supernatant.
18. Resuspend the granulocyte pellet in a single final volume of 500 μL PES.

3.3 Positive Selection of Human CD15+ Cells (See Note 2)

1. Add the antibody panel to the cell suspension and incubate for 10 min at RT.
2. Add the selection beads and incubate for an additional 10 min at RT.
3. Adjust the volume to 6 mL with PES and place the tube on the magnet for 5 min.
4. Carefully discard the supernatant without disturbing the bead-bound cells.
5. Remove the tube from the magnet and add 6.5 mL of PES.
6. Put it back on the magnet for 5 min.
7. Discard the supernatant.
8. Remove the tube from the magnet and add 7 mL of PES.
9. Gently resuspend the cell pellet.
10. Return the tube to the magnet for 5 min.
11. Discard the supernatant.
12. Remove the tube from the magnet.
13. Add 8 mL of PES to wash the cells.
14. Centrifuge at 800× *g* for 8 mn at RT with acceleration 9 and brake 9.
15. Discard the supernatant.
16. Resuspend the neutrophils pellet in a final volume of 500 μL R10 for further applications.

3.4 Neutrophil Culture and Stimulation Protocol

1. Count cells using a cell counting chamber slides.
2. Adjust the concentration of purified neutrophils to 3×10^5 cells in 75 μL with R10.
3. Add G-CSF to achieve a final concentration of 10 ng/mL (add 20 ng/mL for a 2× stock solution).
4. Plate the resuspended neutrophils (75 μL) in 96-well U-bottom plates.

5. Prepare the conditioning or stimulating medium as follows: LPS at 1 μg/mL and TNF-a at 2 ng/mL.
6. Add 75 μL of the stimulating medium to each well containing neutrophils.

Purified neutrophils are plated at a concentration of 2 million cells/mL in 96-well U-bottom plates, with each well containing 3×10^5 cells in 150 μL. Cells are cultured in the presence of G-CSF (10 ng/mL) and either LPS (0.5 μg/mL) or TNF-α (1 ng/mL) under stimulating conditions. As a control, 75 μL of R10 medium is used. Incubation is carried out at 37 °C in a 5% CO_2 environment.

3.5 Neutrophil Characterization

3.5.1 Observation by Microscope

Neutrophil morphology and activation state are assessed through microscopic observation. Cells can be examined using light microscopy to determine characteristics such as cell shape, size, and any signs of activation (Fig. 1).

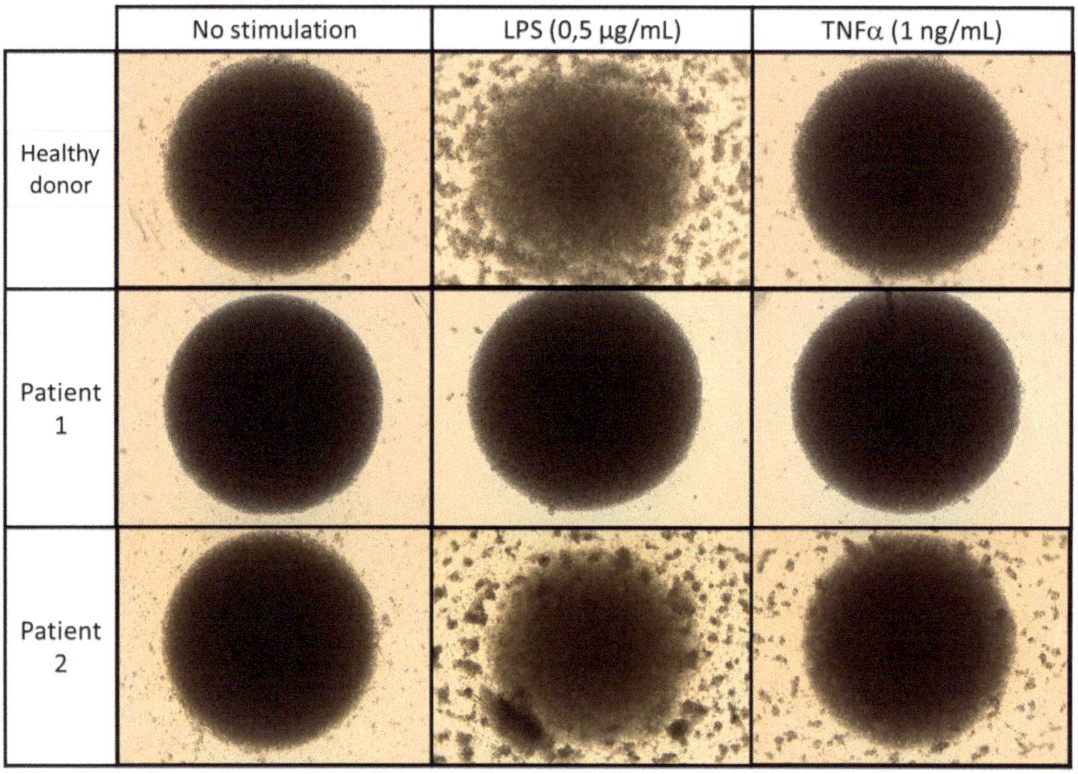

Fig. 1 Image captured using an Olympus CKX53 inverted microscope, equipped with a 4× objective lens and a 10× lens zoom, resulting in a total magnification of 40×. The image was captured with an Olympus EP50 camera for optimal clarity and detail. Shown are purified neutrophils from a healthy donor and two different RA patients after 20 h of culture under stimulating conditions (LPS or TNF-α), compared to the healthy donor

3.5.2 Collection of Supernatant for Cytokine and Chemokine Evaluation

After neutrophil stimulation, the supernatant is collected for further analysis of cytokines and chemokines. These soluble mediators, which include pro-inflammatory cytokines like TNF-α and IL-1β, are measured using techniques like cytokine arrays to assess the neutrophil's secretory response. The supernatant can also be analyzed for the presence of chemokines such as CXCL8 (IL-8) to determine chemotactic signaling involved in neutrophil recruitment and activation.

3.5.3 Phenotype and ROS Characterization (See ***Note 3****)*

1. Flow cytometry is used to characterize the neutrophil phenotype (CD15), assessing the expression of activation markers such as CD11b, CD66b, and CD16 (Table 1).

   These markers help determine the activation status and functional capacity of neutrophils (Fig. 2).

2. The ROS production of neutrophils is evaluated using specific fluorescent probes. For mitochondrial ROS, MitoTracker and MitoSox can be employed, which specifically stain mitochondria and measure mitochondrial oxidative stress. For total ROS measurement, a ROS total kit (such as DCFDA or other general ROS probes) can be used to evaluate the overall ROS production in activated neutrophils. These assays help determine the oxidative burst, a critical function of neutrophils in pathogen defense and immune response. Incubate the cells for 30 min at 37 °C with 5% CO_2 in the incubator (final concentration, *see* Table 2). Thoroughly wash the cells with PES.

4 Notes

1. Double density gradient process typically involves the use of multiple gradients to achieve the desired separation of components based on their density. However, it is noteworthy that this approach can be significantly simplified under certain

Table 1
Flow cytometry analysis of neutrophil phenotype and activation markers

Marker	Fluorescence	Clone	Final concentration
CD45	BV421	HI30	2:1000
CD15	PerCP Cy5.5	HI98	2:1000
CD16	PE Cy7	3G8	0.5:1000
CD11b	AF488	ICRF44	4:1000
CD66b	AF647	G10F5	2:1000

Neutrophils were identified based on CD15 expression and further characterized by the surface expression of activation markers CD11b, CD66b, and CD16

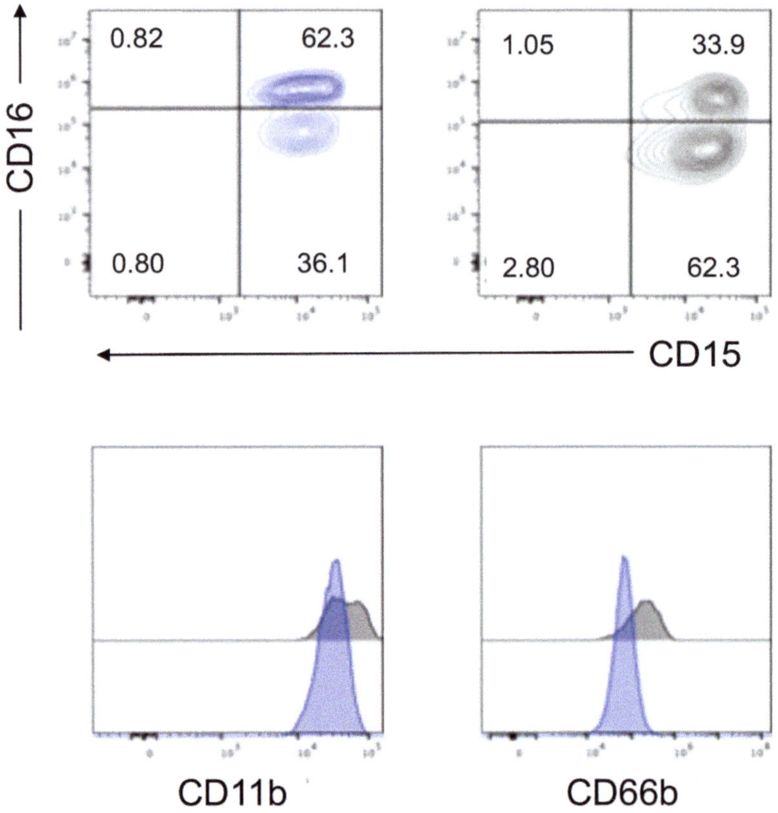

Fig. 2 Flow cytometry plots showing the analysis of cell populations. The singlet gate was applied to exclude doublets and debris. The plots display the staining of CD15 and CD16 markers on the cell populations. The CD15 staining is shown on the x-axis, and CD16 staining is shown on the y-axis, allowing for the identification and characterization of the respective cell subsets based on their expression of these surface markers (CD11b and CD66b) before (blue) and after (black) 20 h stimulation

Table 2
Evaluation of neutrophil ROS production using fluorescent probes

Marker	Fluorescence	Stock	Final concentration
MitoTracker DeepRed	APC	1 mM	50 nM
MitoSox (ROS mitochondria)	PE	5 mM	5 μM
Total ROS assay kit	GFP	500×	1×

MitoTracker and MitoSOX were used to stain mitochondria and assess mitochondrial ROS levels

conditions. Specifically, when low-density gradient (LDG) layers are not a consideration, the procedure can be streamlined by utilizing H1119 directly. This modification reduces the complexity of the gradient system, making it more efficient without compromising the resolution of separation.

2. Neutrophil Purification: Positive vs. Negative Selection. In the purification of neutrophils, the method outlined here primarily describes the use of positive selection techniques. This approach is widely utilized to isolate neutrophils by targeting specific surface markers, ensuring a high level of purity in the final cell population. However, it is important to note that a more recent alternative has emerged—negative selection. This "untouched" method involves the depletion of non-neutrophil cells, thereby preserving neutrophils without the need for surface marker binding. Negative selection has gained attention for its ability to maintain the integrity and functionality of neutrophils, minimizing potential activation or alterations during the isolation process. Both methods have their advantages depending on the experimental goals, and researchers should consider the specific needs of their study when choosing between positive and negative selection.
3. ROS staining and mitochondrial probe. In this protocol, we specify a concentration for TMRE and MitoSOX probes; however, it is important to emphasize that this concentration may need to be adjusted based on the specific flow cytometer used. Variability in instrument sensitivity, laser power, and detector settings can influence fluorescence intensity and optimal probe performance. Therefore, users should perform preliminary titration experiments to determine the most suitable concentration for their cytometer, ensuring accurate and reproducible results while avoiding signal saturation or excessive background fluorescence.

Acknowledgment

We acknowledge the support of Immun4Cure IHU "Institute for Innovative Immunotherapies in Autoimmune Diseases" (France 2030/ANR-23-IHUA-0009). Quiterie Labaye was a recipient of an IHU PhD fellowship.
This work was supported by grants from ARTHRITIS R&D and Foundation ARTHRITIS (#81698), Société Française de Rhumatologie (SFR#3394 and #6188) and CHU Montpellier (AO Tremplin NeutroARTherapy) and INSERM state funding granted to M. Naranjo-Gomez (U1183NAR). This work was also supported by the consortium ACT4COVIDCellnex (220110FF, RAK200019FFA). Daouda Abba Moussa and M. Naranjo-Gomez are members of the "MabImprove Labex," a public grant overseen by the French National Research Agency (ANR) as part of the

"Investments for the future" program (reference: ANR-10-LABX-53-01) and also supported this work.

References

1. Amulic B, Cazalet C, Hayes GL, Metzler KD, Zychlinsky A (2012) Neutrophil function: from mechanisms to disease. Annu Rev Immunol 30:459–489. https://doi.org/10.1146/annurev-immunol-020711-074942
2. Doz-Deblauwe E, Bounab B, Carreras F, Fahel JS, Oliveira SC, Lamkanfi M, Le Vern Y, Germon P, Pichon J, Kempf F, Paget C, Remot A, Winter N (2024) Dual neutrophil subsets exacerbate or suppress inflammation in tuberculosis via IL-1β or PD-L1. Life Sci Alliance 7(7):e202402623
3. Lemaitre J, Desjardins D, Gallouët AS, Gomez-Pacheco M, Bourgeois C, Favier B, Sáez-Cirión A, Le Grand R, Lambotte O (2022) Expansion of immature neutrophils during SIV infection is associated with their capacity to modulate T-cell function. Front Immunol 13:781356. https://doi.org/10.3389/fimmu.2022.781356
4. Marsile-Medun S, Souchard M, Abba Moussa D, Reynaud É, Tuaillon E, Naranjo-Gomez M, Pelegrin M (2024) Fc receptors are key discriminatory markers of granulocytes subsets in people living with HIV-1. Front Immunol 15:1345422. https://doi.org/10.3389/fimmu.2024.1345422
5. Crainiciuc G, Palomino-Segura M, Molina-Moreno M, Sicilia J, Aragones DG, Li JLY, Madurga R, Adrover JM, Aroca-Crevillén A, Martin-Salamanca S, Del Valle AS, Castillo SD, Welch HCE, Soehnlein O, Graupera M, Sánchez-Cabo F, Zarbock A, Smithgall TE, Di Pilato M, Mempel TR, Tharaux PL, González SF, Ayuso-Sacido A, Ng LG, Calvo GF, González-Díaz I, Díaz-de-María F, Hidalgo A (2022) Behavioural immune landscapes of inflammation. Nature 601(7893):415–421. https://doi.org/10.1038/s41586-021-04263-y. Epub 2022 Jan 5
6. Aroca-Crevillén A, Vicanolo T, Ovadia S, Hidalgo A (2024) Neutrophils in physiology and pathology. Annu Rev Pathol 19:227–259. https://doi.org/10.1146/annurev-pathmechdis-051222-015009
7. Soehnlein O, Steffens S, Hidalgo A, Weber C (2017) Neutrophils as protagonists and targets in chronic inflammation. Nat Rev Immunol 17(4):248–261. https://doi.org/10.1038/nri.2017.10. Epub 2017 Mar 13
8. Hidalgo A, Libby P, Soehnlein O, Aramburu IV, Papayannopoulos V, Silvestre-Roig C (2022) Neutrophil extracellular traps: from physiology to pathology. Cardiovasc Res 118(13):2737–2753. https://doi.org/10.1093/cvr/cvab329
9. Manoj H, Gomes SM, Thimmappa PY, Nagareddy PR, Jamora C, Joshi MB (2025) Cytokine signalling in formation of neutrophil extracellular traps: implications for health and diseases. Cytokine Growth Factor Rev 81:27–39. https://doi.org/10.1016/j.cytogfr.2024.12.001. Epub 2024 Dec 9
10. Wise AD, TenBarge EG, Mendonça JDC, Mennen EC, McDaniel SR, Reber CP, Holder BE, Bunch ML, Belevska E, Marshall MG, Vaccaro NM, Blakely CR, Wellawa DH, Ferris J, Sheldon JR, Bieber JD, Johnson JG, Burcham LR, Monteith AJ (2025) Mitochondria sense bacterial lactate and drive release of neutrophil extracellular traps. Cell Host Microbe 33(3):341–357.e9
11. Wang Y, Yang Q, Dong Y, Wang L, Zhang Z, Niu R, Wang Y, Bi Y, Liu G (2025) Piezo1-directed neutrophil extracellular traps regulate macrophage differentiation during influenza virus infection. Cell Death Dis 16(1):60. https://doi.org/10.1038/s41419-025-07395-5
12. Ackermann M, Anders HJ, Bilyy R, Bowlin GL, Daniel C, De Lorenzo R, Egeblad M, Henneck T, Hidalgo A, Hoffmann M, Hohberger B, Kanthi Y, Kaplan MJ, Knight JS, Knopf J, Kolaczkowska E, Kubes P, Leppkes M, Mahajan A, Manfredi AA, Maueröder C, Maugeri N, Mitroulis I, Muñoz LE, Narasaraju T, Naschberger E, Neeli I, Ng LG, Radic MZ, Ritis K, Rovere-Querini P, Schapher M, Schauer C, Simon HU, Singh J, Skendros P, Stark K, Stürzl M, van der Vlag J, Vandenabeele P, Vitkov L, von Köckritz-Blickwede M, Yanginlar C, Yousefi S, Zarbock A, Schett G, Herrmann M (2021) Patients with COVID-19: in the dark-NETs of neutrophils. Cell Death Differ 28(11):3125–3139. https://doi.org/10.1038/s41418-021-00805-z. Epub 2021 May 24
13. Mistry P, Nakabo S, O'Neil L, Goel RR, Jiang K, Carmona-Rivera C, Gupta S, Chan DW, Carlucci PM, Wang X, Naz F, Manna Z, Dey A, Mehta NN, Hasni S, Dell'Orso S, Gutierrez-Cruz G, Sun HW, Kaplan MJ (2019) Transcriptomic, epigenetic, and functional analyses implicate neutrophil diversity in

the pathogenesis of systemic lupus erythematosus. Proc Natl Acad Sci USA 116(50): 25222–25228. https://doi.org/10.1073/pnas.1908576116. Epub 2019 Nov 21

14. Bashant KR, Aponte AM, Randazzo D, Rezvan Sangsari P, Wood AJ, Bibby JA, West EE, Vassallo A, Manna ZG, Playford MP, Jordan N, Hasni S, Gucek M, Kemper C, Conway Morris A, Morgan NY, Toepfner N, Guck J, Mehta NN, Chilvers ER, Summers C, Kaplan MJ (2021) Proteomic, biomechanical and functional analyses define neutrophil heterogeneity in systemic lupus erythematosus. Ann Rheum Dis 80(2):209–218. https://doi.org/10.1136/annrheumdis-2020-218338. Epub 2020 Sep 28
15. Henning S, Reimers T, Abdulahad W, Fierro JJ, Doornbos-van der Meer B, Bootsma H, Horvath B, de Leeuw K, Westra J (2025) Low-density granulocytes and neutrophil extracellular trap formation are increased in incomplete systemic lupus erythematosus. Rheumatology (Oxford) 64(3):1234–1242. https://doi.org/10.1093/rheumatology/keae300
16. Fresneda Alarcon M, McLaren Z, Wright HL (2021) Neutrophils in the pathogenesis of rheumatoid arthritis and systemic lupus erythematosus: same foe different M.O. Front Immunol 12:649693. https://doi.org/10.3389/fimmu.2021.649693. eCollection 2021. PMID: 33746988
17. Karmakar U, Vermeren S (2021) Crosstalk between B cells and neutrophils in rheumatoid arthritis. Immunology 164(4):689–700. https://doi.org/10.1111/imm.13412. Epub 2021 Sep 8
18. O'Neil LJ, Kaplan MJ (2019) Neutrophils in rheumatoid arthritis: breaking immune tolerance and fueling disease. Trends Mol Med 25(3):215–227. https://doi.org/10.1016/j.molmed.2018.12.008
19. Wright HL, Moots RJ, Edwards SW (2014) The multifactorial role of neutrophils in rheumatoid arthritis. Nat Rev Rheumatol 10(10): 593–601
20. Ribon M, Seninet S, Mussard J, Sebbag M, Clavel C, Serre G, Boissier MC, Semerano L, Decker P (2019) Neutrophil extracellular traps exert both pro- and anti-inflammatory actions in rheumatoid arthritis that are modulated by C1q and LL-37. J Autoimmun 98:122–131. https://doi.org/10.1016/j.jaut.2019.01.003
21. Aymonnier K, Amsler J, Lamprecht P, Salama A, Witko-Sarsat V (2023) The neutrophil: a key resourceful agent in immune-mediated vasculitis. Immunol Rev 314(1): 326–356. https://doi.org/10.1111/imr.13170. Epub 2022 Nov 21
22. Arnold S, Kitching AR, Witko-Sarsat V, Wiech T, Specks U, Klapa S, Comdühr S, Stähle A, Müller A, Lamprecht P (2024 May) Myeloperoxidase-specific antineutrophil cytoplasmic antibody-associated vasculitis. Lancet Rheumatol 6(5):e300–e313
23. Thieblemont N, Wright HL, Edwards SW, Witko-Sarsat V (2016) Human neutrophils in auto-immunity. Semin Immunol 28(2): 159–173. https://doi.org/10.1016/j.smim.2016.03.004. Epub 2016 Mar 29

Chapter 8

Neutrophil Transmigration Analyses for Multiplexed Drug Screening and In Vivo Like Application Testing Using an Organ on a Chip Model

Clara M. Reichardt and Luis E. Muñoz

Abstract

Neutrophils are the most abundant type of innate immune cell in humans and make up for 40–60% of leukocytes in humans. As a part of the innate immunity and due to their high frequency, they form the first line of defense against harmful invaders. While neutrophils are crucial for combating infections, dysregulated neutrophil activity is a hallmark of chronic inflammation and autoimmune disorders. Here, we describe two methods to evaluate neutrophil migration, an in vitro transwell migration assay and an ex vivo organ-on-chip model, which allows quantification of the migration of neutrophils under flow and through human umbilical vein endothelial cells to best mimic conditions found in the human body.

Key words Organ-on-a-chip, Neutrophil, Migration, Perfusion, Flow, HUVECs, Transwell

1 Introduction

Neutrophils are short-lived leukocytes with multiple specific effector functions that contribute to the pathogenesis of many chronic and acute diseases [1]. Studies on neutrophil response to pathogens, inflammatory signaling, or other stimuli, however, often miss a very important step ahead of the neutrophil-mediated immunity, migration [2]. Transwell migration assays have been published before; however, the quantification of the transmigrated neutrophils remains challenging. Among other reasons, the short life span and the potential release of neutrophil extracellular traps, which result in the death of the cell, can be challenging [3]. Well-established classic transwell migration assays as we describe it in this chapter enable the screening of the transmigration and chemotactic activities against various stimuli in a very fast and easy manner. Neutrophils are in constant movement throughout the organism patrolling tissues and must pass through capillary endothelial cell layers [4]. Two very important environmental factors, blood flow

Rami Bechara (ed.), *Autoimmunity: Methods and Protocols*, Methods in Molecular Biology, vol. 2990,
https://doi.org/10.1007/978-1-0716-4997-8_8,

and capillary endothelial cells, are missed in classic transwell settings. Organ-on-a-chip models with two channels separated by a membrane offer the opportunity to mimic vasculature in one channel with or without an organ tissue on the other side of the membrane. As in the human body, neutrophils must perform transendothelial migration, out of the flow, towards stimuli released from the organ, thus mimicking in vivo conditions [5, 6]. Together, both assay systems offer the ability to pre-select stimuli with the classic transwell migration assay and test them under in vivo-like conditions with the organ-on-a-chip transmigration model.

2 Materials

2.1 Neutrophil Isolation

1. Aqua ad injectabilia.
2. Dulbecco's phosphate buffer saline (DPBS).
3. PBS (10×): 80 g NaCl, 14.4 g Na_2HPO_4, 2 g KCl, 2 g KH_2PO_4, fill up to 1000 mL dH_2O, pH 7.5, filter sterile.
4. EDTA or heparin blood collection tube and blood draw equipment.
5. Density centrifugation medium, density: 1.077 g/mL.

2.2 Transwell Migration Assay

1. 5 mM CFSE.
2. Transwell plate and receiver plate, 96-well, pore size: 5 μm, membrane diameter: 4.26 mm, polycarbonate.
3. Fluorescence plate reader (capable to measure fluorescence of CFSE with excitation: 455 nm and emission: 520 nm).

2.3 Organ on a Chip Transmigration Model

1. EtOH (70%).
2. Aqua ad injectabilia.
3. DPBS.
4. DPBS with Ca^{2+}& Mg^{2+}.
5. RPMI 1640 without phenolred (R0).
6. Peristaltic pump, accepted tubing size 1.9 mm.
7. BioChip with two cavities, pore size: 8 μm (Dynamic42 GmbH).
8. Combifix adapter (male-male).
9. Original perfusor tubing, 75 cm, inner diameter 0.9 mm, outer diameter 1.9 mm.
10. Mobicol Gravi Classic with Luer-Lock and screw cap, bottom plug, without filters.
11. Collagen A.

12. Human umbilical vein endothelial cells and appropriate growth medium.
13. Disinfection fleece wipes.
14. FACS buffer: 500 mL DPBS, 10% FBS, 5 mM EDTA.
15. Fluorescent-labelled antibodies against: CD11b, CD15, CD16, CD62L, CD66b.
16. 96-well, v-bottom plate.

3 Methods

3.1 Neutrophil Isolation by Density Centrifugation

1. Pour 18 mL of anticoagulated blood from a healthy human donor (or disease group) to a 50 mL centrifugation tube.
2. Rinse the blood tubes with 20 mL DPBS and add it to the falcon.
3. Prepare a 50 mL centrifugation tube with 15 mL density centrifugation medium (DCM).
4. Carefully overlay the DCM with the blood (*see* **Note 1**) and centrifuge the tube at 398× *g* for 30 min at RT with the lowest acceleration possible and without brake to separate Plasma, PBMCs, DCM, PMNs, and Erythrocytes (Fig. 1).
5. Discard the plasma, PBMC, and the DCM layers. Aspirate the PMN layer (directly above the erythrocyte layer) with a serological pipette and transfer to a fresh 50 mL centrifugation tube.
6. Perform erythrocyte lysis by adding 36 mL cold, sterile water to the tube. Close and invert the tube for 20 s and add 4 mL PBS (10×). Centrifuge at 300× *g* for 5 min at room temperature. Discard the supernatant and resuspend the cell pellet in 1 mL DPBS.

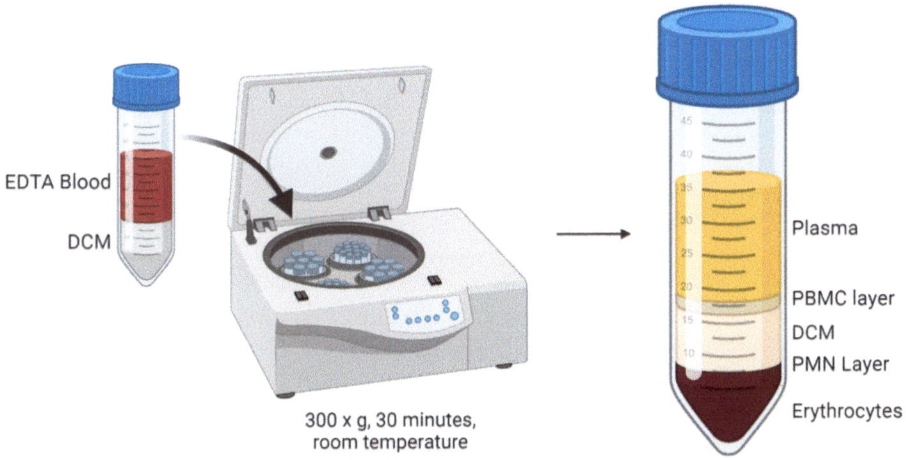

Fig. 1 Density centrifugation with DCM. (Created with BioRender.com)

7. Repeat **Step 6**.
8. Resuspend the cells in 1 mL DPBS and determine the concentration with a Neubauer hemocytometer chamber or an automated cell counting system.

3.2 Transwell Migration Assay

1. Isolate neutrophils according to Subheading 3.1.
2. Adjust the concentration to 2×10^6 cells/mL with DPBS in 50 mL reaction tube.
3. Add 1 μL CFSE (5 mM in DMSO) to the cell suspension and incubate at 37 °C. Fill up the tube with DPBS and centrifuge at 300× *g* for 5 min at room temperature. Determine the after-staining cell count and re-adjust to 2×10^6 cells/mL (*see* **Note 2**).
4. Add medium containing stimuli at a final volume of 150 μL/well to the receiver plate of the 96-well transwell system (*see* **Note 3**). Add unconditioned medium containing 300 pg/mL IL-8 as positive control in DPBS as well as a suitable negative control (e.g., pure medium when working with conditioned medium) in triplicates.
5. *Optional*: Prepare a standard curve to determine absolute numbers of migrating neutrophils by adding a dilution series of the CFSE-stained neutrophil cell suspension to the receiver plate starting with 100.000 cells/well. Fill up the corresponding transwell inserts with DPBS.
6. Set the transwell insert into the plate and add 50 μL of CFSE-stained neutrophil suspension per insert. Incubate the transwell system for 90 min at 37 °C with 5% CO_2 (Fig. 2).
7. Retrieve the transwell system from the incubator, remove the transwell insert, and measure the fluorescent signal (excitation: 488 nm, emission: 520 nm) per well with a fluorescence plate reader (*see* **Note 4**).
8. *Optional*: Prepare a standard curve with the signal from the neutrophil standard curve and calculate neutrophil numbers using a linear best fit function (Figs. 3 and 4).

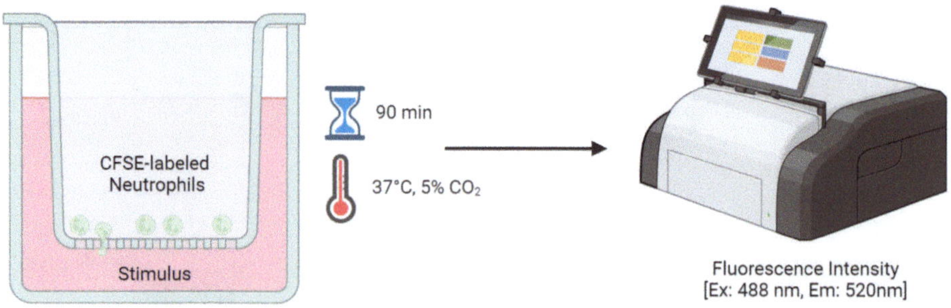

Fig. 2 Transwell Migration Assay. (Created with BioRender.com)

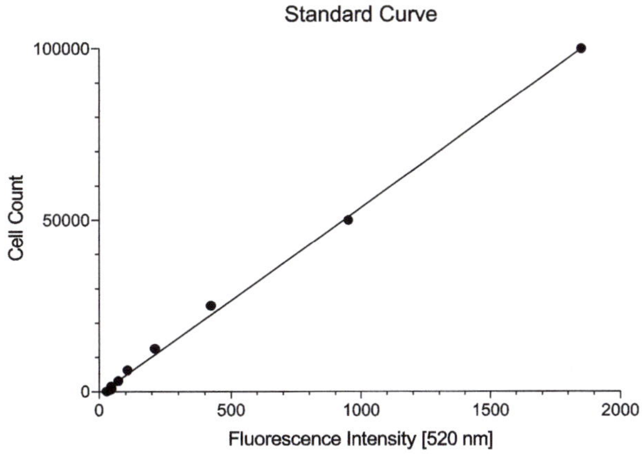

Fig. 3 Standard curve to determine absolute neutrophil numbers by fluorescence intensity [520 nm] values

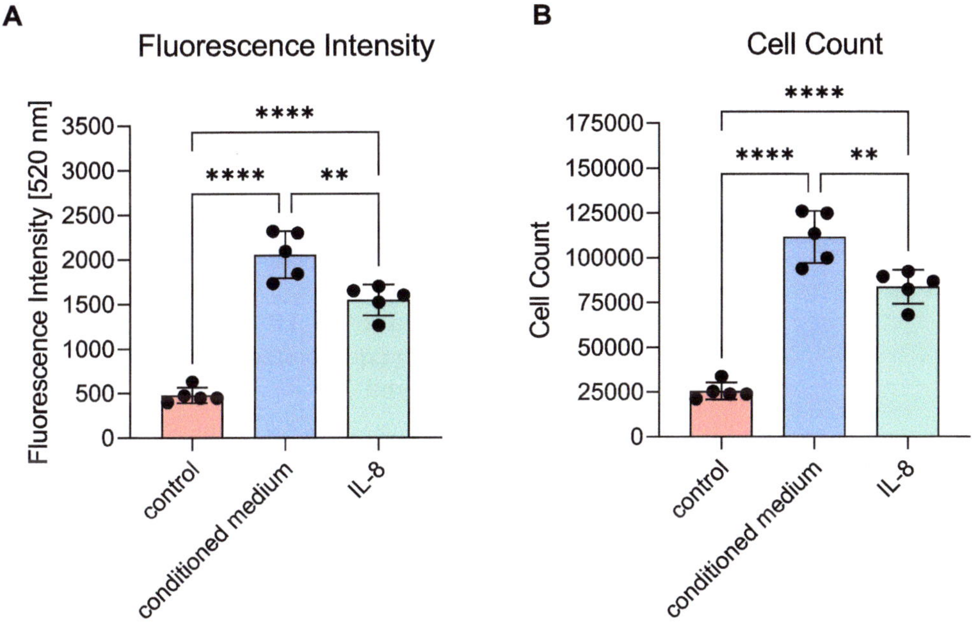

Fig. 4 Neutrophil transmigration readout for three conditions. Control (unconditioned medium), conditioned medium (tumor cell secretome), and IL-8 (300 pg/mL in unconditioned medium). (**a**) Fluorescence intensity [520 nm] of CFSE stained neutrophils in the receiver plate per condition. (**b**) Absolute cell counts of transmigrated cells determined by the standard curve (Fig. 3). Statistics were calculated using an ordinary one-way ANOVA with correction for multiple comparison testing (Tukey test). Significance values: *: $p \leq 0.05$, **: $p \leq 0.01$, ***: $p \leq 0.001$, $n = 5$

3.3 Organ on a Chip Transmigration Model

1. Perform sterilization of the BioChip by submerging the chip in EtOH (70%). Flush the cavities by reverse pipetting the surrounding EtOH through the inlets and outlets of the BioChip. Incubate 30 min under sterile conditions at room temperature.

2. Exchange the surrounding EtOH with ddH_2O and flush the cavities with ddH2O. Repeat the process once with fresh ddH_2O.
3. Coat the chip membrane with collagen by flushing the lower channel of the chip with DPBS with Ca^{2+}& Mg^{2+} followed by 130 μL Collagen A (1:100 in DPBS with Ca^{2+}& Mg^{2+}). Incubate the BioChip at 37 °C for 1–24 h (*see* **Note 5**).
4. Remove the collagen from the chip by flushing with 200 μL DPBS with Ca^{2+} and Mg^{2+} followed by 200 μL human umbilical vein endothelial cells growth medium (*see* **Note 5**).
5. Seed human umbilical vein endothelial cells (HUVEC) to the lower channel of the BioChip by flushing the cavity with cell-media suspension (*see* **Notes 5** and **6**).
6. Incubate the BioChip upside down at 37 °C, 5% CO_2, change medium every 24 h until the cells reach confluence (*see* **Notes 6** and 7).
7. Start perfusion of the lower channel with growth medium by connecting the perfusor tubing and the reservoir (*see* **Note 8**) to the BioChip (Fig. 5).
8. Perfuse the lower channel with 25 μL/min for 24 h at 37 °C, 5% CO_2 prior to the experiment.
9. At the day of the experiment, isolate neutrophils according to Subheading 3.1.
10. Adjust the concentration to 2×10^6 cells/ mL in RPMI 1640 without phenolred (R0) equilibrated at 37 °C, 5% CO_2 for at least 24 h (*see* **Notes 9** and **10**).
11. Stop perfusion of the lower channel. Remove the reservoir and exchange with a fresh reservoir (*see* **Note 11**).

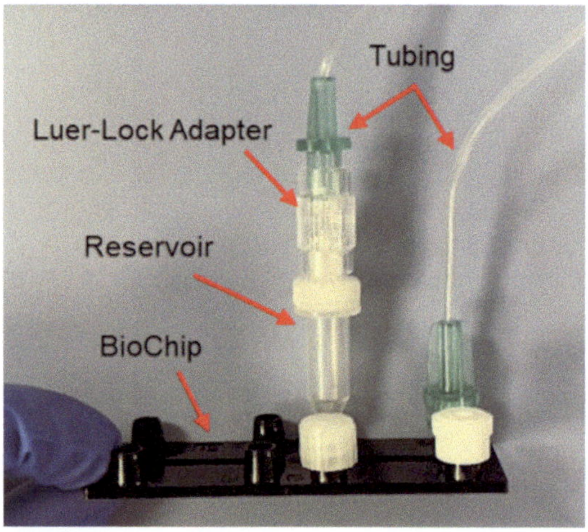

Fig. 5 BioChip with perfusor tubing and reservoir connected to one cavity

12. Add 1.5 mL of the neutrophils suspension to the reservoir. Connect the R0 containing tubing to the reservoir.
13. Add the stimuli of interest to the upper channel. Use IL-8 at a concentration of 300 pg/mL in R0 as positive control (*see* **Note 5**).
14. Start perfusion with 25 μL/min at 37 °C, 5% CO_2 for 2.5 h.
15. Stop perfusion. Flush the upper channel with 1000 μL fresh R0 and collect the suspension in a reaction tube (*see* **Note 12**).
16. Prepare staining solution by diluting the antibodies against CD11b, CD15, CD16, CD62L, CD66b according to the manufactures instructions in FACS buffer.
17. Transfer the cell suspension to a 96-well v-bottom plate and centrifuge at 300× *g* for 5 min. Remove supernatant and resuspend in staining solution.
18. Incubate cells and antibodies 30 min at 4 °C in the dark. Centrifuge at 300× *g* for 5 min. Remove supernatant and resuspend cells in 150 μL FACS Buffer.
19. Analyze specimens with a flow cytometer.

4 Notes

1. To ease overlaying of blood and DCM use 10 mL serological pipettes to pipette the blood-PBS mix and tilt the falcon with the density centrifugation medium to 50°. Blood will mix less with the DCM and separation of the cell types is improved.
2. If the neutrophil viability is significantly decreased by CFSE staining, reduce the CFSE concentration for staining.
3. For other well plate sizes, reconsider the working volumes according to the manufacturer's instructions.
4. If the OD measurements are out of range for your plate reader, reduce the CFSE concentration for staining.
5. The protocol displays the working volumes for the BC005 BioChip. For other BioChip models, adjust working volumes according to the manufactures instructions.
6. To reach confluence within 3 days, approximately 0.5×10^6 cells/mL should be used.
7. Start perfusion when cells reach confluence of 80% and start the experiment when cells have reached 100% confluence.
8. To prevent spilling of medium or cells, flush the perfusor tubing with medium using a syringe. Connect the tubing to the outlet (right chimney) of the BioChip channel used with the syringe on the other end of the tubing connected. Add the empty reservoir to the inlet (left chimney) and fill up with

medium. Close the lid of the reservoir, remove the syringe from the tubing, and connect the tubing to the reservoir lid (Fig. 5).

9. R0 should be equilibrated with the appropriate CO_2 conditions prior to use. Prepare a conventional cell culture flask with R0 and incubate at least 24 h at 37 °C, 5% CO_2. The R0 can be stored in the incubator for several days.
10. In case of low neutrophil transmigration, the neutrophil count can be adjusted up to 7×10^6 cells/mL.
11. To exchange reservoirs, work under sterile conditions and remove the tubing from the reservoir lid and attach a syringe with growth medium. Remove the lid and empty the chip-coupled reservoir with a 1000 μL pipette. Quickly remove the old reservoir and add a fresh one (without lid) to the chip. Carefully flush the tubing with the medium from the syringe. Empty the reservoir before adding the cell suspension.
12. To collect the cell suspension form the chip channel, stack and hold an empty unused 1000 μL pipette tip in the inlet of the cavity and flush fresh medium through the outlet using a 1000 μL pipette and pipette tip. The medium will end up in the empty pipette tip. Close the tip with the finger (in gloves) and transfer to a reaction tube.
13. To analyze the fraction of neutrophils that did not migrate through the endothelial cell layer, flush the lower channel similar to the upper once after perfusion. Collect the cells from the reservoir and the tubing by flushing with a sterile syringe. Stain and acquire with a flow cytometer.

Acknowledgments

We acknowledge the detailed instructions to the BioChip at the Dynamic 42 Academy and further outstanding support by the company.

References

1. Shafqat A, Khan JA, Alkachem AY et al (2023) How neutrophils shape the immune response: reassessing their multifaceted role in health and disease. Int J Mol Sci 24(24):17583
2. de Oliveira S, Rosowski EE, Huttenlocher A (2016) Neutrophil migration in infection and wound repair: going forward in reverse. Nat Rev Immunol 16(6):378–391
3. Lauwers M, Verschelden G, Boero C et al (2022) Optimization of the Transwell(R) assay for the analysis of neutrophil chemotaxis using flow cytometry to refine the clinical investigation of immunodeficient patients. Clin Immunol 238:108994
4. Filippi MD (2019) Neutrophil transendothelial migration: updates and new perspectives. Blood 133(20):2149–2158
5. Feile A, Wegner VD, Raasch M et al (2024) Immunocompetent intestine-on-chip model for analyzing gut mucosal immune responses. J Vis Exp 207:e66603
6. Koceva H, Amiratashani M, Rennert K et al (2024) Immunocompetent alveolus-on-chip model for studying alveolar mucosal immune responses. J Vis Exp 207:e66602

Chapter 9

Induction, Isolation, and Characterization of Human Neutrophil Extracellular Traps (NETs) to Be Used in Functional Assays

Quang Khai Doan, Cindy Qiu, and Patrice Decker

Abstract

Polymorphonuclear neutrophils (PMNs) represent ~60% of circulating leukocytes in humans. Once activated, PMNs can release neutrophil extracellular traps (NETs), molecular complexes composed of DNA and proteins from granules. Although NETs play a protective role in fighting microbes, they may become harmful under certain circumstances, especially in some inflammatory autoimmune diseases.

Because NETs can be either beneficial or deleterious depending on the microenvironment, a deeper characterization of NETs and NET-inducing stimuli as well as a better understanding of the different properties and activities of NETs is required, in both physiological and pathological contexts.

In the present chapter, we describe tools not only to induce and isolate NETs, but also required methods to characterize NETs via biochemical approaches and imaging, in order to prove that obtained structures conform to the definition of NETs exhibiting typical extracellular fibers. NETs produced with our protocol are suitable for detailed assessment of their biological properties in functional studies.

Key words Neutrophils, Neutrophil extracellular traps (NETs), Inflammation, Autoimmunity, Neutrophil isolation, NET induction and isolation, Immunofluorescence, Biochemical characterization

1 Introduction

Polymorphonuclear neutrophils (PMNs) are the most abundant blood leukocytes. They are among the first cells recruited at inflammation sites. Classically described as cells with a short lifespan (few hours), PMN survival is increased in pro-inflammatory environments and can reach up to 5 days in the circulation [1]. Traditionally, PMNs were known to work through phagocytosis, degranulation, production of reactive oxygen or nitrogen species, as well as secretion of cytokines and chemokines. More recently, a new innate immune mechanism involving PMNs has been

Quang Khai Doan and Cindy Qiu contributed equally with all other contributors.

Rami Bechara (ed.), *Autoimmunity: Methods and Protocols*, Methods in Molecular Biology, vol. 2990,
https://doi.org/10.1007/978-1-0716-4997-8_9,

described, namely, formation of neutrophil extracellular traps (NETs) [2]. Although often defined as pro-inflammatory cells, PMNs also display immuno-regulatory properties. In fact, PMN subpopulations have been described, explaining the different properties reported.

In response to certain stimuli, PMNs expel NETs, which are by definition complexes made of DNA and proteins from granules, but devoid of membrane. Additional proteins are present in NETs. Originally described as a defense mechanism against bacteria [2], NETs are also produced in response to viruses, fungi, or parasites. In addition, NET formation is triggered during sterile inflammation in response to, e.g., cytokines or danger-associated molecular patterns (DAMPs). NET formation can occur via the release of nuclear DNA accompanied by the death of PMNs, a process known as (suicidal) NETosis [2]. Alternatively, NET formation occurs without PMN death via the release of mitochondrial DNA [3] or the release of nuclear DNA through vesicles [4]. The correct name for the latter two mechanisms is NET formation or vital NETosis.

NET formation can be triggered by a variety of stimuli, either physiological or chemical ones. However, phorbol myristate acetate (PMA) is the most described NET-inducing stimulus so far and the most used in vitro; it triggers potent NET formation and its mode of action is well characterized. PMA is a typical stimulus to induce classical suicidal NETosis. Therefore, NET induction by PMA can be used as a model. Indeed, PMA and physiological stimuli like *Candida albicans* and Group B Streptococcus use related pathways to induce NETs [5].

Although beneficial during infections, NET formation can become deleterious in some circumstances, e.g., in inflammatory autoimmune diseases like systemic lupus erythematosus (SLE) and rheumatoid arthritis (RA), due to excessive NET production/impaired NET clearance or altered NET composition. Thus, NETs may become pro-inflammatory and even pathogenic [6], triggering, e.g., the activation of macrophages and PMNs [7] as well as fibroblasts [8, 9]. Moreover, NETs represent a source of autoantigens, are recognized by autoantibodies in RA and SLE patients, and might be involved in the break of tolerance following B/T-cell activation. Therefore, inappropriate activation of PMNs and uncontrolled NET formation may lead to a pro-inflammatory milieu and may favor autoimmunization. In contrast, NETs may also display anti-inflammatory activities through inhibition of cytokine secretion by macrophages [7] or proteolysis of secreted cytokines [10].

The pathogenic stimuli involved in inducing NET formation as well as those modulating the pro-inflammatory activity and antigenicity of NETs in autoimmune diseases are still not characterized. Moreover, depending on the type of stimuli [11, 12] and the source of PMNs (healthy individuals versus patients) [13], the composition of NET varies. Likewise, and as mentioned above, NET formation occurs via different mechanisms and pathways, with or without the involvement of reactive oxygen species [5]. Therefore, although typical NETs are DNA filaments decorated with proteins—many of which originate from granules—the composition may differ (containing histones or not, associated or not with histone H3 citrullination, containing nuclear versus mitochondrial DNA). Particularly, NETs may also contain DAMPs. It is thus essential to precisely characterize NETs in each individual study. Specifically, studies should always confirm NET formation by visualizing NET fibers with imaging/microscopy, especially when testing potential new or non-classical NET-inducing stimuli or when setting up NET protocols.

For all the reasons mentioned above, a better understanding of NET functions and activities is required, especially in inflammatory autoimmune diseases. Particularly, it is necessary: (1) to identify disease-associated stimuli for NET formation; (2) to characterize their impact on NET composition; and (3) to determine the consequences on NET activity and how NETs influence other immune cells or target cells such as stromal cells. Therefore, a precise characterization of NETs and high-quality analytic tools are required.

Different protocols or approaches to detect and measure NETs have been reported, but some lack accuracy. For instance, measuring only extracellular DNA, sometimes named (circulating) cell-free DNA, in biological fluids or cell culture supernatants is not optimal to evaluate the presence of NETs, as DNA can be released by different processes, e.g., necrosis, and cannot be used to estimate NET formation, especially without detection of associated proteins. Likewise, citrullinated histone H3 is often used as a NET marker; however, it is not specific to NET formation and is not clearly observed in response to all NET-inducing stimuli. Finally, as a surrogate NET marker, ELISAs have been developed to detect complexes made of DNA and neutrophil elastase or of DNA and myeloperoxidase, but these assays are merely indicative and do not provide a direct evidence of NET structures. In all cases, NET formation should be confirmed by imaging, demonstrating NET fibers made of extracellular DNA and proteins.

Here, we describe methods to induce, detect, and isolate (soluble) NETs from primary human blood PMNs in vitro as well as methods to analyze and characterize NET preparations. Experimental controls that should be included are suggested. We also describe how to characterize NET fibers, a necessary step to prove that the used stimulus triggers NETs. This control should be included, at least when setting up the protocol or when testing novel NET inducers. We also recommend using freshly isolated primary PMNs and avoiding cell lines. Our protocol can be used to follow NET induction by PMA or, after adjustment, other

NET-inducing stimuli, including physiological ones, and especially those associated with autoimmune diseases such as cytokines, DAMPs, or autoantigens [14]. It is especially adapted for preparing sterile soluble NETs in order to assess their biological activity in vitro, e.g., on immune cells.

2 Materials

2.1 Neutrophil Isolation

1. Fresh, human peripheral whole blood collected in ethylenediaminetetraacetic acid (EDTA)-treated (e.g., BD Vacutainer) tubes.
2. Biosafety cabinet.
3. PolymorphPrep™ dextran density gradient medium from Axis-Shield. Store at room temperature (RT) protected from light.
4. Sterile disposable pipettes.
5. 15 mL conical tubes.
6. 50 mL conical tubes.
7. Centrifuge.
8. Pasteur Pipette/Filter pipette tips.
9. Sterile phosphate-buffered saline (PBS) 1X without Ca^{2+}/Mg^{2+}. Store at 4 °C.
10. Cold ammonium-chloride-potassium (ACK) buffer: 155 mM NH_4Cl, 10 mM $KHCO_3$, 0.1 mM Na_2EDTA. Store at 4 °C.
11. A timer.
12. Roswell Park Memorial Institute (RPMI)-1640 GlutaMAX™ medium supplemented with 50 μg/mL Penicillin, 50 μg/mL Streptomycin, 100 μg/mL Neomycin (PSN), 25 mM HEPES, and 50 μM β-mercaptoethanol, with or without serum (*see* **Note 1**). Store at 4 °C.

2.2 Neutrophil Phenotypic Characterization

1. FACS tubes.
2. Filter pipette tips.
3. Centrifuge.
4. Staining buffer: PBS supplemented with 5% fetal calf serum (FCS), 0.02% sodium azide, 100 μg/mL of human gamma globulin. Store at 4 °C.
5. Fridge.
6. Fluorochrome 1-conjugated anti-human CD66b (clone G10F5, mouse IgM, κ) antibody.
7. Fluorochrome 1-conjugated isotype control (mouse IgM, κ).

8. Fluorochrome 2-conjugated anti-human CD11b (clone ICRF44, mouse IgG1, κ) antibody. Verify first that fluorochromes 1 and 2 are compatible (e.g., FITC and APC).
9. Fluorochrome 2-conjugated isotype control (mouse IgG1, κ).
10. Flow Cytometer.
11. Flow cytometry software for result acquisition and analysis.

2.3 Soluble NET Induction and Isolation

1. Human peripheral blood PMNs isolated as described in Subheading 3.1.
2. 1-well Lab-Tek™ II chamber slides (Nunc).
3. Humidified CO_2 (5%) incubator, 37 °C.
4. Biosafety cabinet.
5. Micro high-speed centrifuge.
6. Sterile pipettes and tips.
7. Sterile 50 mL conical tubes.
8. Sterile 15 mL conical tubes.
9. Nuclease-free and pyrogen-free 1.5 mL microcentrifuge tubes.
10. Nuclease-free and pyrogen-free 0.6 mL microcentrifuge tubes.
11. Sterile PBS 1X, pH 7.4 without Ca^{2+}/Mg^{2+}. Store at RT.
12. Sterile PBS 1X, pH 7.4 with Ca^{2+}/Mg^{2+}. Store at RT.
13. NET induction medium: RPMI-1640 GlutaMAX™ medium supplemented with 25 mM HEPES, 50 μg/mL Penicillin, 50 μg/mL Streptomycin, 100 μg/mL Neomycin (PSN), 50 μM β-mercaptoethanol, without serum (*see* **Note 1**). Store at 4 °C.
14. Sterile 0.001% poly-L-lysine solution in PBS 1X, pH 7.4 without Ca^{2+}/Mg^{2+}. Store at 4 °C.
15. 0.2 M EDTA, pH 8.0 solution. Store at 4 °C.
16. 100 μg/mL Phorbol 12-myristate 13-acetate (PMA) (*see* **Note 2**). Store at −20 °C.
17. 3000 Units/mL Deoxyribonuclease (DNase) I. Store at −20 °C.

2.4 NET DNA Quantification

1. Quant-iT™ PicoGreen™ dsDNA Assay Kit (Invitrogen). Store at 4 °C.
2. Fluorescence microplate reader.
3. 96-well black polystyrene plate.

2.5 NET Immunofluorescence Staining

2.5.1 Induction of NET Formation

1. Isolated PMNs.
2. 8-well Lab-Tek II chamber slides (Nunc).
3. 0.001% Poly-L-Lysine solution in PBS 1X, pH 7.4 without Ca^{2+}/Mg^{2+}.
4. PBS 1X, pH 7.4 without Ca^{2+}/Mg^{2+}.
5. PMA.
6. Culture medium (RPMI-1640 GlutaMAX™ supplemented with 25 mM HEPES, 50 μg/mL Penicillin, 50 μg/mL Streptomycin, 100 μg/mL Neomycin (PSN), 50 μM β-mercaptoethanol) with 10% FCS.
7. Humidified CO_2 (5%) incubator, 37 °C.
8. Biosafety cabinet.

2.5.2 Immunofluorescent Staining

1. PBS 1X, pH 7.4 without Ca^{2+}/Mg^{2+}.
2. 4% Paraformaldehyde (PFA) in PBS.
3. Blocking buffer (PBS, 2% bovine serum albumin (BSA), 2% 76 °C inactivated goat serum, 0.2% Triton X-100).
4. Staining buffer (PBS, 2% BSA, 0.2% Triton X-100).
5. Primary antibodies (*see* **Note 3**).
6. Fluorochrome-conjugated secondary antibodies.
7. SYTOX™ Green (Invitrogen).
8. Mounting medium without DAPI or Hoechst 33342.
9. 60 mm Glass coverslips.
10. Chemical hood.
11. Fluorescence microscope and image analysis software.

3 Methods

3.1 Neutrophil Isolation

1. Under a biosafety cabinet, strongly mix and add 5 mL of PolymorphPrep to each 15 mL conical tube.
2. Carefully add 5 mL of undiluted blood on top of the PolymorphPrep, without mixing them (*see* **Note 4**).
3. Centrifuge at 500 g for 40 min at RT without brake in a swing-out rotor (*see* **Note 5**).
4. Remove the tube carefully from the centrifuge, without mixing the different phases. Two cell bands should be visible, while the pellet contains red blood cells. The upper ring contains mononuclear cells (PBMC), while the lower ring contains PMNs.
5. With the help of a Pasteur pipette or a tip, collect the PMN layer carefully without touching the PBMC layer and transfer it to a new 50 mL conical tube (*see* **Note 6**).

6. Add PBS up to 50 mL.
7. Centrifuge at 300 g for 8 min at RT.
8. Discard the supernatant.
9. Break the cell pellet.
10. Add 5 mL of ACK buffer and incubate for 3 min to lyse the remaining red blood cells.
11. Add PBS up to 50 mL to neutralize the ACK.
12. Centrifuge at 300 g for 8 min at RT.
13. Discard the supernatant.
14. Break the cell pellet.
15. Resuspend the cells in the appropriate culture medium (either medium for cell culture or medium for soluble NET induction).
16. PMNs are isolated and ready to be used for cell counting and purity check before further experimentation.

3.2 Neutrophil Phenotypic Characterization

1. Transfer 100,000 cells in 4 FACS tubes for: (a) purity staining, (b) fluorescence-minus-one (FMO) isotype controls for CD66b, (c) for CD11b, and (d) unstained control.
2. Wash twice to remove the culture medium:

   Centrifuge at 300 g for 5 min.

   Discard the supernatant to wash out the medium.

   Break the cell pellet and resuspend it in 150 μL staining buffer.
3. Blocking step: incubate the cells in 150 μL staining buffer for 30 min at 4 °C.
4. Wash the cells to remove the staining buffer.
5. Antibody staining (indicated antibody volumes are for antibody formats corresponding to 20 μL/test, where one test is defined by 10^6 cells; to be verified for each manufacturer; *see* **Note** 7):

   Purity staining tube: 2 μL anti-CD66b + 2 μL anti-CD11b antibodies +146 μL staining buffer.

   CD66b FMO isotype control tube: 2 μL CD66b isotype control +2 μL anti-CD11b antibody +146 μL staining buffer.

   CD11b FMO isotype control tube: 2 μL anti-CD66b antibody +2 μL CD11b isotype control +146 μL staining buffer.

   Unstained control tube: 150 μL staining buffer.
6. Incubate the cells for 20–30 min at 4 °C, protected from light.
7. Wash to remove excess of antibodies.
8. Add 150 μL of staining buffer in each tube.

9. The cells are now stained and ready to be acquired on the cytometer (*see* **Note 8**).
10. Analyze neutrophil purity and phenotype using flow cytometry software (*see* **Note 9**).

3.3 Soluble NET Induction and Isolation

1. Under the biosafety cabinet, add 2 mL of 0.001% poly-L-lysine solution per 1-well Lab-Tek™ II chamber slide.
2. Put the chamber slides in a humidified CO_2 incubator at 37 °C for 15 min.
3. Discard the poly-L-lysine and wash with 2 mL of PBS 1X, pH 7.4 without Ca^{2+}/Mg^{2+} per chamber slide.
4. Remove the PBS.
5. Resuspend the PMNs in the NET induction medium at the concentration of 1.5×10^6 cells/mL.
6. Seed 2 mL of PMN suspension per chamber slide (*see* **Note 10**).
7. Place the chamber slides back to the humidified CO_2 incubator at 37 °C for 30 min (PMNs settle).
8. Carefully retrieve the chamber slides from the humidified CO_2 incubator.
9. Add 0.62 μL of PMA 100 μg/mL per chamber slide (50 nM final concentration).
10. Carefully place the chamber slides in the humidified CO_2 incubator at 37 °C for 4 h (*see* **Note 11**).
11. Gently aspirate the supernatant and carefully add 2 mL of PBS 1X, pH 7.4 without Ca^{2+}/Mg^{2+} per chamber slide (*see* **Note 12**). Repeat this step once.
12. Carefully discard the PBS.
13. Prepare DNase I 25 Units/mL in a sterile 15 mL conical tube by diluting DNase I 3000 Units/mL 120 times in PBS 1X, pH 7.4 with Ca^{2+}/Mg^{2+}.
14. Add 400 μL of DNase I 25 Units/mL per chamber slide.
15. Place the chamber slides in the humidified CO_2 incubator for 30 min at 37 °C.
16. Add 6 μL of EDTA 0.2 M, pH 8.0 per chamber slide (*see* **Note 13**).
17. Using a P1000 pipette, aspirate the supernatant and then release it strongly into the chamber slide. Repeat this step thoroughly until all NET fibers are detached.
18. Collect all the liquid from the chamber slides into nuclease-free and pyrogen-free 1.5 mL microcentrifuge tubes.
19. Centrifuge the tubes at 300 g for 10 min at 4 °C.

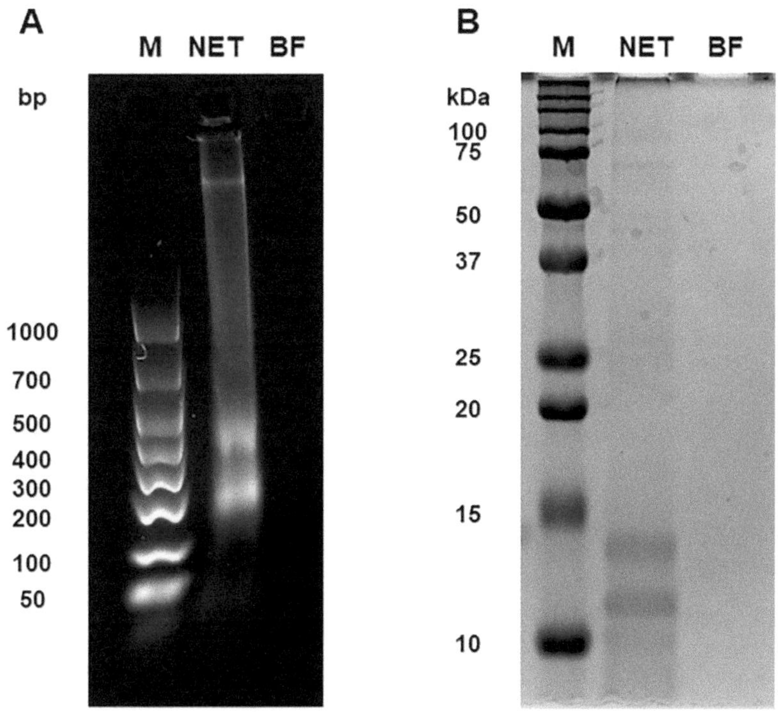

Fig. 1 Biochemical characterization of soluble NETs. Purified NETs were analyzed by 1.5% agarose gel (**a**) and 16% SDS-PAGE (**b**) following classical protocols. In (**a**), DNA fragments presenting as multiple of ~200 bp are observed after DNase I digestion. In (**b**), protein bands typical of histones are seen between 10 and 15 kDa. The identity of bands of interest can be confirmed by Western blot. BF NET purification buffer (all the NET isolation steps were followed but without PMNs), bp base pairs, kDa kilodalton, M molecular weight markers

20. Carefully collect supernatants in fresh nuclease-free and pyrogen-free 1.5 mL microcentrifuge tubes.
21. Centrifuge the tubes at 16,000 g for 10 min at 4 °C.
22. Collect 80 μL of the supernatant in nuclease-free and pyrogen-free 0.6 mL microcentrifuge tubes (*see* **Note 14**).
23. Carefully transfer the remaining supernatants to new nuclease-free and pyrogen-free 1.5 mL microcentrifuge tubes (*see* **Note 15**).
24. Perform a biochemical characterization of NETs (*see* **Note 14**). This step is *required*. Typical results are presented in Fig. 1.

3.4 NET DNA Quantification

1. Under the biosafety cabinet, use Quant-iT™ PicoGreen™ dsDNA Assay Kit to quantify NET DNA (*see* **Note 16**).
2. Take 5 μL from the 80 μL aliquots (obtained from Subheading 3.3, **step 22**) and add it to a 96-well black polystyrene plate.

3. Add 45 μL of Tris-EDTA (TE) 1X working solution to the wells containing NET samples. Also prepare the purified DNA for the standard curve as recommended (in the same final volume).
4. Add 50 μL of aqueous working solution of the Quant-iT™ PicoGreen™ dsDNA Reagent to each sample.
5. Incubate the plate for 2 min at RT, protected from light.
6. Read the plate using a fluorescence microplate reader (excitation at 480 nm; emission at 520 nm).
7. Calculate NET DNA concentrations according to the DNA standard from the kit. Concentrations of soluble NET preparations usually range from 10 μg/mL to 20 μg/mL DNA.

3.5 NET Immunofluorescence Staining

3.5.1 Induction of NET Formation

1. Under the biosafety cabinet, add 200 μL of 0.001% poly-L-lysine solution diluted in sterile PBS per well of a Lab-Tek II chamber slide.
2. Put the chamber slide in a humidified CO_2 incubator at 37 °C for 15 min.
3. Discard the poly-L-lysine and rinse with 500 μL/well of PBS 1X, pH 7.4 without Ca^{2+}/Mg^{2+}.
4. Add PMNs at 50,000 cells per well in complete medium with FCS (final volume = 200 μL).
5. Let the PMNs settle in the Lab-Tek II for 30 min in the humidified CO_2 incubator at 37 °C.
6. Add PMA (50 nM final) in wells dedicated to NET induction. As a negative control, use wells without (PMA or other) stimulation. We recommend performing duplicates for each condition.
7. Incubate 2 h in the humidified CO_2 incubator at 37 °C.

3.5.2 Immunofluorescent Staining

1. Wash twice with RT PBS (500 μL/well) very carefully.
2. Add 200 μL 4% PFA/well for 15 min (not more) at RT in a chemical hood.
3. Wash twice with 500 μL/well RT PBS.
4. Add 200 μL/well of blocking buffer for 20 min at RT.
5. Wash twice with 500 μL/well RT PBS.
6. Incubate overnight at 4 °C the primary selected antibodies (200 μL/well) diluted in PBS 2% BSA 0.2% Triton.
7. Wash twice with 500 μL/well RT PBS.
8. Incubate 40 min at RT in the dark the secondary antibodies (200 μL/well) diluted in PBS 2% BSA 0.2% Triton, conjugated to chosen and compatible fluorochromes (*see* **Note 17**).

Fig. 2 Fluorescence imaging of NETs in non-activated (non-NETting) neutrophils (left) and in PMA-activated neutrophils (right). NETs were labeled by immunofluorescence with antibodies specific for neutrophil elastase (NE, red) and total histones (purple) followed by DNA staining with SYTOX (green). Under basal conditions, NE and histone localization is intracellular, while NE and histones are both detected on extracellular DNA after PMA stimulation. Typical NET fibers are composed of DNA, NE, and histones (white arrow). Nuclei of non-activated PMNs are visible

9. Wash twice with 500 μL/well RT PBS.
10. Incubate 5 min at RT with 200 μL/well SYTOX™ Green (1/20,000) diluted in complete RPMI (no serum) in the dark to stain extracellular NET DNA.
11. Wash twice with 500 μL/well complete RPMI and once with 500 μL/well PBS.
12. Disconnect the plastic chamber from the glass slide.
13. Add a small drop of mounting medium per well and add a coverslip on the slide.
14. Wait at least 15 min and keep the slide in an opaque box at 4 °C before observation under a fluorescence microscope.
15. Analyze the different fluorescence channels recorded using an appropriate image analysis software (*see* **Note 18**). Compare resting to activated PMNs. A representative image is depicted in Fig. 2.

4 Notes

1. Use serum-free medium to avoid nuclease-mediated NET digestion.
2. PMA is used here as an example, but other NET-inducing stimuli may be tested.
3. Classical antibodies to characterize NET fibers (in addition to DNA staining), are those specific for neutrophil elastase or histones, depending on the stimulus. This combination is adapted to PMA-induced NET. However, for example, histones are not present in NET made of mitochondrial DNA. Antibodies against citrullinated histone H3 are often used. Although often claimed, citrullinated H3 is not a NET marker. Moreover, in PMA-induced NET (i.e., NETosis), citrullinated H3 can be detected but only at a low level.

4. For optimal separation, blood samples should be handled at ~20 °C.
5. If the separation is not optimal, perform an additional 5–10 min centrifugation.
6. This protocol can be used to isolate PBMCs simultaneously with PMNs.
7. Antibody volumes should be titrated.
8. Stained cells can be fixed if not acquired immediately.
9. CD66b is a granulocyte marker, while CD11b intensity is used to estimate PMN activation. PMNs are $CD66b^{+}CD11b^{+}$.
10. As a negative control, especially for functional assays, we recommend following the same protocol but without PMNs in order to produce the NET purification buffer.
11. Incubation time depends on the source of PMNs (e.g., human vs mouse).
12. NETs are very fragile.

    Some authors use those supernatants as a source of NETs. Although they may contain NETs, they also contain many factors released by activated PMNs, e.g., cytokines.
13. EDTA is used to stop DNase I activity. EDTA concentration should be titrated when NETs are used in functional assays. Indeed, EDTA may interfere depending on the assay and the NET activity studied.
14. The aliquot is dedicated to NET characterization (i.e., DNA quantification, agarose gel electrophoresis and SDS-PAGE) to avoid repeated thawing of the prepared stock. Western blot may also be useful. In that case, freeze more than 80 μL.

    You may in addition measure optical densities at 260 nm and 280 nm to estimate both DNA and protein contents.

    Depending on the studies (testing new NET-inducing stimuli, looking for a specific NET-associated protein), a deeper analysis of NET composition should be performed by Western blot.
15. When transferring the supernatants, do not disturb the precipitated debris. The obtained soluble NETs can be stored at −20 °C for future use.
16. Follow Quant-iT™ PicoGreen™ dsDNA Assay Kit instructions, which include the preparation of the TE 1X working solution, standard, and aqueous working solution of the Quant-iT™ PicoGreen™ dsDNA Reagent.

    PicoGreen assay is only used to quantify NET DNA. It is not used to detect NETs. NET formation should be confirmed by imaging.

17. For staining with two primary antibodies, we use AlexaFluor546- and AlexaFluor647-conjugated secondary antibodies. Those fluorochromes should also be compatible with SYTOX™ Green used for nuclear DNA staining/extracellular NET DNA staining.

18. By definition, NETs are composed of DNA and proteins from granules, regardless of the mechanisms or pathways involved in NET formation. Depending on the stimulus inducing NETs, and therefore the mechanisms/pathways triggered, other proteins are present in NETs.

    To identify NET fibers, it is therefore necessary to stain extracellular DNA AND at least one, but ideally two, different proteins associated with DNA. To prove observed structures are NETs, an overlay of the three signals should be obtained.

    If the aim is, for example, to analyze NET composition in more details or to study another associated protein of interest, an *additional* staining is required. If technically it is not possible to use four different fluorescence channels, omit one staining of NET protein. In that case, it is necessary to stain DNA, one NET protein and the protein of interest.

    With SYTOX™ Green, it will be possible to distinguish nuclear DNA staining (non-NETting neutrophils) from extracellular NET DNA staining (fibers).

    Staining with anti-neutrophil elastase antibody is more specific for NETs than, e.g., anti-myeloperoxidase antibody (MPO), as MPO is also expressed by monocytes. Indeed, depending on the cell source, you may have contaminating cells. Moreover, monocytes also expel extracellular traps.

Acknowledgments

This work was supported by University Sorbonne Paris Nord and INSERM. We thank Dr. Roxane Hervé for her technical help in the immunofluorescence analysis of NETs.
C.Q. was supported by a CIFRE fellowship from Arthritis R&D and ANRT (Association Nationale de la Recherche et la Technologie).

References

1. Pillay J et al (2010) In vivo labeling with 2H2O reveals a human neutrophil lifespan of 5.4 days. Blood 116:625–627
2. Brinkmann V et al (2004) Neutrophil extracellular traps kill bacteria. Science 303:1532–1535
3. Yousefi S et al (2009) Viable neutrophils release mitochondrial DNA to form neutrophil extracellular traps. Cell Death Differ 16:1438–1444
4. Pilsczek FH et al (2010) A novel mechanism of rapid nuclear neutrophil extracellular trap formation in response to Staphylococcus aureus. J Immunol 185:7413–7425

5. Kenny EF et al (2017) Diverse stimuli engage different neutrophil extracellular trap pathways. elife 6:e24437
6. Melbouci D, Haidar AA, Decker P (2023) Neutrophil extracellular traps (NET): not only antimicrobial but also modulators of innate and adaptive immunities in inflammatory autoimmune diseases. RMD Open 9: e003104
7. Ribon M et al (2019) Neutrophil extracellular traps exert both pro- and anti-inflammatory actions in rheumatoid arthritis that are modulated by C1q and LL-37. J Autoimmun 98: 122–131
8. Khandpur R et al (2013) NETs are a source of citrullinated autoantigens and stimulate inflammatory responses in rheumatoid arthritis. Sci Transl Med 5:178ra40
9. Carmona-Rivera C et al (2017) Synovial fibroblast-neutrophil interactions promote pathogenic adaptive immunity in rheumatoid arthritis. Sci Immunol 2:eaag3358
10. Schauer C et al (2014) Aggregated neutrophil extracellular traps limit inflammation by degrading cytokines and chemokines. Nat Med 20:511–517
11. Chapman EA et al (2019) Caught in a trap? Proteomic analysis of neutrophil extracellular traps in rheumatoid arthritis and systemic lupus erythematosus. Front Immunol 10:423
12. Petretto A et al (2019) Neutrophil extracellular traps (NET) induced by different stimuli: a comparative proteomic analysis. PLoS One 14:e0218946
13. Bruschi M et al (2019) Neutrophil extracellular traps protein composition is specific for patients with lupus nephritis and includes methyl-oxidized alphaenolase (methionine sulfoxide 93). Sci Rep 9:7934
14. Lindau D et al (2014) TLR9 independent interferon alpha production by neutrophils on NETosis in response to circulating chromatin, a key lupus autoantigen. Ann Rheum Dis 73: 2199–2207

Chapter 10

Analysis of Cellular Metabolism Using Flow Cytometry

Christos Aronis and Matteo Villa

Abstract

The understanding that cellular metabolism underlines the differentiation, activation, and function of immune cells has opened new avenues to modulate immunity. Here, we describe a method to analyze cellular metabolism using the quintessential immunological technique: flow cytometry. We analyze single cell suspensions, by combining the staining of surface immune cell markers and nutrient transporters, with the staining using metabolic dyes, to readout surrogates of metabolic pathway utilization with cell subset resolution.

Key words Cellular metabolism, Flow cytometry, Single cell analysis, Surface nutrient transporters, Metabolic dyes

1 Introduction

Cellular metabolism is a key modulator of cell fate and function. Nutrient availability and expression of metabolic enzymes regulate how a cell partitions nutrients in different metabolic pathways, to support its functions. How certain key metabolic pathways such as glycolysis and the tricarboxylic acid (TCA) cycle modulate cell behavior have been extensively studied, whereas the role of lesser known pathways is being (re)discovered [1–4]. Metabolic pathways are intertwined and intermediates of one pathway constitute the substrates for another one. Also, cellular metabolism fulfils multiple tasks, from energy production, to controlling redox balance, to generating biomolecules [5]. The integration of metabolic pathways allows a given substrate (e.g., glucose) to feed into different pathways (e.g., glycolysis and pentose phosphate pathway) that produce different end products (e.g., ATP and nucleotides) that ultimately converge to support cell function (e.g., proliferation).

Mass spectrometry is the golden standard used to describe the metabolic landscape of cells [6], but it is not efficient in analyzing the metabolic complexity of samples containing different cell types, such as tissues. Instead, flow cytometry captures with single cell

Rami Bechara (ed.), *Autoimmunity: Methods and Protocols*, Methods in Molecular Biology, vol. 2990,
https://doi.org/10.1007/978-1-0716-4997-8_10,

resolution the complexity of a given sample, and it has already been cleverly used to quantify certain aspects of cellular metabolism [7].

In this book chapter, we describe a strategy to use flow cytometry to analyze cellular metabolism. We stain cell surface markers using antibodies to identify specific immune cell populations within complex cell suspensions, such as blood. In addition, we use metabolic dyes or antibody staining of nutrient transporters to quantify the potential of cells to use a given metabolic pathway. Of note, we do not use our protocol to quantify cellular metabolism, but to quantify *proxies of metabolic pathways* that could instruct further cell-targeted validation approaches.

2 Materials

Sterility is not required while performing this protocol, but the use of a safety cabinet should be considered to protect operators during the handling of hazardous materials such as patient-derived cell suspensions.

2.1 Consumables

1. Pipette and pipette tips.
2. 1.5 mL tubes.
3. 5 mL tubes.
4. 96 wells V-bottom plate.

2.2 Instrumentation

1. Vortex.
2. Cell counter.
3. Refrigerated centrifuge, capable of spinning plates up to 500 g.
4. Cell culture incubator, with 5% CO_2, at 37 °C.
5. Flow cytometer, equipped with at least a 488 nm laser (used for excitation of MitoTracker Green FM and tetramethylrhodamine, methyl ester, TMRM).

2.3 Solutions and Reagents

1. Staining buffer: 2 mM ethylenediaminetetraacetic acid (EDTA), 3% fetal bovine serum (FBS), 0.05% sodium azide (NaN_3) in 1X phosphate-buffered saline (PBS) solution, pH ~7.4. Keep at 4 °C (*see* **Note 1**).
2. Cell culture medium: 10% FBS, 2 mM glutamine, 50 Units/mL penicillin/50 μg/mL streptomycin solution, in RPMI 1640. Keep at 37 °C (*see* **Note 2**).
3. Fluorophore-conjugated antibody staining of surface markers is necessary to identify specific populations in complex cell suspensions using flow cytometry. The reader should choose the appropriate antibody combination to use, to focus the analysis on the cell types of interest.

In addition to the identification of specific cell subsets, antibody staining to quantify the expression of the nutrient transporters CD36, CD71, and CD98 is needed. CD36 is quantified as a proxy of the ability of cells to take up long chain fatty acids that are then used as fuels of fatty acid oxidation [8]. To quantify the ability of cells to sustain the activity of mTORC1 (mammalian target of rapamycin complex 1), the engagement of glycolysis and proliferation, use CD71 and CD98 [9].

Prepare a solution containing the surface antibodies using staining buffer and keep the solution at 4 °C, protected from light, until use (*See* **Note 3**).

4. Metabolic dyes MitoTracker Green FM, TMRM, and CellROX Deep Red are needed to quantify, respectively, mitochondrial content, mitochondrial membrane potential, and cellular reactive oxygen species (*See* **Note 4**).

   Prepare a solution containing the metabolic dyes using cell culture medium and keep the solution at 37 °C, protected from light and in an air-tight tube, until use (*See* **Note 5**).

3 Methods

1. Prepare the solution containing the metabolic dyes in cell culture medium, vortex to mix, and keep it at 37 °C. Use the metabolic dyes at the concentration recommended by the Manufacturer, or the one identified upon titration. Our final concentrations were 20 nM MitoTracker Green FM, 10 nM TMRM, and 5 μM CellROX Deep Red.

   Perform the staining of 0.5×10^6 cells with metabolic dyes in 100 μL of metabolic dyes solution, in a 96-well plate. Prepare a volume of metabolic dyes solution accordingly (*See* **Note 6**)
2. Prepare the solution containing the surface antibodies in staining buffer, pipette to mix, and keep at 4 °C. Use surface antibodies at concentrations identified upon titration. Perform the staining of 0.5×10^6 cells with surface antibodies in 50 μL, in a 96-well plate. Prepare a volume of surface antibodies solution accordingly (*See* **Note 7**).
3. Seed 0.5×10^6 cells in the well of a 96-well plate (*See* **Note 8**).
4. Top up each well to 200 μL, using the same medium used to resuspend the cells.
5. Centrifuge the plate, and after centrifugation discard the supernatant (*See* **Note 9**).

6. Resuspend the cell pellet in 100 μL of metabolic dyes solution, and incubate for 30 min, at 37 °C, in a cell culture incubator (*See* **Note 10**).
7. Move the cells to 4 °C, add 100 μL of staining buffer previously kept at 4 °C, and centrifuge as before. After centrifugation, discard the supernatant.
8. Wash the cells by resuspending the pellet in 200 μL of staining buffer at 4 °C, centrifuge as before, and discard the supernatant.
9. Resuspend the cell pellet in 50 μL of surface antibody solution, and incubate for 20 min, at 4 °C, protected from light (*See* **Note 11**).
10. Add 150 μL of staining buffer previously kept at 4 °C to each well and centrifuge as before. After centrifugation, discard the supernatant.
11. Wash the cells by resuspending the pellet in 200 μL of staining buffer at 4 °C, centrifuge as before, and discard the supernatant.
12. Resuspend the cell pellets in 150 μL of staining buffer, transfer the suspensions in 5 mL tubes, and keep at 4 °C until ready to analyze them on a flow cytometer (*See* **Note 12**).

4 Notes

1. EDTA chelates Ca^{2+} and Mg^{2+} ions, reducing cell-to-cell adhesion.

   Proteins in FBS reduce non-specific binding of antibodies to surface molecules, reduce cell adhesion to polystyrene in 96-well plates and in 5 mL tubes, and prevent cell apoptosis [10]. However, high concentration of FBS may contribute to autofluorescence signal in flow cytometry. The final FBS concentration should be tested beforehand. In alternative, FBS can be substituted with bovine serum albumin.

   NaN_3 prevents bacterial growth and photobleaching of fluorochromes [11]. It also inhibits some cellular enzymes and may affect cell viability.
2. Cell culture medium can be any medium used to keep the cells of interest metabolically viable. In general, a medium containing glucose, glutamine, and nutrients typically found in FBS should be used, as many cells require them to support their basic metabolic activity.
3. Protect fluorochrome-conjugated antibodies from light exposure, as fluorophores are photosensitive. It is recommended that the surface antibody solution is prepared shortly before

the staining step (within 30 min), but it may remain stable at 4 °C for up to 12 h, depending on the fluorochromes used.

4. We used MitoTracker Green FM to quantify mitochondrial content in mixed cell suspensions. Previous work has shown that, in certain cell types, MitoTracker dyes claimed to measure mitochondrial content, actually load into mitochondria dependently on membrane potential, thus making them suboptimal to purely measure mitochondrial content [12]. Other work showed that MitoTracker Green may be actively pumped out of mitochondria by the efflux pumps encoded, in the mouse, by *Bcrp1* and *Mdr1a/b* [13]. We recommend testing the applicability of MitoTracker Green FM to measure mitochondrial content independently of membrane potential in each specific setting, by performing key technical controls. For instance, treatment of cells with Verapamil inhibits the activity of the efflux pumps. Drugs like carbonyl cyanide-*p*-trifluoromethoxyphenylhydrazone (FCCP) dissipate mitochondrial membrane potential, thus allowing to test for MitoTracker Green FM loading behavior. We also recommend performing technical controls for the stainings with TMRM and CellROX dyes. Finally, we recommend corroborating the analysis of cellular metabolism by flow cytometry with alternative techniques such as biochemical or mass spectrometry-based assays.
5. Prepare the solution containing metabolic dyes immediately before the staining step, as those dyes are sensitive to light and oxygen exposure (i.e., CellROX Deep Red). Because of such sensitivity, it is advised to prepare single use aliquots of the stock solutions to avoid multiple freezing and thawing cycles.
6. Depending on the number of samples to stain, you can scale up the final volume of the metabolic dyes and surface antibodies solutions. Pipetting error should also be considered when calculating the final volume.
7. Do not vortex, in order to mix the surface antibody solution, as it may affect fluorochromes. Use pipette mixing instead.
8. Cells should always be kept at 4 °C, unless indicated otherwise, in order to preserve their metabolic status.

   Before resuspending the cells in metabolic dyes solution, consider preparing an unstained control to use as a negative control in flow cytometry. Given that every cell contains mitochondria, MitoTracker Green FM will load into every cell, leaving no negative control after staining with metabolic dyes.
9. When discarding the supernatant from a 96-well plate, do so by flicking the plate in a quick, continuous, and sharp movement, to avoid cross-contamination of neighboring wells and disturbance of pellet.

Choose the centrifugation speed and time according to the cell type of interest. A centrifugation speed of 400–450 g, for 5 min, at 4 °C is usually recommended for primary immune cells like T cells, B cells, and monocytes. Cells that are more prone to activation such as neutrophils may be centrifuged for 300 g, for 5 min, at 4 °C.

10. To prepare the unstained negative control (*see* **Note 8**), instead of using the metabolic dyes solution, resuspend cells in complete medium and treat them with the same procedures.
11. Certain surface antigens (e.g., cytokine and chemokine receptors) require to be stained at room temperature or at 37 °C, due to their dynamic trafficking through the plasma membrane. Unless staining of such molecules is necessary, always carry out surface antibody staining at 4 °C, to minimize non-specific binding of antibodies and preserve the metabolic status of cells.
12. Acquire the stained samples as soon as possible after the procedure is complete. If this is not possible, you can keep stained samples shortly (maximum 4 h) at 4 °C, protected from light.

    Fixation of stained flow cytometry samples with 1–4% formaldehyde solution is often used to preserve samples and postpone acquisition on the flow cytometer for up to 24 h, without substantial loss of fluorescence signal. We do not recommend the fixation of samples processed following our protocol, since MitoTracker Green FM and TMRM are non-fixable dyes. If fixation is a necessary step, consider using fixable alternatives.

Acknowledgments

This work was supported by a grant from the Medical University of Graz to M.V.

References

1. Buck MD, Sowell RT, Kaech SM et al (2017) Metabolic instruction of immunity. Cell 169(4):570–586. https://doi.org/10.1016/j.cell.2017.04.004
2. Finley LWS (2023) What is cancer metabolism? Cell 186(8):1670–1688. https://doi.org/10.1016/j.cell.2023.01.038
3. Puleston DJ, Villa M, Pearce EL (2017) Ancillary activity: beyond core metabolism in immune cells. Cell Metab 26(1):131–141. https://doi.org/10.1016/j.cmet.2017.06.019
4. Zhu J, Thompson CB (2019) Metabolic regulation of cell growth and proliferation. Nat Rev Mol Cell Biol 20(7):436–450. https://doi.org/10.1038/s41580-019-0123-5
5. Spinelli JB, Haigis MC (2018) The multifaceted contributions of mitochondria to cellular metabolism. Nat Cell Biol 20(7):745–754. https://doi.org/10.1038/s41556-018-0124-1
6. Jang C, Chen L, Rabinowitz JD (2018) Metabolomics and isotope tracing. Cell 173(4):822–837. https://doi.org/10.1016/j.cell.2018.03.055

7. Arguello RJ, Combes AJ, Char R et al (2020) SCENITH: a flow cytometry-based method to functionally profile energy metabolism with single-cell resolution. Cell Metab 32(6): 1063–1075. e1067. https://doi.org/10.1016/j.cmet.2020.11.007
8. Chen Y, Zhang J, Cui W et al (2022) CD36, a signaling receptor and fatty acid transporter that regulates immune cell metabolism and fate. J Exp Med 219(6). https://doi.org/10.1084/jem.20211314
9. Sinclair LV, Rolf J, Emslie E et al (2013) Control of amino-acid transport by antigen receptors coordinates the metabolic reprogramming essential for T cell differentiation. Nat Immunol 14(5):500–508. https://doi.org/10.1038/ni.2556
10. Song X, Wang J, Li Y et al (2021) Improved strategy for jet-in-air cell sorting with high purity, yield, viability, and genome stability. FEBS Open Bio 11(9):2453–2467. https://doi.org/10.1002/2211-5463.13248
11. Miyoshi N, Arai T, Tokumaru K et al (1997) Photobleaching of fluorescent probe in microfluorometry and detection of active oxygen species. Acta Histochem Cytochem 30(2): 189–191. https://doi.org/10.1267/ahc.30.189
12. Neikirk K, Marshall AG, Kula B et al (2023) MitoTracker: a useful tool in need of better alternatives. Eur J Cell Biol 102(4):151371. https://doi.org/10.1016/j.ejcb.2023.151371
13. de Almeida MJ, Luchsinger LL, Corrigan DJ et al (2017) Dye-independent methods reveal elevated mitochondrial mass in hematopoietic stem cells. Cell Stem Cell 21(6):725–729. e724. https://doi.org/10.1016/j.stem.2017.11.002

Chapter 11

Quantifying Mitochondrial Metabolism and Metabolic Fluxes in Soft Agar Cultures

Birte Dowerg, Fangfang Chen, and Thekla Cordes

Abstract

Anchorage-independent cultures provide insights into cell proliferation, differentiation, and tumorigenesis beyond traditional two-dimensional models by mimicking parts of the extracellular matrix (ECM). The soft agar colony formation assay enables cells to proliferate in a three-dimensional manner resulting in metabolic phenotypes that are distinct from traditional monolayer cultures. Here, we established a soft agar colony formation assay with subsequent cell isolation to analyze mitochondrial metabolism, metabolic fluxes, morphology, and gene expression within the same sample. We applied mass spectrometry and tracing approaches to decipher carbon utilization for tricarboxylic acid (TCA) cycle metabolism. We also quantified the alteration of immune-related genes in response to inflammatory stimuli in soft agar cultures that might be relevant to autoimmune diseases, which are frequently associated with inflammatory environments and may contribute insights into chronic inflammation and immune cell survival that parallel tumorigenic processes. Our methodology offers a robust model to better understand cell metabolism and function of anchorage-independent cultures that may contribute to the development of new treatment strategies.

Key words Extracellular matrix, Soft agar, Mitochondria, Stable isotope tracer, Metabolite extraction, Metabolic flux, Mass spectrometry, Anchorage-independent cultures, Metabolism

1 Introduction

Anchorage-independent cell culture models have emerged in the fields of cell differentiation, proliferation, and cancer to better mimic the physiological conditions encountered by cells in circulation or within the extracellular matrix (ECM). Culture systems that enable cell growth independent of cell attachments, such as spheroids and soft agar colony formation, are common methodologies for anchorage-independent culture systems. These culture systems closely mimic the architecture and microenvironment of tissues, resulting in gradients of nutrients and oxygen that influence metabolism, cellular processes, growth, and survival [1–3]. Autoimmunity often develops in response to chronic inflammation, where immune cells face environmental stress [4]. To better understand the

Rami Bechara (ed.), *Autoimmunity: Methods and Protocols*, Methods in Molecular Biology, vol. 2990,
https://doi.org/10.1007/978-1-0716-4997-8_11,

biological processes and metabolic reprogramming, anchorage-independent cell culture systems provide a more complex network of cell-cell and cell-matrix interactions.

In liquid anchorage-independent spheroid cultures, cells aggregate in culture media without attachment to an ECM and reprogram metabolism compared to traditional two-dimensional culture methods [5, 6]. For instance, isotopic tracing approaches identified that spheroid cultures have increased reductive carboxylation and pyruvate dehydrogenase (PDH) flux, suggesting that spheroid cultures reprogram mitochondrial metabolism and substrate utilization [7, 8]. Further, spheroid cultures have altered amino acid metabolism influencing lipid composition and cell growth [9, 10]. Compared to spheroids, metabolic quantification of colonies from soft agar requires additional steps to isolate colonies for subsequent analysis [6, 9–11]. Due to the lack of robust colony isolation methods, little is known about the metabolic reprogramming of different cell types grown in soft agar, and functional metabolic assays are limited.

Here, we provide a detailed workflow for the isolation of colonies from soft agar for subsequent analysis of cell metabolism and function. Our protocol combines the quantification of metabolic and functional parameters from one sample including metabolite levels, metabolic fluxes, gene expression, morphology, and growth (Fig. 1). As a case study, we applied uniformly labeled [U-$^{13}C_6$]glucose to leukemia-like RAW 264.7 macrophages cultured in soft agar to quantify metabolite levels and carbon utilization for mitochondrial TCA cycle metabolism. We also quantified levels of immune-related genes including immunoresponsive gene 1 (*Irg1*) and interleukin-6 (*Il-6*), in response to lipopolysaccharide (LPS) treatments in soft agar cultures. Our isolation protocol provides a valuable tool for studying metabolic reprogramming in anchorage-independent cultures grown in soft agar to better understand biological processes.

2 Materials

2.1 List of Chemicals

1. Isotopic tracer (e.g., [U-$^{13}C_6$]glucose).
2. DMEM without glucose and glutamine for tracing studies (Sigma-Aldrich).
3. Fetal Bovine Serum (FBS).
4. High glucose Dulbecco's Modified Eagle Medium (DMEM).
5. Interferon-γ (IFN-γ).
6. Internal standard for metabolite analysis.
7. ITaq Universal SYBR-Green Supermix.
8. Low-gelling temperature agarose.

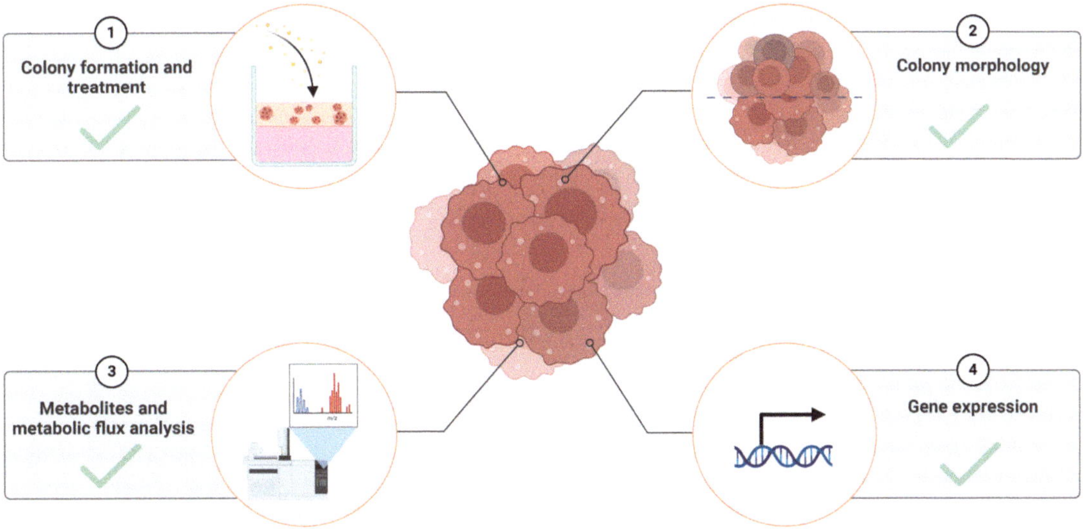

Fig. 1 Quantification of cell metabolism and function in an anchorage-independent soft agar culture system. The methodology provides quantification of metabolism and cellular processes of cells grown in soft agar: (1) Colony count and treatment, including colony activation by external inflammatory stimuli (2) Cell morphology: cell shapes visualized via microscopy (3) Metabolites: including metabolite extraction and quantification and metabolic flux analysis with stable isotope tracer via gas chromatography combined with mass spectrometry (GC-MS) (4) Gene expression analysis of isolated RNA from interface via qRT-PCR

Table 1
Primer used in this study

Gene target	Forward/reverse	Sequence (5′-3′)	Supplier
Il-6	Forward	TCTATACCACTTCACAAGTCGGA	Eurogentec
Il-6	Reverse	GAATTGCCATTGCACAACTCTTT	Eurogentec
Irg1	Forward	GCAACATGATGCTCAAGTCTG	Invitrogen
Irg1	Reverse	TGCTCCTCCGAATGATACCA	Invitrogen
Rpl27	Forward	GCGATCCAAGATCAAGTCCTTTG	Eurofins genomics
Rpl27	Reverse	TCAAAGCTGGGTCCCTGAACAC	Eurofins genomics

The table depicts primer pairs used in this study including sequences (5′-3′) and supplier information. *Il-6* interleukin-6, *Irg1* immunoresponsive gene 1, *Rpl27* mitochondrial ribosomal protein gene

9. Lipopolysaccharide (LPS).
10. Methoxyamine hydrochloride (MOX).
11. Milli-Q-purified water or equivalent pure water.
12. N-Tertbutyldimethylsilyl-N-methyltrifluoroacetamide (MTBSTFA) with 1% tert-butyldimethylchlorosilane (TBDMS).

13. Penicillin (stock 10,0000 U/mL)/Streptomycin (stock: 10,0000 μg/mL).
14. Primer pairs (*see* Table 1).
15. NaCl (0.9%).
16. Methanol (99.8%, HPLC grade) equilibrated to −20 °C.
17. Chloroform (HPLC grade) equilibrated to −20 °C.

2.2 List of Instruments and Materials

1. 15 mL reaction tube.
2. 1.5 mL reaction tube (e.g., Eppendorf, Safe-Lock) (*see* **Note 1**).
3. 96-well cell culture plate.
4. Analytical platform (e.g., gas chromatograph coupled to mass spectrometry (GC-MS)).
5. Rotary vacuum evaporator, e.g., CentriVap Concentrator (Labconco) cooled to 4 °C.
6. Autoclave.
7. Ball mill (e.g., Retsch MM 400).
8. CDNA reverse transcription kit (e.g., High Capacity cDNA Reverse Transcriptase Kit).
9. Cell culture plates (e.g., 6-well culture dishes).
10. Centrifuges (prewarmed to 37 °C, and cooled to 4 °C).
11. Microscope (e.g., Axcio Vert.A1 microscope coupled with Axiocam 105 color).
12. Magnet for removing the metal balls.
13. Microwave.
14. Protein assay kit.
15. qPCR.
16. RNA extraction kit.
17. Tecan Infinite® 200 Spark® reader.
18. Water bath at 55 °C.

2.3 Reagents and Solutions

1. 3% (w/v) agarose in Milli-Q-purified water: weigh 0.9 g agarose and add 30 mL Milli-Q-purified water in a 100 mL glass bottle. Microwave the solution for 30–45 s intervals until the agarose is dissolved. Autoclave the agarose solution at 121 °C for 15 min. Store at room temperature until use.
2. LPS: Prepare a 1:200 dilution from 1 mg/mL stock for a final working stock of 5000 ng/mL.
3. IFN-γ: Prepare a 1:40 dilution from stock (100 μg/mL) for a final working stock of 2500 ng/mL.

4. 0.9% (w/v) NaCl: Dissolve 4.5 g NaCl in 500 mL Milli-Q-purified water. Prewarm to 55 °C.
5. 5 μg/mL norvaline: Water with 5 μg/mL norvaline as an internal standard for metabolite extraction: Mix 200 μL 1 mg/mL norvaline stock with 40 mL Milli-Q-purified water. Store at 4 °C.
6. Growth medium containing DMEM supplemented with 10% FBS and 0.5% P/S.
7. Methoxyamine hydrochloride (MOX) reagent 2% (w/v): Prepare a 2% MOX solution in pyridine by dissolving 20 mg MOX in 1 mL pyridine.

2.4 List of Software Programs

1. Data visualization software (e.g., GraphPad Prism 9).
2. Imaging software (e.g., Fiji open-source software, ImageJ 2.9.0/1.53 t Java 1.8.0_322, 64-bit).
3. Mass spectrometry analysis software (e.g., OpenChrom open-source software, Lablicate).
4. Software for qPCR analysis (e.g., QuantStudio Design and Analysis Software v1.5.2, Thermo Fisher Scientific).

3 Methods

A detailed workflow for quantifying metabolism and metabolic fluxes in soft agar cultures is depicted in Fig. 2. We provide protocols for colony formation assay (*see* Subheading 3.1), colony treatments (*see* Subheading 3.2), isotopic tracing (*see* Subheading 3.3), quantifying colony growth and morphology (*see* Subheading 3.4), colony isolation from soft agar (*see* Subheading 3.5), subsequent metabolite extraction (*see* Subheading 3.6), and metabolic analysis using GC-MS (*see* Subheadings 3.7 and 3.8). Further, transcriptomic approaches provide insights into the molecular dynamics of gene expressions [12] (*see* Subheading 3.9). As a case study, we provide protocols for mammalian cell cultures (RAW 264.7 cells, ATCC, TIB-71) grown on a 6-well tissue culture plate, and volumes are calculated for 6-well tissue culture plates. Volumes may be adjusted accordingly if different culture formats are used (*see* **Note 2**).

3.1 Cell Seeding and Colony Formation Assay

The protocol describes the preparation of the soft agar colony formation assay in a 6-well cell culture plate as previously applied in Cordes et al. 2022 [10]. The lower layer represents a 2 mL hard layer containing 0.6% of low-gelling temperature agarose in DMEM. The upper layer was made of a 1 mL soft agar layer with a final concentration of 0.3% agarose in DMEM which contains the cells. Conduct all steps under the clean bench unless otherwise

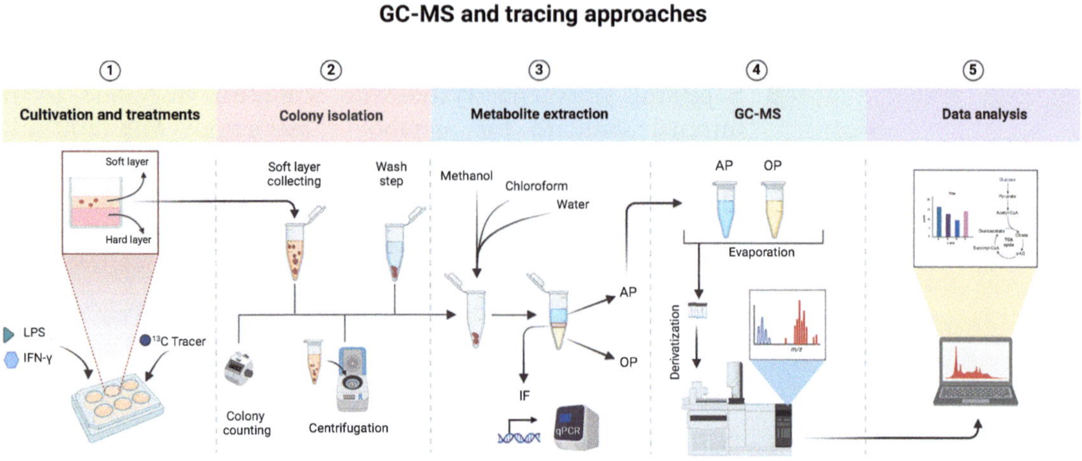

Fig. 2 Schematic overview of the experimental procedure using GC-MS and tracing approaches in anchorage-independent soft agar cultures. Schematic overview of the experimental workflow that comprises five parts: (1) Cultivation and treatments including colony formation assay and treatments with stable isotope tracer and inflammatory stimuli; (2) colony isolation including colony counting, soft-layer collection and colony isolation; (3) metabolite extraction from the aqueous phase (AP) containing hydrophilic metabolites and the organic phase (OP) including hydrophobic metabolites as well as gene expression analysis from, the interface (IF) (4) GC-MS for quantifying metabolite abundances and fluxes; (5) data analysis, including raw data preparation, statistical tests, and data visualization. (The figure was created with Biorender.com)

noticed. Volume is calculated for a total of six culture wells with three wells for control and three wells for treatment conditions (*see* **Note 3**).

3.1.1 Preparation of the Lower Hard Layer with 0.6% Agarose (15 mL)

1. Heat the previously made 3% agarose solution in the microwave until the agarose is entirely dissolved and store at 55 °C (*see* **Note 4**).
2. Mix 10.425 mL DMEM, 1.5 mL FBS (final concentration of 10% for 15 mL), and 75 μL P/S (final concentration of 0.5% for 15 mL). Prewarm to 37 °C.
3. Add 3 mL of prewarmed 3% agarose (final concentration of 0.6%) to the 12 mL medium prepared in **Step 2** for a final volume of 15 mL.
4. Carefully mix the solution by pipetting up and down without inducing air bubbles using a serological pipette (*see* **Note 5**).
5. Pipette 2 mL to each 6-well of the culture plate.
6. Allow the solution to solidify for 10–15 min at room temperature.
7. Keep the 3% agarose at 55 °C (*see* **Note 4**).

3.1.2 Preparation of the Upper Soft Layer with 0.3% Agarose (8 mL) Containing Cells

1. Count the cells and prepare a 5 mL cell suspension containing 12,000 cells/mL in growth medium (*see* **Note 6**).
2. Prepare 5 mL agarose portion by mixing 3.475 mL DMEM with 0.5 mL FBS (10% final in 5 mL), 25 μL P/S (0.5% final in 5 mL), and 1 mL of 3% agarose (0.6% final) (*see* **Note** 7).
3. Mix 4 mL of the cell suspension (*see* **step 1**) with 4 mL of the agarose portion (*see* **step 2**) (1:1) and mix thoroughly by pipetting up and down to obtain 8 mL with a final cell concentration of 6000 cells/mL in 0.3% agarose.
4. Pipette 1 mL of the cell suspension agarose mixture containing a total of 6000 cells slowly from the edge into each well.
5. Avoid or remove bubbles with a needle or pipette tip.
6. Put the plate for 5–10 min at 4 °C until the agar is solidified. This step improves uniform cell distribution within the soft agar layer to ensure colony formation from single cells.
7. Check the plate under the microscope for equal distribution and single cell seeding.
8. Culture the plate at 37 °C and 5% CO_2 in a humidified cell culture incubator.
9. Depending on the cell type, colonies are formed within 5–7 days (Fig. 3) (*see* **Note 8**).
10. Depending on the experimental design, colonies can be counted, treated (*see* Subheading 3.2), traced (*see* Subheading 3.3), or isolated for metabolite quantification (*see* Subheading 3.5).

3.2 Treatment of Soft Agar Cultures

To determine the impact of alteration of the microenvironment, soft agar colonies can be exposed to external treatment, such as inflammatory stimuli LPS. If treatments should be combined with metabolic tracing approaches, tracing media as outlined in Subheading 3.3. can be applied and combined with the treatment. Depending on the experimental design, the timing for treatments and tracing may be adjusted.

1. Prepare 4 mL of treatment medium containing a final concentration of 100 ng/mL LPS (20 μL stock 5000 ng/mL) and 50 ng/mL IFN-γ (20 μL stock 2500 ng/mL) in growth DMEM with 10% FBS and 1% P/S.
2. Prepare 4 mL of control DMEM containing 10% FBS and 1% P/S (*see* **Note 9**).
3. Slowly add 1 mL treatment or control medium from the edge into each culture well (*see* **Note 10**).
4. Incubate the culture plate at 37 °C and 5% CO_2 in a humidified cell culture incubator for 48 h (*see* **Note 11**).

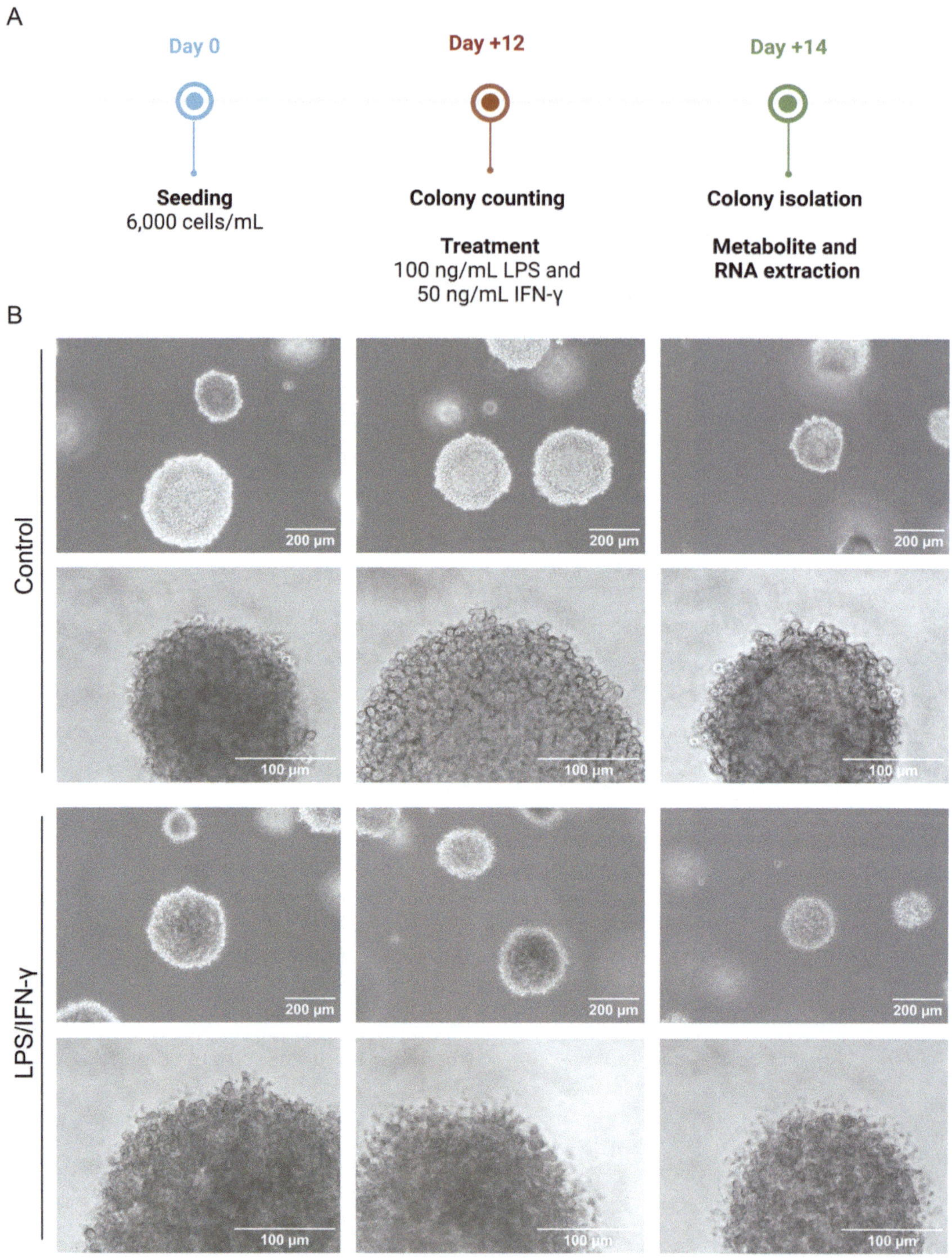

Fig. 3 Colony formation assay of RAW 264.7 cells grown in soft agar with inflammatory stimuli. (**a**) Schematic timeline of the experimental procedure for RAW 264.7 soft agar cultures. (**b**) Representative microscopy images of RAW 264.7 cells after 12 days and 14 days in culture. 6000 cells were seeded per culture well in 0.3% agarose on top of a 0.6% hard layer. After 12 days in culture, cells were treated with 100 ng/mL LPS and 50 ng/mL IFN-γ for 48 h. Images depict colony shapes of the control and after treatment with LPS and IFN-γ. Microscopy images represent colonies of three cellular replicates. The scale bar is 200 μm or 100 μm, as

3.3 Metabolic Tracing of Soft Agar Cultures

A wide variety of stable isotope tracers are available to trace into metabolic pathways, including ^{15}N, ^{13}C, and ^{2}H [13]. In our case study, we cultured soft agar colonies with [U-$^{13}C_6$]glucose to quantify ^{13}C carbon utilization for TCA cycle metabolism in the presence of LPS and IFN-γ. For detailed preparation of the tracing media, we refer to our previous protocol Cordes and Metallo 2019 [14].

1. Prepare 8 mL [U-$^{13}C_6$]glucose tracer medium with 25 mM [U-$^{13}C_6$]glucose tracer and add 4 mM ^{12}C glutamine. Supplement the medium with 10% FBS and 0.5% P/S.
2. The tracing will be combined with treatment strategies (*see* Subheading 3.2). Use 4 mL tracer media for treatment and 4 mL for control conditions.
3. Slowly add 1 mL treatment or control ^{13}C tracing medium from the edge into each culture well. The medium will reach the cells by diffusing through the soft layer (*see* **Note 10**).
4. Incubate the culture plate at 37 °C and 5% CO_2 in a humidified cell culture incubator for 48 h (*see* **Note 11**).

3.4 Quantifying Colony Growth and Morphology

Soft agar assays are widely used to provide insights into growth, and morphological changes are reliably determined with microscopy techniques [15]. Quantify the size and number of colonies before metabolite extraction using an imaging platform. Imaging may also reveal changes in colony morphology introduced by treatment strategies. In our case study, we visualize colonies with Fiji open-source software as depicted in Fig. 3.

3.5 Colony Isolation from Soft Agar

This protocol describes the isolation of colonies from soft agar for subsequent metabolite extraction to quantify metabolite levels, metabolic fluxes, and gene expression from the same sample (Fig. 4).

1. Prewarm 0.9% sodium chloride solution (NaCl) to 55 °C to avoid solidification of the agar (*see* **Note 4**).
2. Add 1 mL of 55 °C prewarmed NaCl to the soft layer of each culture well (*see* **Note 12**).
3. Carefully solve the upper soft layer with the NaCl or scrape it off using a cell scraper.
4. Mix the NaCl with the upper 0.3% agarose soft layer using a 1000 μL pipette tip with the end cutoff to minimize sheer stress

Fig. 3 (continued) indicated in the images. Images depicting colonies were visualized and uncolored with Fiji open-source software (ImageJ 2.9.0/1.53 t Java 1.8.0_322 (64-bit)). Experiments were independently repeated three times with similar results. (Figure **a** was created with Biorender.com)

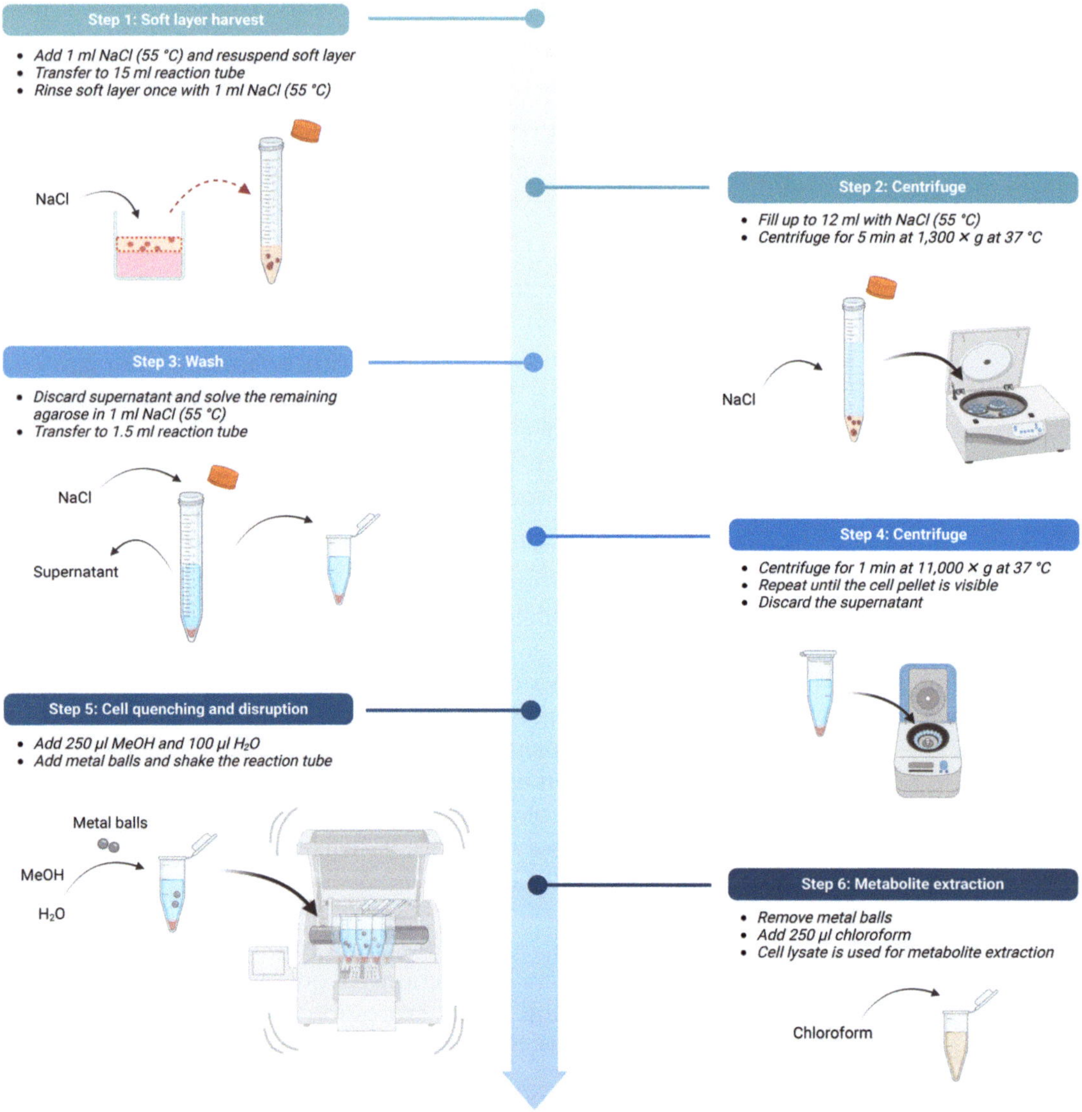

Fig. 4 Protocol for colony isolation from soft agar. The illustration depicts a schematic workflow of the colony isolation protocol, divided into four steps: (1) Soft layer harvest by resuspending the soft layer in 1 mL NaCl and washing the well with 1 mL NaCl (2) Centrifugation step to remove leftover agarose by building a cell pellet (3) Wash step with 1 mL NaCl and transfer to a 1.5 mL reaction tube (4) Centrifuge step to remove remaining agarose (5) Cell quenching and lysis: adding methanol, water with internal standard, and metal balls to disrupt the cells via mechanical shaking (6) Metabolite extraction: adding chloroform for further metabolite extraction. (The figure was created with BioRender.com)

and formation of air bubbles, taking care not to disturb the lower hard layer (*see* **Note 13**).

5. Transfer the cell-agar solution into a 15 mL reaction tube.

6. Wash the leftover 0.6% agarose hard layer once with prewarmed 1 mL NaCl and collect the remaining colonies to the 15 mL reaction tube.
7. Fill up the reaction tubes to 12 mL with prewarmed NaCl.
8. Centrifuge the tubes at 37 °C for 5 min at 1300× *g* in a prewarmed centrifuge (*see* **Note 14**).
9. Remove the supernatant using a serological pipette.
10. Resuspend the cell pellet containing cell colonies with 1 mL prewarmed NaCl by pipetting up and down (*see* **Note 15**).
11. Transfer the solution to a 1.5 mL reaction tube.
12. Centrifuge at 37 °C for 1 min at 11,000× *g*.
13. Remove the remaining supernatant using a serological pipette.
14. Repeat **steps 10–13** until colonies are free from agar (*see* **Note 16**).
15. The pellet contains isolated colonies that can be used for metabolite extraction (*see* Subheading 3.6) or other biological assays.

3.6 Metabolite Extraction from Isolated Soft Agar Colonies

A wide range of metabolite extraction protocols are available for different metabolites and mass spectrometry approaches. Here, we employ a modified Bligh and Dyer method using methanol/water/chloroform [16] for subsequent analysis using a GC-MS analytical platform (*see* Subheading 3.7).

1. Add 0.25 mL −20 °C cold methanol to the reaction tubes containing the cell pellets to quench metabolism.
2. Add 0.1 mL 4 °C cold water containing 5 μg/mL of the internal standard norvaline (*see* **Note 17**).
3. Add two metal balls to each reaction tube (*see* **Note 18**).
4. Disrupt the cells in a ball mill for 1 min at a frequency of 30/s within the frozen Retsch adapter (*see* **Note 19**).
5. Remove the metal balls with a magnet (*see* **Note 20**).
6. Take 25 ul of the methanol/water phase for protein quantification (*see* Subheading 3.8).
7. Add 0.25 mL of −20 °C chloroform to the extracts (*see* **Note 21**).
8. Vortex the cell suspension at 14,000 rpm for 10 min at 4 °C.
9. Centrifuge the reaction tubes at 17,000× *g* for 5 min at 4 °C.
10. Collect 200 μL of the upper aqueous phase (AP, containing hydrophilic metabolites) in an analytical sample vial (Fig. 2).
11. Collect 200 μL of the lower organic phase (OP, hydrophobic metabolites) into an analytical sample vial.

12. Wash the interface with 100 μL methanol and centrifuge at 17,000× *g* for 5 min at 4 °C.
13. Continue with the RNA isolation protocol or store the interface (IF) at −80 °C until use.

3.7 Metabolic Analysis Using Gas Chromatograph-Mass Spectrometry (GC-MS)

Mass spectrometry-based platforms are commonly used for the quantification of metabolite levels and fluxes. In our case study, we analyzed our samples on a gas chromatograph coupled to mass spectrometry (GC-MS) with prior derivatization. All steps are described in detail in our previous protocol including GC-MS method, metabolite fragment ions, and data analysis [14].

1. Evaporate the aqueous upper phase in an analytical sample vial under vacuum in a rotary vacuum evaporator at 4 °C until each sample dried entirely.
2. Derivatize samples using 15 μL MOX reagent (Subheading 2.3) (incubate at 45 °C for 60 min) and 15 μL MTBSTFA reagent (incubate at 45 °C for 30 min).
3. Analyze derivatized samples on GC-MS analytical platform using e.g. an Agilent 7890A gas chromatograph (GC) equipped with a DB-35MS UI (Agilent J&W Scientific, 122-3832 UI) and interfaced with e.g. an Agilent 5977 mass spectrometer (MS). For MTBSTFA derivatized samples, the GC oven is held at 100 °C for 2 min, increased to 300 °C at 10 °C/min, and held at 325 °C for 3 min.
4. Identify metabolites and analyze data using mass spectrometry analysis.
5. Determine mass isotopomer distributions and total metabolite abundances by integrating mass fragments with corrections for natural isotope abundances.
6. Normalize metabolite levels, e.g., to the sum of metabolites or protein content (*see* Subheading 3.8) (Fig. 5).

3.8 Normalization of Metabolite Abundances

Metabolite abundances might be normalized to account for the variation of colony size and number within the culture wells. Metabolite abundances can be normalized for instance to the sum of the metabolite levels (Fig. 5) or the protein content. Here, we provide a protocol for the quantification of the protein content for each sample.

1. Add 25 μL of the methanol/water phase from metabolite extraction (*see* Subheading 3.6, **step 6**) to a 96-well plate.
2. Dry the plate within a chemical fume hood.
3. Add 25 μL protein extraction buffer (e.g., RIPA) to each sample and shake the plate to lyse each sample.
4. Quantify protein content using a protein assay.

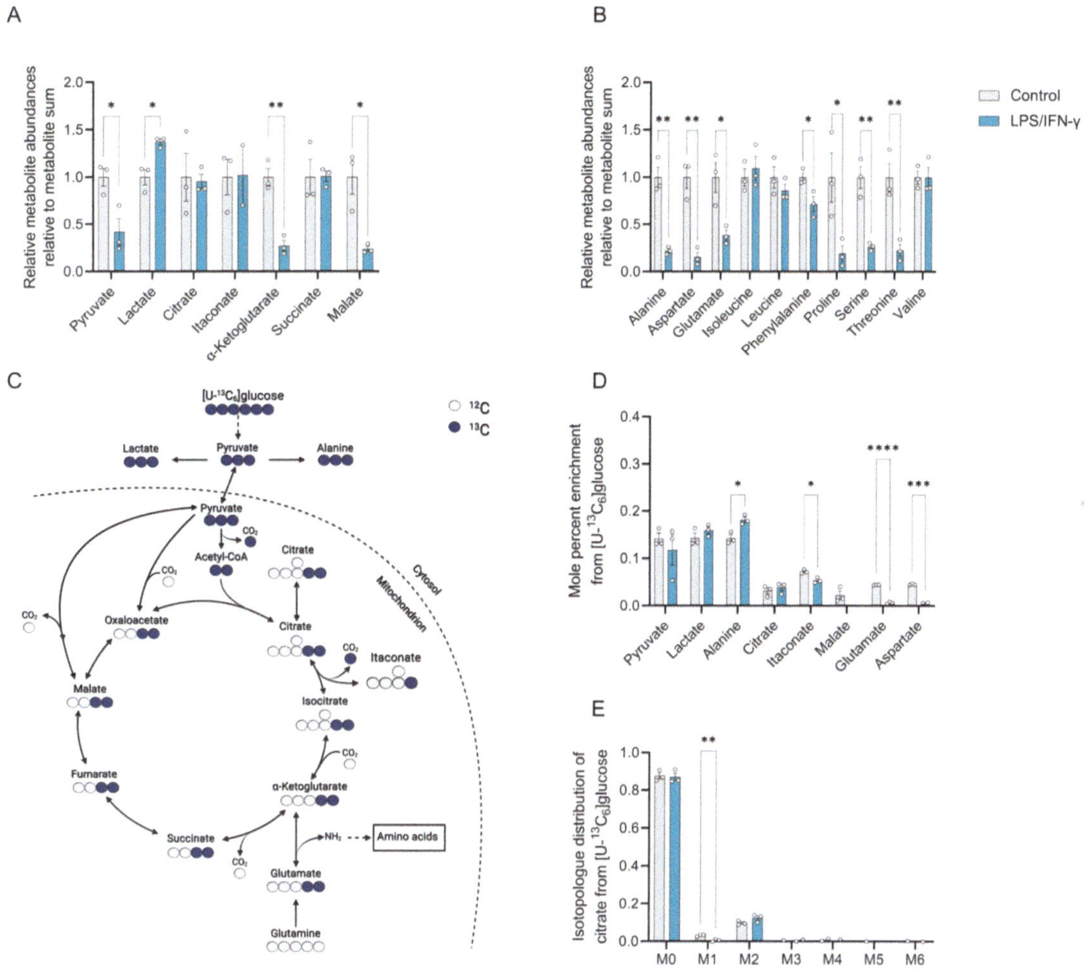

Fig. 5 Metabolite quantification and metabolic flux analysis using stable isotope labeling approaches. Relative metabolite abundances relative to metabolite sum in Arbitrary Units (AU) of (**a**) TCA cycle intermediates and (**b**) amino acids of RAW 264.7 soft agar cultures. 6000 cells per culture well were seeded in 0.3% agarose and colonies were treated after 12 days in culture with 100 ng/mL LPS and 50 ng/mL IFN-γ for 48 h. (**c**) Metabolic map depicting atom transitions from [U-$^{13}C_6$]glucose. ^{13}C carbons are indicated in blue, and ^{12}C carbons are indicated in grey. (**d**) Mole percent enrichment of TCA cycle intermediates and metabolites from [U-$^{13}C_6$] glucose. (**e**) Isotopologue distribution of citrate labeled from [U-$^{13}C_6$]glucose. Data is depicted as barplots ± SEM from three cellular replicates and experiments were independently repeated three times with similar results. *P*-values performed with multiple unpaired *t*-test: $*p < 0.05$, $**p < 0.01$, $***p < 0.001$, $****p < 0.0001$. (Figure **c** was created with BioRender.com)

5. Normalize metabolite abundance to the protein content for each sample.

3.9 Quantification of Gene Expression from Isolated Soft Agar Colonies

A wide variety of RNA isolation protocols are available. In this present study, the RNA was isolated from the interface (IF) obtained during the metabolite extraction (*see* Subheading 3.6).

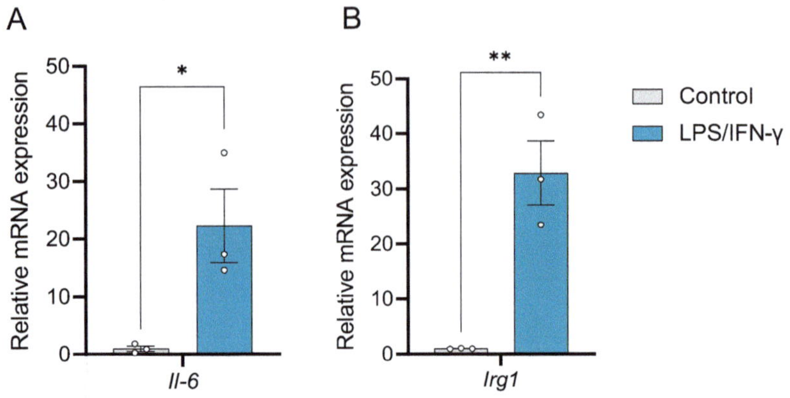

Fig. 6 Quantification of immune-related genes in soft agar cultures. Gene expression of (**a**) *Il-6* and (**b**) *Irg1* relative to housekeeping gene *Rpl27* expression in RAW 264.7 cells. Colonies were treated after 12 days in culture with 100 ng/mL LPS and 50 ng/mL IFN-γ for 48 h. Data is depicted as barplots ± SEM obtained from three cellular replicates each with two technical replicates. The experiment was independently repeated three times with similar results. *P*-values performed with multiple unpaired *t*-test: $*p < 0.05$, $**p < 0.01$

1. Wash the interface (*see* Subheading 3.6, **step 11**) with 100 μL methanol and centrifuge at 17,000× *g* for 5 min at 4 °C.
2. Remove the entire liquid and use the pellet for RNA isolation.
3. Isolate RNA according to the manufacturer's instructions.
4. Quantify RNA concentration by using a microplate reader.
5. Reverse transcribe 10 μL RNA (50 ng/μL) per sample into cDNA.
6. Pipette 2 μL cDNA (5 ng/μL) in the middle of each well of the reaction plate (*see* **Note 22**).
7. Prepare one master mix for each primer pair targeting the genes of interest consisting of 10 μM of reverse and forward primer prepared with ITaq Universal SYBR-Green Supermix. The primer list is depicted in Table 1. The master mix for each sample includes 5 μL SYBR-Green Supermix 1.5 μL forward primer (10 μM) and 1.5 μL reverse primer (10 μM).
8. Add 8 μL of the master mix into each 96 well containing 2 μL of cDNA (*see* **step 6**).
9. Cover the 96-well plate with a foil.
10. Centrifuge the plate at 300× *g* for 30 s.
11. Quantify gene expression and normalize to the housekeeping gene (e.g., *Rpl27*) (Fig. 6).

4 Notes

1. We recommend safe lock tubes to withstand centrifugation forces and to protect against unintentional opening during sample handling.
2. The volume of hard and soft layers can be scaled up and down according to the used culture plate and well size.
3. We recommend working with cellular triplicates for metabolic studies.
4. The agarose glass bottle can be kept in a 55 °C water bath or a beaker filled with 55 °C warm water to avoid solidification until use.
5. Prewarm the serological pipettes to 37 °C to minimize solidification of the agar.
6. The double amount of the cells (12,000 cells/mL) was calculated since the mixture will dilute 1:1 after adding the agarose portion for a final seeding concentration of 6000 cells/mL. Pipette the cell suspension up and down several times to reach a single-cell suspension. The number of cells may be adjusted depending on the cell type and expected growth to observe single colony formation.
7. The double agarose concentration was calculated since the mixture will dilute 1:1 after adding the cell suspension. Treatments such as substrates can be added directly to the agarose portion, while the concentration needs to be doubled accordingly.
8. Colony size, number, and morphology can be determined via bright-field microscopy (Fig. 3). Most cell types will form colonies within 7 days. If colony formation requires more than 7 days, we recommend a feeding layer containing 1 mL of upper 0.3% agarose layer in normal DMEM without cells (*see* Subheading 3.1.2, **steps 2–3**).
9. Some additives may influence the pH of culture media. Sterile filter and adjust all culture media to pH = 7.3.
10. To reduce the risk of disrupting the soft layer, slowly add the medium along the well's edge.
11. The liquid medium may separate the soft layer from the hard layer. Move the plate gently in and out of the incubator. The time for treatments and tracing may be adjusted depending on the experimental design.
12. The colony isolation should be performed as fast as possible to minimize metabolic changes introduced by the extraction procedure. We recommend using a 1000 μL pipette to remove the supernatant as an automatic aspirator may remove the colonies.

13. Pipetting up and down can easily disrupt the soft layer. Cutting the 1000 μL pipette tips to increase the opening or using a cell scraper can also be helpful to scrape off the soft layer from the hard layer. The separation of the hard and soft layers is visible.
14. This step is important to release the colonies from the surrounding agarose. After removing the supernatant, repeat the washing step to remove residual agarose after each centrifugation step.
15. Avoid vortexing as it may destroy colonies and lead to foam, and the colonies will not be visible anymore.
16. Repeat this step until the cell pellet is visible.
17. From this step on, the extracts should be kept at 4 °C.
18. Type and number of metal balls may be adjusted depending on the cell pellet.
19. Use a frozen Retsch adapter for 1.5 mL reaction tubes to keep the extracts cold. Next to normalizing the metabolites to the internal standard norvaline, the metabolites are normalized further to the total metabolite sum. After this step, 25 μL of the cell extract can be transferred to a 96-well plate to determine protein quantity for alternative normalization.
20. Use a magnet to remove the magnet balls before centrifugation of the tubes.
21. We recommend working inside a chemical fume hood for steps where chloroform is involved.
22. We recommend biological and technical replicates.

Acknowledgments

Figures were illustrated with BioRender.com. This work was supported by Cambridge Isotope Laboratories, Inc.

Author Contributions T.C. designed and conceptualized the study, supervised, and secured funding; T.C. and B.D. wrote the manuscript, analyzed data, and prepared figures; T.C., B.D., and F.C. performed the experiments.

References

1. Khare S, Kim LC, Lobel G et al (2021) ASS1 and ASL suppress growth in clear cell renal cell carcinoma via altered nitrogen metabolism. Cancer Metab 9(1):40. https://doi.org/10.1186/s40170-021-00271-8
2. Schafer ZT, Grassian AR, Song L et al (2009) Antioxidant and oncogene rescue of metabolic defects caused by loss of matrix attachment. Nature 461(7260):109–113. https://doi.org/10.1038/nature08268

3. Grimes DR, Kelly C, Bloch K et al (2014) A method for estimating the oxygen consumption rate in multicellular tumour spheroids. J R Soc Interface 11(92):20131124. https://doi.org/10.1098/rsif.2013.1124
4. Sun L, Wang X, Kaplan DL (2011) A 3D cartilage–inflammatory cell culture system for the modeling of human osteoarthritis. Biomaterials 32(24):5581–5589. https://doi.org/10.1016/j.biomaterials.2011.04.028
5. Baker BM, Chen CS (2012) Deconstructing the third dimension: how 3D culture microenvironments alter cellular cues. J Cell Sci 125 (Pt 13):3015–3024. https://doi.org/10.1242/jcs.079509
6. Elia I, Broekaert D, Christen S et al (2017) Proline metabolism supports metastasis formation and could be inhibited to selectively target metastasizing cancer cells. Nat Commun 8: 15267. https://doi.org/10.1038/ncomms15267
7. Parker SJ, Svensson RU, Divakaruni AS et al (2017) LKB1 promotes metabolic flexibility in response to energy stress. Metab Eng 43:208–217. https://doi.org/10.1016/j.ymben.2016.12.010
8. Jiang L, Shestov AA, Swain P et al (2016) Reductive carboxylation supports redox homeostasis during anchorage-independent growth. Nature 532(7598):255–258. https://doi.org/10.1038/nature17393
9. Muthusamy T, Cordes T, Handzlik MK et al (2020) Serine restriction alters sphingolipid diversity to constrain tumour growth. Nature 586(7831):790–795. https://doi.org/10.1038/s41586-020-2609-x
10. Cordes T, Kuna RS, McGregor GH et al (2022) 1-Deoxysphingolipid synthesis compromises anchorage-independent growth and plasma membrane endocytosis in cancer cells. J Lipid Res 63(10):100281. https://doi.org/10.1016/j.jlr.2022.100281
11. van Gorsel M, Elia I, Fendt S-M (2019) 13C tracer analysis and metabolomics in 3D cultured cancer cells. Methods Mol Biol 1862:53–66. https://doi.org/10.1007/978-1-4939-8769-6_4
12. Nakamura D (2023) The evaluation of tumorigenicity and characterization of colonies in a soft agar colony formation assay using polymerase chain reaction. Sci Rep 13(1):5405. https://doi.org/10.1038/s41598-023-32442-6
13. Cordes T, Metallo CM (2016) Tracing insights into human metabolism using chemical engineering approaches. Curr Opin Chem Eng 14: 72–81. https://doi.org/10.1016/j.coche.2016.08.019
14. Cordes T, Metallo CM (2019) Quantifying intermediary metabolism and lipogenesis in cultured mammalian cells using stable isotope tracing and mass spectrometry. Methods Mol Biol 1978:219–241. https://doi.org/10.1007/978-1-4939-9236-2_14
15. Ivascu A, Kubbies M (2006) Rapid generation of single-tumor spheroids for high-throughput cell function and toxicity analysis. J Biomol Screen 11(8):922–932. https://doi.org/10.1177/1087057106292763
16. Bligh EG, Dyer WJ (1959) A rapid method of total lipid extraction and purification. Can J Biochem Physiol 37(8):911–917. https://doi.org/10.1139/o59-099

Chapter 12

Assessing Mitochondrial Respiration of In Vitro Models for the Study of Immune-Related Diseases Using a Seahorse Analyzer

Chloé Chivé and Arnaud Tête

Abstract

Energy metabolism, particularly mitochondrial function, has emerged as a key regulator of the immune response and plays a significant role in the pathophysiology of inflammatory diseases. Studying mitochondrial parameters is therefore crucial in understanding these pathologies. This protocol outlines the evaluation of mitochondrial respiration parameters using a Seahorse Analyzer. With the growing use of innovative models that better mimic human physiology, such as cell cultures in inserts, this chapter provides a detailed guide to Seahorse analysis for this specific type of culture.

Key words Mitochondria, OXPHOS, ETC, Metabolism, Cell culture models

1 Introduction

Inflammatory diseases such as systemic autoimmune diseases are characterized by inappropriately aberrant autoreactive immune responses against its own healthy cells, which ultimately resulting in tissue damage. Energy metabolism is emerging as a key mechanism in the regulation of immune cells for either their activation, differentiation, or function [1, 2]. In particular, it is becoming increasingly evident that mitochondria are the organelles that play a pivotal role in functionality and responsiveness of immune cells [2, 3]. Moreover, mitochondria are also involved in epithelial cells and their implication in the immune response. Mitochondrial dysfunction has been demonstrated to play a role in the response to inflammatory stimuli in surrounding parenchymal epithelial cells [4, 5]. Besides, an alteration of epithelial cell mitochondrial function could lead to activation or modulation of immune cells by releasing mitochondrial constituents or metabolic byproduct then can act as mitochondrial damage-associated molecular patterns (mtDAMPs) [6]. It is also known that epithelial cells are important

Rami Bechara (ed.), *Autoimmunity: Methods and Protocols*, Methods in Molecular Biology, vol. 2990,
https://doi.org/10.1007/978-1-0716-4997-8_12,

for the establishment of the immune response to viral infection and that mitochondria are involved in this antiviral response. Infection by RNA viruses is recognized by pattern recognition receptors (PRRs), mainly the viral sensor retinoic acid-inducible gene-I (RIG-I), whose activation involved mitochondria, and leads to the release of cytokines and interferons required for the recruitment of immune cells to the site of infection [7, 8]. The antiviral response is dependent on the function and quality of the mitochondrial network, as studies have shown the influence of reactive oxygen species, mitochondrial membrane potential, and autophagic flux in inhibiting RIG-I-mediated antiviral signaling [9–11]. It is therefore essential to take an interest in energy metabolism and especially mitochondrial function when studying inflammatory diseases. Mitochondria has multiple roles in numerous metabolic pathways including the generation of adenosine triphosphate (ATP) through oxidative phosphorylation (OXPHOS). This process is essential for providing the energy required for a multitude of physiological processes and for synthesizing intermediates. Glucose that enters the cells is converted into pyruvate and lactate during the glycolysis. Pyruvate is then imported in the mitochondria, where it is used as a substrate for the tricarboxylic acid cycle (TCA cycle), also referred to as the citric acid cycle or the Krebs cycle. The TCA cycle is closely linked to OXPHOS, by generating the reducing equivalents NADH and $FADH_2$ which are required for the transfer of electrons through the electron transport chain (ETC), also known as the mitochondrial respiratory chain, located in the inner mitochondrial membrane. The electrons transfer through the various complexes (complexes I to IV) of the ETC resulting in the generation of a mitochondrial membrane potential, which is then used to produce ATP by complex V. This process necessitates the consumption of oxygen and is referred as OXPHOS [12]. Furthermore, lactate, a byproduct of glycolysis, is exported from cells along with protons through specific monocarboxylate transporters (MCTs), resulting in an extracellular acidification. One common method for analyzing energy metabolism is to measure glycolytic and mitochondrial respiration capacities with an assay using a Seahorse analyzer (Agilent Technologies). Among these assays, the Seahorse XF Cell Mito Stress Test is a rapid, effective, and useful tool for evaluating multiple parameters of mitochondrial function, including basal respiration, maximal respiration, and ATP-linked respiration. These different parameters are obtained by measuring oxygen consumption rate (OCR), which reflect OXPHOS, using sequential modulators of the mitochondrial ETC. This protocol provided a step-by-step guidance of the rapid and easy implementation of this technology, particularly the preparation of the Seahorse XF Cell Mito Stress Test, from cell seeding to results analysis. In addition, the distinctive feature of this protocol is the delineation of methodologies for measuring mitochondrial respiration, not only in adherent or non-adherent cells cultivated on culture plates, but also in cells grown on cell culture inserts for which this protocol has been optimized and presented here. Indeed, this type of cell culture can be employed, for example, for the study of cellular communication in co-culture or for tissue engineering.

2 Materials

2.1 Solutions, Supplies, and Equipment Required

1. Seahorse analyzer. These protocols are optimizing using a Seahorse XFe24 analyzer (i.e., the 24-well plate format) from Agilent Technologies.
2. Cell assay medium (*see* **Note 1**).
3. Cell media supplements: 1 M glucose, 100 mM pyruvate, and 200 mM glutamine.
4. Seahorse XF Calibrant solution.
5. Seahorse XFe24 sensor cartridges (Agilent Technologies).
6. Seahorse XF24 cell culture plates and XF24 Islet Capture Microplates (Agilent Technologies).
7. Seahorse Islet Capture Screen Insert Tool (Agilent Technologies).
8. Cell culture inserts.
9. Biopsy punch Ø 3 mm.
10. Cell Mito Stress Test inhibitors: 1 mM oligomycin, 1 mM FCCP (carbonyl cyanide-4(trifluoromethoxy)-phenylhydrazone), 1 mM antimycin A, and 1 mM rotenone.
11. Cell surface coating reagent, e.g., Poly-L-lysine 0.01% (w/v) in H_2O, Cell-Tak, or polyethyleneimine (stock solution: 20 mg/mL in 150 mM NaCl).
12. Phosphate-buffered saline (PBS).
13. 20 mM Hoechst 33342 for nuclear staining.
14. RIPA (25 mM Tris–HCl (pH 7.6), 150 mM NaCl, 1% NP-40, 1% sodium deoxycholate, 0.1% sodium dodecyl sulfate) lysis buffer supplemented with protease inhibitors for cell lysis and protein extraction.
15. Pierce™ BCA Protein assay kit.
16. Automated Cell Imaging System.
17. 37 °C non-CO_2 incubator.
18. 37 °C water bath.

Table 1
Cell assay medium preparation (usual concentrations)

Supplements	Stock solution concentration	Final concentration
Glucose	1 M (100X)	10 mM
Glutamine	200 mM (100X)	2 mM
Pyruvate	100 mM (100X)	1 mM

Table 2
Preparation of drugs solutions used in the Cell Mito Stress Tests (mitochondrial respiration modulators and nuclear staining)

Compounds	Stock solution concentration (mM)	Stock solution volume (μL)	Cell assay medium volume (μL)	Final concentration in cartridge port: at 10X (μM)
Oligomycin	1	30	1970	15
FCCP	1	20	1980	10
Rotenone/antimycin A/Hoechst 33342	1/1/20	10/10/3	1977	5/5/30

FCCP carbonyl cyanide-4(trifluoromethoxy)-phenylhydrazone

2.2 Preparation of the Cell Culture Media and the Mitochondrial Respiration Modulators and Nuclear Staining Solutions

1. Prepare the cell assay medium (*see* **Note 1**) by the supplementation of the Seahorse XF assay medium buffered to pH 7.4 with glucose, pyruvate, and glutamine. The typical concentrations of glucose, glutamine, and pyruvate in most cells are 10 mM, 2 mM, and 1 mM, respectively (Table 1), but can be adapted to suit the cell type or the scientific question being asked. This medium can be prepared 1 day before the experiment and store at 4 °C and pre-heat to 37 °C before use.

Prepare the mitochondrial respiration modulators and Hoechst 33342 working solutions as described in Table 2. The concentrations of the drug solutions should be adapted in accordance with the cells employed. To facilitate the selection of appropriate concentrations, *see* **Note 2**, which provides instructions on guidance on the use of the publications database. These drug solutions will be filled in sensor cartridge port at a 10X concentration and injected into the cell assay medium during the assay.

3 Methods

This section outlines the experimental procedure, which begins with the preparation of the cell cultures to be tested. Three distinct procedures are described in this protocol, depending on whether the cells are adherent, non-adherent, or if they are cultivated cell culture inserts. Thereafter, the workflow of the Seahorse Cell Mito Stress Test is then described in detail.

3.1 Placing Cells in Seahorse XF24 Cell Culture Plates

If the cell type of interest (either if it is a cell line, primary cell or induced pluripotent cell) is grown attached on the surface of culture plates, it can be seeded and cultivated on Seahorse XF24 cell culture plates before the assay or can alternatively be grown on other support and be seeded a few hours prior to running the assay. Importantly, the first consideration is to choose the optimal cell-seeding density (*see* **Note 2**).

3.1.1 Adherent Cells

1. Trypsin, harvest and count the cells.
2. Re-suspend the cell to the previously defined final concentration to seed in 100 μL of culture medium. Typically, it will be around 1.10^4 and 1.10^5 cells per well if the cells are seeded the day before the assay (*see* **Note 2**).
3. Seed the cells on the Seahorse XF24 cell culture plates with the 100 μL of cell suspension. Do not add any cells to the wells designated for the background (culture medium only). It is recommended to leave the plate at room temperature in the tissue culture hood for 1 h. This may prove beneficial in promoting even cell distribution and in reducing the edge effect.
4. Once the cells are adhered well, add 150 μL of cell medium (add it gradually to the sides of the well, taking care not to disrupt the recently attached cells).

3.1.2 Non-adherent Cells

Cells that grow in suspension and do not naturally adhere to bottom of the well such as lymphocytes need to be attached to plate using coating factor. Several choices are possible, with CellTak and poly-lysine being the most commonly used. This protocol details the procedure using polyethyleneimine (PEI) as an attachment factor. PEI is an inert hydrophobic polymer with adhesive properties for cell culture [13] and a less expensive alternative than other coating factor.

1. Prepare the PEI working solution by diluting PEI stock solution to a final concentration of 25 μg/mL in 150 mM NaCl. Sterile filter the diluted PEI working solution on 0.22 μm membrane filter.
2. Coating the Seahorse XF24 cell culture plates by adding 100 μL of PEI at 25 μg/mL in each well and incubate for 30 min at room temperature.
3. Remove the PEI solution and wash three times with PBS. PEI solution can be used up to three times.

4. Plate the cells on the PEI-coated Seahorse XF24 cell culture plates as described in Subheading 3.1.1 for adherent cells.

3.1.3 Cells on Culture Inserts

As cells cultivated on culture inserts cannot be grown directly on Seahorse XF24 cell culture plates, the protocol must be optimized. The membrane must be cut and transferred to the bottom of the Seahorse XF24 cell culture plate (Fig. 1a). This procedure requires the use of Seahorse XF24 Islet Capture Microplates, which were initially developed to measure the mitochondrial respiration of pancreatic islet cells, to hold the insert piece at the bottom of the well, and to prevent it from moving during analysis and touching the Seahorse probes, which can damage cells and alter measurements (Fig. 1a).

1. Rinse apical and basolateral compartments of cell insert with PBS.
2. Remove the insert containing the membrane with the cell from the plate of culture and placed them on a glass slide, as illustrated in Fig. 1a.
3. Using a biopsy punch (Ø 3 mm), excise a piece of membrane. Place the membrane carefully in a bottom of the Seahorse XF24 Islet Capture Microplates with the cells facing upwards, using tweezers. For 12-well inserts (area of 1.12 cm^2), the same membrane can be used for up to three excisions, allowing technical replicates to be carried out (Fig. 1b).
4. Pre-wetting the islet capture grid in Seahorse assay medium (*see* Subheading 3.2.1). This step is important to minimize the formation of bubbles which can alter the measurements.
5. Use the Islet Capture Screen Insert Tool to take the pre-wetting grid and carefully place it at the bottom of the wells, just above the membranes with cells (Fig. 2). The plate is ready for the assay to be run.

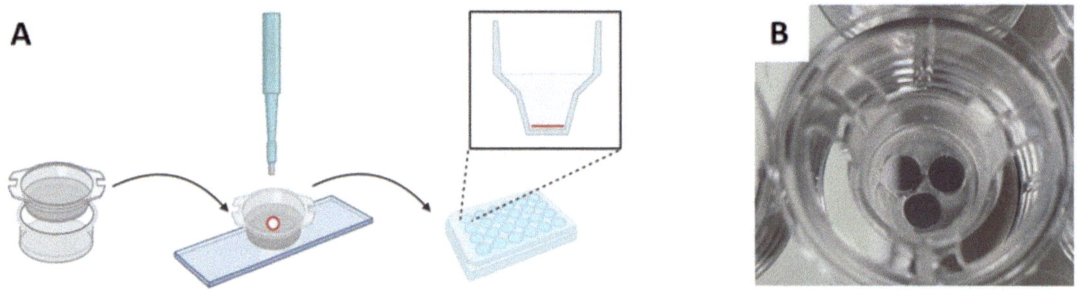

Fig. 1 Protocol used to perform Seahorse analysis on cells grown in culture inserts. Transfer process of a piece of insert cultivated with cells to the wells of the Seahorse XF24 Islet Capture microplate (**a**). Picture of one 12-well insert where three samples were recovered using a biopsy punch (**b**). (Created with BioRender.com)

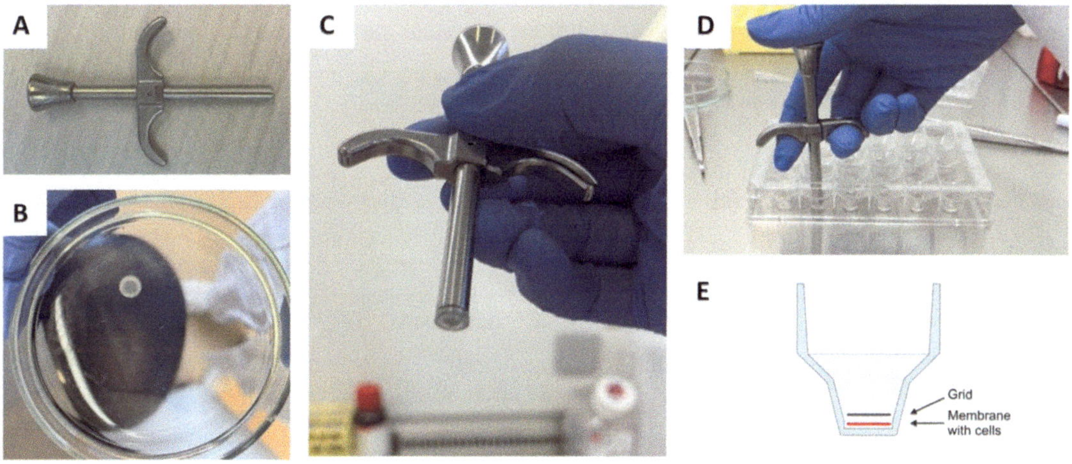

Fig. 2 Preparation of Seahorse XF24 Islet Capture Microplates. Seahorse Islet Capture Screen Insert Tool (**a**) and islet capture grid (**b**) are used to hold the punch at the bottom of the well. Transfer of the grid on the bottom of Seahorse XF24 Islet Capture Microplate wells (**c**, **d**). Illustration of the membrane and grid at the bottom of the wells (**e**). (Created with BioRender.com)

3.2 Assessing Mitochondrial Respiration Using Seahorse Mito Stress Test Assay

The methodology for the preparation and execution of the experiment is identical, regardless of the type of cell culture employed. This assay allows parameters of mitochondrial function to be monitored by measuring the OCR of live cells directly on the Seahorse XFe24 analyzer in real time. The day before running this assay, it is first necessary to hydrate the sensor cartridge and warm up the seahorse analyzer.

3.2.1 Hydration of the Sensor Cartridge and Instrument Warming Up

Hydration of the sensor cartridge must be done 1 day before the assay, according to this method:

1. Open the Agilent Seahorse XFe24 Extracellular Flux Assay kit that contains the Seahorse sensor cartridge and a utility plate.
2. Remove the Seahorse sensor cartridge upside down on the bench.
3. Fill each well of the utility 24-well plate with 1 mL of Seahorse XF Calibrant solution.
4. Re-place the cartridge on the utility plate and place them in a 37 °C non-CO_2 incubator overnight.
5. Turn on the Seahorse analyzer in order to warm the machine overnight to ensure a constant and uniform temperature (minimum 5 h). It is feasible to pre-fill the plate template and protocol in the Wave software (Agilent Technologies).

1. Prepare the cell assay medium, the mitochondrial respiration modulators, and nuclear staining solutions as described in Subheading 2.2.

Table 3
Loading of drug solutions in sensor cartridge ports

Sensor cartridge port	Drug solution	Volume
A	Oligomycin	56 μL
B	FCCP	62 μL
C	Rotenone/antimycin A/Hoechst 33342	69 μL

3.2.2 Assessing Mitochondrial Respiration Using Mito Stress Test Protocol

2. Heat the cell assay medium to 37 °C in a water bath.
3. For cell preparation, refer to the preceding paragraphs on the specific requirement depending of the type of cell culture.
4. Wash the cells with 500 μL of the supplemented cell assay medium.
5. Add 500 μL of the supplemented cell assay medium and place the plate at 37 °C in a non-CO_2 incubator for 45 min to 1 h prior to the assay.
6. During this time, load the drug solutions into the ports of the hydrated sensor cartridge, as indicated in Table 3. Port D does not need to be filled in for the Mito Stress assay.
7. Start the program on Wave Software to initiate the equilibration of the sensor cartridge.
8. Place the sensor cartridge and the utility plate into the Seahorse XFe24 Analyzer with care to discard the lid and the pink hydro booster.
9. After equilibration of the sensor cartridge (approximatively 20 min), discard the utility plate and replace it by the cell culture plate previously prepared.
10. Run the assay.

Following the completion of the test, the resulting data are expressed as OCR, with measurements taken in each individual well. It is necessary to normalize these results in accordance with the cell number per well as variations may occur during plate preparation. These variations may be either due to the effect of a treatment that induces cell proliferation or death or technical bias during sample preparation. However, it is important to note that these differences will influence the OCR results and require normalization. It is nevertheless necessary to carry out technical replicates (minimum 3) per condition. A straightforward and pertinent approach to normalizing these results is to divide the OCR by the number of cells counted after nuclear staining with Hoechst 33342. Hoechst 33342 is a widely used cell-permeable DNA-binding fluorescent probe that emits blue fluorescence, which can be detected with a DAPI filter.

3.2.3 Cell Counting or Total Protein Estimation for Normalization

For this purpose, Hoechst 33342 is added to sensor cartridge port C, along with rotenone and antimycin A (Table 2). After the assay, the number of individual cells is estimated by identifying the Hoechst-marked nuclei. No additional incubation time is required for labelling. The exposure time of the cells between the injection of port C and the end of the test is sufficient.

1. Load the plate in an automated Cell Imaging System (e.g., Imagexpress Pico from Molecular Devices) and acquire image with DAPI filter (or equivalent).
2. Using a preconfigured protocol, analyze the cell count. This parameter will be used to normalize data as described in Subheading 3.2.4.

If the cell labelling is not feasible, the results can be normalized according to the quantity of total proteins, which serves as an indirect marker of cell quantity. This is particularly relevant for cells grown in inserts, as there may be bubbles and detached cells as a result of the protocol with punches and the use of the grid may potentially interfere with the Hoechst 33342 reading. In such instances, the cells can be lysed in RIPA lysis buffer supplemented with protease inhibitors, and the protein concentration can be determined using a BCA (bicinchoninic acid) assay, as the Pierce™ BCA Protein assay kit from ThermoFisher Scientific.

3.2.4 Results Analysis

As a reminder, with a workflow of utilization of successive mitochondrial respiration modulators deposit in the sensor cartridge port and directly injected in assay media (Fig. 3a), real-time

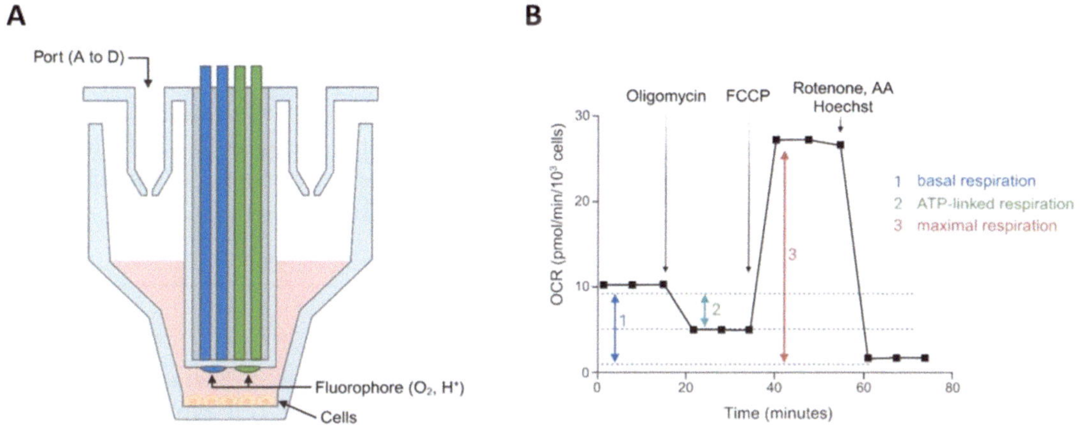

Fig. 3 Schema of the Seahorse and the MitoStress assay methods. Inhibitors are injected sequentially via ports A–D, and oxygen consumption and proton production are measured by probes in wells (**a**). Oxygen consumption rate (OCR) is measured over time after sequential injection of oligomycin, FCCP, and rotenone/antimycin A to calculate basal respiration, respiration related to ATP production, and maximal respiration (**b**). (Created with BioRender.com)

measurement of OCR allows calculation of mitochondrial parameters (Fig. 3b).

The first and most important parameter read is the basal level of respiration, which corresponds to cell respiration in the culture medium. The first drug injected is oligomycin, an inhibitor of complex V of the mitochondrial respiratory chain (i.e., ATP synthase), which caused a decrease in OCR, which is used to calculate the ATP-linked respiration. FCCP is then injected and acts as an uncoupler, inhibiting the flow of electrons through the ETC, causing an increase of the OCR to reach its maximum level, allowing the assessment of maximal mitochondrial respiration and spare capacity (corresponding to maximal respiration minus basal respiration or the ability of cells to increase their energy level via OXPHOS). Then, the final injection is a mixture of antimycin A and rotenone, which inhibit complexes III and I of the mitochondrial respiratory chain, respectively. This results in the cessation of mitochondrial respiration and the measurement of a minimal level of mitochondrial OCR. It should be noted that in the previous detailed protocol, the last injection includes Hoechst 33342 to stain nuclei used for normalization.

Thanks to Wave software, the OCR and these metabolic parameters measured with the Mito Stress Test are automatically calculated for each well from OCR measurement. In the software, it is then necessary to normalize the data as described in Subheading 3.2.4. For cells grown directly in Seahorse plates and for which Hoechst-labeled cell counting is possible, normalization is performed by indicating the estimated number of cells for each well and OCR is expressed as pmol/min/cells. For cells grown on inserts and for which the punch method is used, normalization is performed on the software with the amount of total protein, and the OCR is expressed in pmol/min/μg protein. The software then provides the above parameters. For experiments with punches, it is recommended to make three punches per insert and then average the normalized OCRs to determine the mean OCR for the insert. A summary of all results can be exported as an Excel file using the Seahorse XF Cell Mito Stress Test Report Generators.

4 Notes

1. The cell assay medium is a dedicated Seahorse XF assay medium, which is dependent on the cell type in question. The Seahorse XF assay media are based on the standard culture media DMEM (Dulbecco's Modified Eagle medium) or RPMI (Roswell Park Memorial Institute medium) and are buffered to pH 7.4, with optimization for the different Seahorse assays. Select the appropriate Seahorse XF DMEM medium or RPMI

medium, depending on the medium in which your cells are preferentially grown. These media must be supplemented with glucose, pyruvate, and glutamine (*see* Table 1).

2. The degree of cell confluence at the time the assay is particularly a key criterion. The reading area is relatively limited, and a deficiency in cells can introduce bias into the results. Furthermore, the level of cell confluence can influence the energy level of the cells. Indeed, excessively confluent cells will proliferate less, resulting in a reduction in energy demand and a concomitant decrease in the basal respiratory rate. This can mask obscure any differences between the various conditions. The optimal number of cells to seed varies depending on the cell type, with larger and more spread cells requiring fewer cells to achieve confluence. Ideally, confluence should be around 90% on the day of the assay. A valuable resource for determining the number of cells to seed according to their type is the Agilent Life Science Publication Database, accessible via the Agilent website (www.agilent.com). The Agilent Life Science Publication Database offers a convenient platform for searching scientific publications that cite studies utilizing specific cell types and protocols, including the Cell Mito Stress Test. This search function enables the user to obtain information regarding the experimental conditions employed by other researchers in the conduct of assays that may be of interest. Furthermore, it is useful to adapt the concentration of drugs employed in the assay.

Acknowledgments

The authors would like to thank Cyto2BM platform of BioMedTech Facilities of Université Paris Cité INSERM US36 | CNRS UAR2009 | Université Paris Cité. The optimization of the method for cells grown on inserts was part of a project funded by the French Foundation for Medical Research (FRM), the French Ecological Transition Agency (ADEME), and the major area of interest on air quality in Ile-De-France (DIM Qi^2).

References

1. Patel CH, Leone RD, Horton MR, Powell JD (2019) Targeting metabolism to regulate immune responses in autoimmunity and cancer. Nat Rev Drug Discov 18(9):669–688
2. Blanco LP, Kaplan MJ (2023) Metabolic alterations of the immune system in the pathogenesis of autoimmune diseases. PLoS Biol 21(4): e3002084
3. Mills EL, Kelly B, O'Neill LAJ (2017) Mitochondria are the powerhouses of immunity. Nat Immunol 18(5):488–498
4. Kiss AL (2021) Inflammation in focus: the beginning and the end. Pathol Oncol Res 27: 1610136

5. Kominsky DJ, Campbell EL, Colgan SP (2010) Metabolic shifts in immunity and inflammation. J Immunol 184(8):4062–4068
6. Marchi S, Guilbaud E, Tait SWG, Yamazaki T, Galluzzi L (2023) Mitochondrial control of inflammation. Nat Rev Immunol 23(3): 159–173
7. Seth RB, Sun L, Ea C-K, Chen ZJ (2005) Identification and characterization of MAVS, a mitochondrial antiviral signaling protein that activates NF-kappaB and IRF 3. Cell 122(5):669–682
8. Chen X, Liu S, Goraya MU, Maarouf M, Huang S, Chen J-L (2018) Host immune response to influenza a virus infection. Front Immunol 9:320
9. Tal MC, Sasai M, Lee HK, Yordy B, Shadel GS, Iwasaki A (2009) Absence of autophagy results in reactive oxygen species-dependent amplification of RLR signaling. Proc Natl Acad Sci USA 106(8):2770–2775
10. Koshiba T, Yasukawa K, Yanagi Y, Kawabata S (2011) Mitochondrial membrane potential is required for MAVS-mediated antiviral signaling. Sci Signal 4(158):ra7
11. Villeret B, Dieu A, Straube M, Solhonne B, Miklavc P, Hamadi S, Le Borgne R, Mailleux A, Norel X, Aerts J, Diallo D, Rouzet F, Dietl P, Sallenave J-M, Garcia-Verdugo I (2018) Silver nanoparticles impair retinoic acid-inducible gene I-mediated mitochondrial antiviral immunity by blocking the autophagic flux in lung epithelial cells. ACS Nano 12(2):1188–1202
12. Martínez-Reyes I, Chandel NS (2020) Mitochondrial TCA cycle metabolites control physiology and disease. Nat Commun 11(1):102
13. Vancha AR, Govindaraju S, Parsa KVL, Jasti M, González-García M, Ballestero RP (2004) Use of polyethyleneimine polymer in cell culture as attachment factor and lipofection enhancer. BMC Biotechnol 4:23

Chapter 13

Investigating Dysregulated X Chromosome Inactivation in Human B Cells and Plasmacytoid Dendritic Cells by RNA FISH

Berenice Faz-Lopez, Sophie Allart, Julie Chaumeil, and Jean-Charles Guéry

Abstract

RNA fluorescent in situ hybridization (RNA FISH) has been widely used to monitor the allelic expression of primary transcripts and/or non-coding RNAs in individual cell nuclei. In particular, this method is a gold standard to establish which X-linked genes undergo or escape X chromosome inactivation in female mammalian cells by combining probes specific of the X-linked transcripts of interest with probe specific of the long-non-coding *XIST* RNA that coats the inactive X (Xi) territory. Here we describe the dual RNA FISH protocol we have developed to track (1) the coating of the Xi by the *XIST* RNA and (2) the expression of the X-linked *TLR7* gene on the active and inactive X chromosomes, in human plasmacytoid dendritic cells (pDCs) and in female B lymphocytes induced to differentiate into plasmablasts or pre-Antibody Secreting Cells (pre-ASC) upon activation of either TLR7 or TLR9.

Key words X chromosome inactivation, RNA FISH, TLR7, *XIST* RNA, Human immune cells, B cells, Plasmablasts, Pre-Antibody Secreting Cells, Plasmacytoid dendritic cells

Abbreviations

FISH	Fluorescent In Situ Hybridization
PB	Plasmablast
pDC	Plasmacytoid Dendritic Cells
pre-ASC	Pre-Antibody Secreting Cells
TLR	Toll-Like Receptor
XCI	X Chromosome Inactivation
XIST	X Inactive Specific Transcript

Rami Bechara (ed.), *Autoimmunity: Methods and Protocols*, Methods in Molecular Biology, vol. 2990,
https://doi.org/10.1007/978-1-0716-4997-8_13,

1 Introduction

In female mammals, one of the two X chromosomes undergoes random inactivation to equalize the expression of X-linked genes between XX females and XY males [1]. However, X chromosome inactivation (XCI) is incomplete in humans, and 15–23% of X-linked genes escape XCI, with heterogeneity between cell types, tissues, and XX individuals [2, 3]. The X chromosome carries many immunity-associated genes, including the ssRNA sensing TLRs, *TLR7* and *TLR8*, which have been shown to escape XCI to variable levels in immune cells [4–7]. Because of the strong link between number of X chromosomes and the risk of autoimmunity together with the observations that upregulated expression of the X-linked genes, *TLR7* or *TLR8*, alone is sufficient to induce autoimmunity in otherwise wild-type animals [8, 9], it has been hypothesized that gene-dose effects due to escape from XCI of immune regulatory genes could contribute to the sex bias in systemic rheumatic diseases [10–12]. Indeed, recent studies suggest that dysregulated XCI maintenance, associated with reactivation of immune-regulatory genes, in particular *TLR7*, from the inactive X chromosome (Xi), can promote systemic autoimmunity in females [13–15]. Although expression and coating of the long non-coding *XIST* RNA on the future inactive X chromosome was thought to be essential only for the initiation phase of XCI [1], emerging evidence suggests that maintenance of Xi-gene silencing may also require the accumulation of *XIST* on the Xi (called "*XIST* cloud") and the subsequent recruitment of heterochromatic histone marks [11, 12]. Therefore, investigating whether alterations in XCI may occur in immune cells from patients affected by sex-biased autoimmune diseases is of particular interest [12].

It has been shown that *XIST* RNA localization on the Xi is impaired in primary naive B and T lymphocytes and plasmacytoid dendritic cells (pDCs), both in humans and mice, but comes back after activation/stimulation [16]. As we wished to study *XIST* RNA accumulation on the Xi as well as the allelic expression status of *TLR7* in human pDCs, we used an in vitro culture system with human recombinant IL-3 to promote the *XIST* RNA relocalization to the Xi in a large proportion of cells >60–70% [15]. This allowed an easy detection of the Xi in nuclei to determine the expression of the *TLR7* gene from the Xa and the Xi chromosomes, but also to perform combined RNA and DNA FISH analyses on the same nuclei as described elsewhere [15, 17]. We have also applied the same RNA FISH protocol for the assessment of *TLR7* allelic expression in plasmablasts (PB) generated from $CD27^{neg}$ naïve $CD19^{+}$ B cells upon simulation with different TLR-ligands in the presence of cytokines [4]. Similarly, RNA FISH analysis was also performed on naïve B cells stimulated with TLR-ligands, BCR cross-linking, and IFN-γ to induce "ABC"-like $CD11c^{+}T\text{-}bet^{+}$ pre-antibody secreting cells (pre-ASCs) as described [18]. In this protocol, we detail the experimental procedure to assess *XIST* RNA localization and *TLR7* allelic expression by RNA FISH in these different in vitro cultures of human pDCs, PB, and B cells. The generation of the human *XIST* and *TLR7*-specific probes is also explained. We finally describe the method to quantify the proportions of nuclei with (i) type I to IV *XIST* clouds and with (ii) mono or biallelic *TLR7* primary transcript expression, using the machine learning object classification tool of the Imaris software (Bitplane AG, Switzerland).

2 Materials and Reagents

Prepare all solutions using UltraPure distilled RNAse/DNase-free water and analytical grade reagents. Prepare and store reagents at 4 °C (unless indicated otherwise). Diligently follow all waste disposal regulations when disposing waste materials.

2.1 For Human pDC, Naïve B Cell, PB, and ABCs Preparation

All the materials needed for cell isolation have to be manipulated and prepared under sterile conditions into the hood. The solutions need to be sterile, if not, they have to be filtered with a sterile syringe filter of 0.22 μm for small volumes or with a 500 mL bottle -top vacuum filter system of 0.22 μm for higher volumes.

2.1.1. RPMI-10 % FBS complete medium (R10): RPMI medium, 10% FBS, 2 mM L-glutamine, 50 μM β-mercaptoethanol, 1 mM sodium pyruvate, 1X penicilin-streptomycin, 1X non-essential aminoacids.

2.1.2. PBS-FBS-EDTA buffer: 1X PBS, 1% FBS, 2 mM EDTA.

2.1.3. Kits for cell isolation by negative selection: EasySep™ Human B cell Isolation Kit (StemCell), EasySep™ Human Plasmacytoid DC Enrichment Kit (StemCell).

2.1.4. EasySep™ magnets (StemCell).

2.1.5. 5 mL polystyrene round-bottom tubes.

2.1.6. Cytokines and interferons: IL-3 (Peprotech), IFN-γ (Peprotech), IFN-β (Peprotech), IL- (Peprotech), IL-21 (Peprotech), CL097 (InvivoGen), CpG-B (InvivoGen).

2.1.7. Cell Trace Violet (CTV, Thermo Fisher Scientific).

2.1.8. Antibodies for purity Assessment: For naïve B cells and PB: Anti-Human CD19-PE-Texas (clone LT19, Miltenyi), Anti-Human CD27-APC (clone M-T271, Biolegend). For ABCs:Anti-Human CD19-PE-Texas red (Miltenyi), Anti-Human CD27-BV785 (clone O323, Biolegend), Anti-Human IgD-FITC (clone IA6-2, Biolegend), Anti-

Human CXCR3-APC (clone REA232, Miltenyi), Anti-Human CD11c-BV421 (clone B-ly6, BD Biosciences). For pDCs:Anti-Human BDCA4-APC (clone REA774, Miltenyi), Anti-Human CD123-FITC (clone 6H6, Biolegend).

2.2 For RNA Fluorescent In Situ Hybridization (RNA FISH)

2.2.1. Glass slides coated with poly-L-lysine (Epredia) (Alternative: glass slides incubated with Poly-L-lysine (Sigma-Aldrich) for >15 min at 37 °C, washed twice with 1X PBS at room temperature and air dried).

2.2.2. Cytoskeletal CSK-RVC-X solution:

2.2.2.1. 100 mM PIPES buffer stock solution (stock can be kept in the fridge for several weeks): 1.5 g PIPES dissolved in 40 mL H_2O and 1.5 mL of 10 M NaOH; pH adjusted to 6.8–7 with around 1.35 mL of 5 M HCl, and final volume adjusted to 50 mL with H_2O.

2.2.2.2. Cytoskeletal (CSK) buffer (prepared just before use): 100 mM NaCl, 300 mM sucrose, 10 mM PIPES (use the 100 mM stock solution), 3 mM MgCl2. First, dissolve the sucrose in sterile H_2O, adjust pH to 6.8 if necessary and filter with a 0.22 μm filter.

2.2.2.3. Final CSK-RVC-X buffer (prepared just before use): 2 mM Ribonucleoside Vanadyl Complex (RVC, NEB), 0.5% X-100 Trition, CSK buffer. Prepare a stock solution of 10% Triton X-100 in a 50 mL falcon (can be kept at RT for months **in the dark**, check each time that it still makes foam/ bubbles by shaking the falcon).

2.2.2.4. 2% Formaldehyde: prepared from 16% stock solution (formaldehyde without methanol, ChIP grade) diluted in sterile 1X PBS.

2.2.2.5. 70%, 85%, 95% Ethanol solutions: prepared from absolute Ethanol and ultrapure H_2O.

2.2.3. Probe preparation:

2.2.3.1. *XIST* and *TLR7* (F + R) primers (*see* Tables 1 and 2).

2.2.3.2. Genomic DNA from female PBMCs (gDNA).

2.2.3.3. 10 mM dNTPs.

2.2.3.4. GoTaq G2 HotStart Polymerase enzyme and 5X flexi buffer (Promega).

2.2.3.5. 25 mM $MgCl_2$.

2.2.3.6. 50× Tris Acetate EDTA Buffer (TAE).

Table 1
PCR primer pairs used in the preparation of *XIST* RNA FISH probe

Primer pair	Sequences 5′—3′	Amplimer size
XIST-1F + XIST-1R	AATGCTGGTAAAGCCCACAC TCTGGGGACAAGAACCATTC	516 bp
XIST-2F + XIST-2R	CTGGCATTTGGCACTTTACAC TTCCTTTCGCATTGCTTCT	478 bp
XIST-3F + XIST-3R	GACACCATGGCTACCTGTGA TCCCATTACCCTTTGGATTG	492 bp
XIST-4F + XIST-4R	CAACGAGGAAGCAGGAGTCT CATTTGCTCCTGATCTGTGC	491 bp
XIST-5F + XIST-5R	GGAAAGGTGATGGGGTATGA TCTCATGGACTCCCTTTGCT	512 bp
XIST-6F + XIST-6	ATGGAATGGGCAAAGTGGT CGTCTGGGATCCTATGCACT	477 bp
XIST-7F + XIST-7R	AGTGCATAGGATCCCAGACG TTGCAGTTGTCAATGGTCCT	511 bp
XIST-8F + XIST-8R	GTGTCACCAACCATGCTGTC GAGTAGAATCCCGCTTCCTTG	506 bp
XIST-9F + XIST-9R	TTCATAGTCCCAGGGAAAGGT AGGTCCCTGCATCATCTTTG	508 bp
XIST-10F + XIST-10R	GGTGTGAGCAGTTGGGATCT TGCAATTGTCCAGAGTCCTG	475 bp
XIST-11F + XIST-11R	TCGTTAGCAACATCCCACAG AAGACTACCAGTCGCCCTTG	493 bp
XIST-12F + XIST-12R	GGATTCTCCAGAAGCACAGC TTACCCTTGGCATTGCTGA	501 bp
XIST-13F + XIST-13R	CACCAGACCACAGCAATTCA CCTACCCCTTGGCTAATGCT	508 bp
XIST-14F + XIST-14R	GGGAGGTTCAGACCACACAG GCTGTGTGCTTTTCGTGTTG	509 bp
XIST-15F + XIST-15R	GGTTGGGTTATGCAGCAATC TTGTTGTATCGGGAGGCAGT	525 bp
XIST-16F + XIST-16R	AACACTGCGACAGAACTGGA TTGTGGGTTGTTGCACTCTC	501 bp
XIST-17F + XIST-17R	GACTTCCTCTGCCTGACCTG GGATACCTGCTGATTCCCTTC	514 bp

Table 2
PCR primer pairs used in the preparation of *TLR7* RNA FISH probe

Primer pair	Sequences 5′—3′	Amplimer size
TLR7-1F + TLR7-1R	TTATAAGGAAGTGGGGAGGAGA TCTTGCCAGCTGGTTTCAG	477 bp
TLR7-2F + TLR7-2R	TGCTCTGCTCTCTTCAACCAG TTTTCTTTGACAGCACTCAAGC	513 bp
TLR7-3F + TLR7-3R	AGAGCCCTATGGTCTTTGGTG CTGGAAACCTTGCCCTCA	487 bp
TLR7-4F + TLR7-4R	TCAGCAACTTCCCTTATTCACA GGATTTCCTGCTCTCCTGACT	503 bp
TLR7-5F + TLR7-5R	GCTTGGAGGAACAGCTGAGT AGACAGGACATATGGGAGCAG	482 bp
TLR7-6F + TLR7-6R	TCAGTTGGTAGCCCTGTAGCA TCAGGCCTAAAGGTAACACTGC	525 bp
TLR7-7F + TLR7-7R	ACGTGGTGCTAACCCCTAGAA TGCTGTCAGGGAAATCTATGGT	525 bp
TLR7-8F + TLR7-8R	TTCTCCCTCACCTGCTTGA AGGAATTTTTCCCTGAAGATCA	476 bp
TLR7-9F + TLR7-9R	TTTCACAGCCTCCATTTTCC TGCCAGTGGGTACAGGAAG	536 bp
TLR7-10F + TLR7-10R	GTGTACCAGGGCAGGTTTAGC ACCATCTAGCCCCAAGGAGT	513 bp
TLR7-11F + TLR7-11R	ACTCCTTGGGGCTAGATGGT CAGTTTTGGCCCAGGTAGAG	480 bp
TLR7-12F + TLR7-12R	CCTCCCGCCTAGCTTACAG GGGCACATGCTGAAGAGAGT	484 bp
TLR7-13F + TLR7-13R	AATGCTTTTGATGCGCTGA TTGAGCAGAAGCCAACTTCA	490 bp
TLR7-14F + TLR7-14R	AGTGAAGTTGGCTTCTGCTCA TGGTGGAGGAAGAGATGTCA	520 bp
TLR7-15F + TLR7-15R	ATCTCTTCCTCCACCAGCAG GGACATTTTCTGGGAAGCTG	493 bp
TLR7-16F + TLR7-16R	CAGCTTCCCAGAAAATGTCC TGGTAACCAGTCCCTTTCCTC	511 bp
TLR7-17F + TLR7-17R	CAAACTGGAAGACCCAAGAGA CAAAACACGCTTTTGGTGTG	502 bp

2.2.3.7. Agarose.

2.2.3.8. Ethidium Bromide.

2.2.3.9. QIAquick PCR purification kit (QIAGEN).

2.2.3.10. Molecular weight 1Kb.

2.2.3.11. 6X dye loader.

2.2.4. Nick translation:

2.2.4.1. Vysis Nick Translation kit (Abbott).

2.2.4.2. Aminoallyl-dUTP-ATTO-550 and Aminoallyl-dUTP-ATTO-647 N (Jenna Bioscience) (the new vials of the different Aminoallyl-dUTP-ATTO have to be resuspended in 40 μL of RNAse/DNAse-free water (from the NT kit)).

2.2.5. Probe precipitation.

2.2.5.1. Salmon Sperm (Invitrogen, Cat 15632011V).

2.2.5.2. 3 M Sodium acetate pH 5.2.

2.2.5.3. Isopropanol.

2.2.5.4. Ethanol.

2.2.5.5. Formamide.

2.2.5.6. 20X SSC.

2.2.5.7. UltraPure distilled RNAse/DNase-free water.

2.2.6. Hybridization buffer.

2.2.6.1. 50% Dextran sulfate solution: 50 g of Dextran dissolved in 10 mL H_2O. Once prepared, store the solution at −20 °C, in aliquots of 400 μL, in 1.5 mL Eppendorf tubes.

2.2.6.2. Hybridization buffer: 20% Dextran, 4X SSC, 4 mg/mL BSA, 20 mM RVC (NEB). Once prepared, store it at −20 °C.

2.2.7. DAPI solution at 20 mg/mL. Store at 4 °C or −20 °C.

2.2.8. Image acquisitions and analysis.

2.2.8.1. Confocal microscope.

2.2.8.2. FIJI program.

2.2.8.3. Imaris file converter.

2.2.8.4. IMARIS program—10.1.1 version.

3 Methods

3.1 Cell Culture

3.1.1 Cell Isolation

Frozen vials of Peripheral Blood Mononuclear Cells (PBMCs) were used in this protocol to isolate naive B cells and plasmacytoid Dendritic Cells (pDCs).

3.3.1.1. Heat the R10 complete medium (*see* Subheading 2.1) into a warm bath at 37 °C. Add 25 mL of the pre-warmed R10 medium into 50 mL tubes identified with the corresponding cell type to be isolated.

3.3.1.2. Defrost the vials in a warm bath at 37 °C for 30 s approximately.

3.3.1.3. Resuspend the cells by adding 1 mL of R10 medium to the vials with a 2 mL pipet, gently pipet up and down, and transfer the cells into the 50 mL identified tubes.

3.3.1.4. Centrifuge for 10 min at 800 rpm and discard the supernatant. Refill the tube with another 25 mL of R10 medium and count the cells in a Neubauer chamber. Centrifuge again for 10 min at 800 rpm.

3.3.1.5. Resuspend the cells with PBS-FBS buffer (*see* Subheading 2.1) according to the number of cells obtained following the manufacturer's instructions of the corresponding cell isolation kit (EasySepTM Human B cell Isolation Kit; EasySepTM Human Plasmacytoid DC Enrichment Kit).

3.3.1.6. Take an aliquot of the isolated cells to confirm the purity assessment by flow cytometry with the corresponding fluorochrome-conjugated antibodies.

3.3.1.7. Proceed with one of the following stimulation systems:

3.1.2 In Vitro IFNβ Stimulation

3.1.2.1. After purification and cell counting, centrifuge the cells for 5 min at 300 g, resuspend them in R10 complete medium (500 μL/well in 24-w/plate at 1×10^6 cells/mL), and incubate them during 2 h at 37 °C, 5% CO_2, with human recombinant IFNβ (2 ng/mL), to induce *TLR7* transcription (*see* **Note 1**).

3.1.2.2. Recover the cells in a 1.5 mL Eppendorf tube and resuspend with 1 mL of sterile 1X PBS to wash the cells, centrifuge for 5 min at 300 g aspirate the supernatant, and resuspend with 20 μL of 1X PBS for each 2×10^5 cells.

3.1.3 pDC-IL-3 Culture System (See **Note 2**)

3.1.3.1. Culture purified pDCs (3.1.1.) at 1×10^5 cells per well in 96-well plates (200 μL/well) in R10 complete medium with IL-3 (5 ng/mL) at 37 °C in 5% $C0_2$ incubator for 48 h.

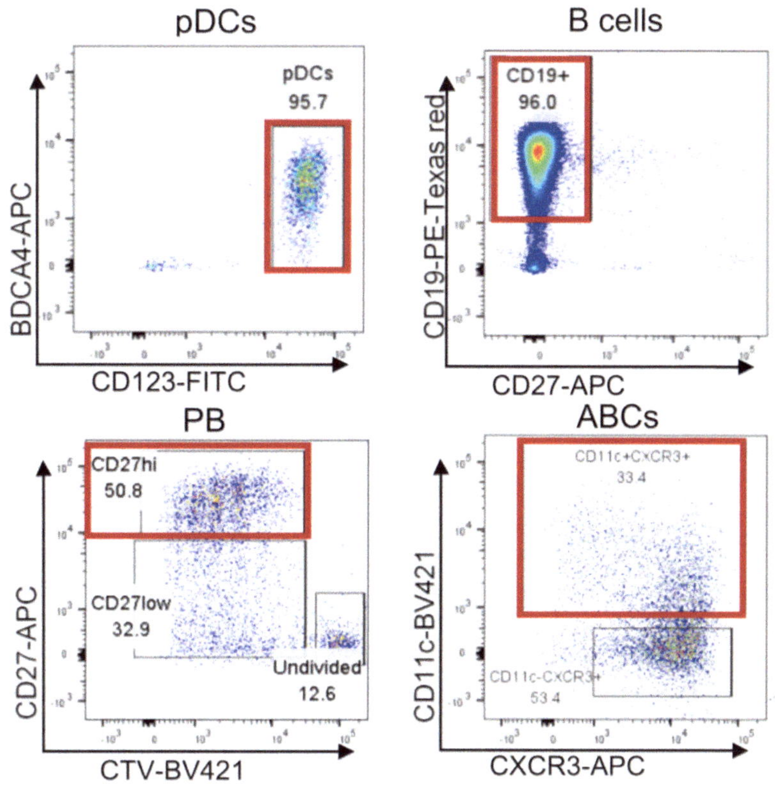

Fig. 1 Representative Dot plot profiles of enriched pDCs, naïve B cells, PB and ABCs. CD123+ BDCA4+ pDCs and naïve CD19 IgD+ CD27neg B cells are selected using the untouched negative selection kits from StemCells. The purity of the isolated pDCs (upper-left panel), B cells (upper-right panel) was typically above 90%. PBs are FACS-sorted as CD27hi CTVlow and ABCs as CD11c+ CXCR3+/– upon in vitro activation as indicated in red

3.1.3.2. Add human recombinant IFN-β (1 ng/mL) for the last 2 h of culture (*see* **Note 1 & 2**).

3.1.3.3. Collect and wash the cells once in 1X PBS, count, and then resuspend with 20 μL of 1X PBS for each 2 × 10^5 cells.

3.1.4 TLR7/9-Driven B Cell Activation Models

Two culture systems can be used to activate the enriched $CD19^+CD27^{neg}$ B cells purified in Subheading 3.1.1 (Fig. 1).

The First Model Using $CD19^+$B Cells (As Described in [4])

3.1.4.1.1. Label enriched B cells (Fig. 1) with Cell Trace Violet (CTV), according to the manufacturer recommendation, and then stimulate them at 1 × 10^5 cell/well in 96-U plates with either CL097 (0.3 μg/mL) or CpG-B (0.3 μg/mL) in the presence of human recombinant IL2 + IL21 (10 ng/mL) in R10 complete medium.

3.1.4.1.2. At day 4, collect, wash once with RPMI and then stain with anti-human CD19, CD27 for cell sorting. FACS-sort CD19^{+} cells according to the expression of CTV^{low} and CD27^{hi} profiles, which label plasmablasts (PB).

3.1.4.1.3. Incubate sorted PB with IFNβ for 2 h as in Subheading 3.1.2 to induce *TLR7* transcription.

3.1.4.1.4. Collect and wash the cells once in 1X PBS, count, and then resuspend with 20 μL of 1X PBS for each 2 × 10^{5} cells.

The Second Culture System Used to Generate Age-Associated Like CD11c^{+} T-bet^{+} B Cells or Pre-ASCs (As Described in [18])

3.1.4.2.1. Stimulate the CD27^{neg} naïve B cells (1 × 10^{5}/well) purified from Subheading 3.1.1 in 200 μL of R10 complete medium supplemented with IL-2 (10 ng/mL), IFN-γ (10 ng/mL), goat anti-human IgA/G/M F(ab′)2 antibodies (2 μg/mL), and CpG-B oligodeoxynucleotides (1 μg/mL) or CL097 (1 μg/mL) in 96-well plates for 3 days.

3.1.4.2.2. At day 3, collect and stain the cells with anti-CD19, anti-IgD, anti-CD2, anti-CD11c, and anti-CXCR3 antibodies. Gate CD19^{+}CD27^{-} B cells and sort CD11c^{+}CXCR3^{+} pre-ASCs ("ABC-like" cells) (see Fig. 1).

3.1.4.2.3. Culture the cells for 2 h with IFNβ (1 ng/mL) in R10 complete medium at 37 °C, then wash the cells once in 1X PBS, count, and then resuspend with 20 μL of PBS for each 2 × 10^{5} cells. All the antibodies used for flow cytometry are used at 1:100 concentration except for IgD-FITC at 1:200.

3.2 Cell Preparation, Permeabilization, and Fixation

3.2.1 Spotting of the Cells on Glass Slides

3.2.1.1. Circle with a diamond pen in the back side of a poly-L-lysine coated slide to delimitate the spot area.

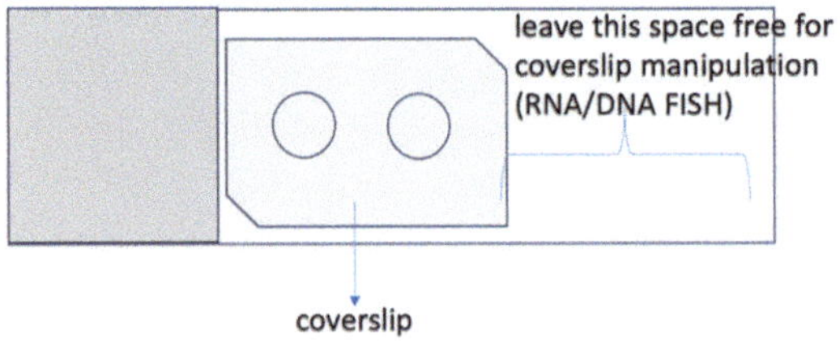

3.2.1.2. Then, in the front side, spot 2 × 10^{5} cells (resuspended in 20 μL of 1×PBS) inside each circled area.

3.2.1.3. Incubate the slides for 10 min at 37 °C.

3.2.1.4. Check at the microscope if the cells are well attached by moving gently the slide.

3.2.2 Cell Permeabilization

3.2.2.1. Incubate slides in ice-cold CSK-RVC-X buffer (*see* Subheading 2.2) in a coplin jar, for 7–10 min on ice. Note that the length of incubation should be adjusted for each cell type: the bigger the cytoplasm is, the longer the permeabilization may be. Note that **RVC is crucial for RNA FISH as it is an RNase inhibitor.**

3.2.2.2. Gently rinse the slides once in 1X PBS.

3.2.3 Cell Fixation

3.2.3.1. Incubate the slides in 2% paraformaldehyde (PFA) diluted in 1X PBS for 10 min at room temperature.

3.2.3.2. Rinse twice in 2X SSC.

3.2.3.3. For subsequent RNA FISH proceed to Subheading 3.4, for slide storage, proceed to Subheading 3.2.4.

3.2.4 Slide Storage

3.2.4.1. Transfer the slides into another Coplin jar with a watertight lid containing 70% ethanol and wash two times for 5 min at room temperature.

3.2.4.2. Refill the Coplin jar to the top with 70% ethanol and store at −20 °C for several weeks. Note that after several weeks, RNAs may be degraded. Before proceeding to RNA FISH, incubate the slides in one 85%, one 95%, and two 100% ethanol baths (5 min each) and proceed to Subheading 3.4.

3.3 Probe Preparation for RNA-FISH

3.3.1 PCR Amplification

The probes can be prepared several days before starting the protocol. It is recommended to prepare aliquots to avoid the degradation of the PCR products. Then they can be stored at −20 for several months. Defrost all the reagents and PCR products at room temperature, mix, and spin down before starting.

3.3.1.1. Prepare the mix for 20 PCRs containing 20 ng of female gDNA in a final volume of 46 μL (*see* Table 3 below).

Table 3
PCR master mix composition for 20 reactions (46 μL each) with 20 ng female gDNA per reaction

gDNA (20 ng/μL)	1 μL × 20 = 20 μL
dNTPs 10 mM	1 μL × 20 = 20 μL
5×Gotaq flexi buffer	10 μL × 20 = 200 μL
$MgCl_2$ 25 mM	4 μL × 20 = 80 μL
GoTaq G2 hotStart polymerase enzyme	0.25 μL × 20 = 5 μL
H_2O (q.s.p 50 μL)	29.75 μL × 20 = 595 μL
Vol total	46 μL × 20 = 920 μL

Prepare the primer mix stock solution at 2.5 μM by adding 2.5 μL of each forward and reverse primers (stock at 100 μM) to 95 μL of ultrapure H_2O (can be stored at −20 °C). Add 4 μL of each primer mix ($n = 17$ *see* Tables 1 and 2 for *XIST* and *TLR7* primer sequences) to 46 μL of PCR mix. Duplicate the PCRs for each primer mix in order to generate more material.

Close the plate with a film, do a quick Spin-down, and then run the PCR in a final volume of 50 μL/tube in the thermocycler according to the program below:

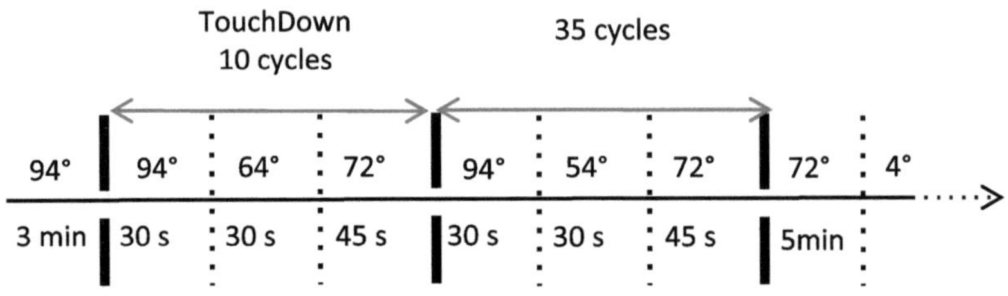

When the PCR is done, store samples at −20 °C.

3.3.1.2. Run an electrophoresis gel to confirm the efficacy of the PCRs. Prepare the agarose gel at 2% in TAE and add 1 μL of ethidium bromide per every 10 mL of TAE (i.e., 10 μL for a 100 mL gel).

Meanwhile, in separate tubes, add 5 μL of PCR product and 1 μL of the 6× dye loader and keep the remaining PCR products at −20 °C until its purification.

Once the gel is ready, place it in the electrophoresis chamber with 1X TAE covering entirely the gel.

Add the 0.1-1 kb molecular weight ladder (5 μL per well) in the first and last wells, and the PCR products in order from wells 1 to 17 plus the negative control (*see* Fig. 2 for example of the *XIST* RNA PCR products). Run the gel for ~45 min at 100 V.

If all of the PCR products are well amplified, then proceed with the DNA purification.

3.3.2 DNA Purification

Use the QIAquick PCR purification kit, and follow the manufacturer's instructions (briefly described below).

Mix the 17 DNA samples obtained (46 μL × 17 × 2 = 1564 μL), including the duplicates, and add 5 times the total volume of binding buffer (7820 μL). Load 750 μL per column, 13 columns are needed. Elute the PCR products with 30 μL of elution buffer per column. NB: be sure to add the buffer in the middle of the column not on the walls, then incubate for 1 min at room temperature and centrifuge for 1 min at 15,000 g.

Collect and pool the PCR products from the Eppendorf tubes, and check the concentration with a Nanodrop. The concentration has to be higher than 65 ng/μL to proceed to the fluorescent label

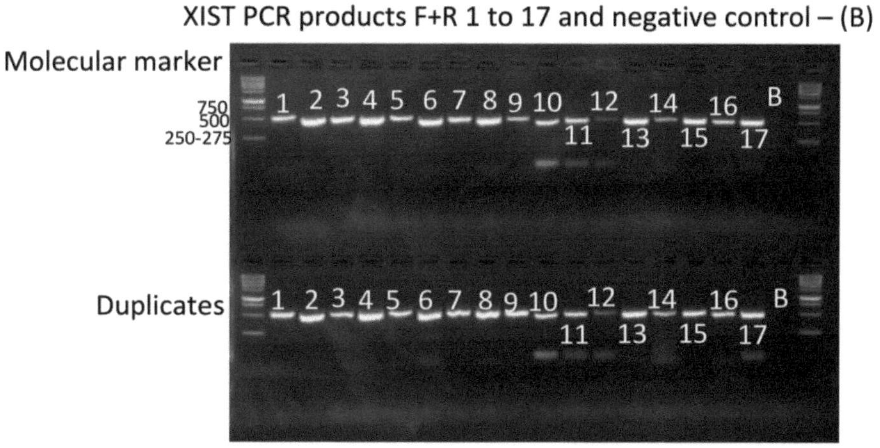

Fig. 2 Representative image of electrophoresis gel of XIST PCR products (amplification between 475 and 525 pb). Arrangement of samples in the electrophoresis gel: XIST PCR products in wells 1–17 and the negative control (B)

of the probe by nick translation. If the concentration is lower, you can concentrate the products with a speedvac in order to reduce the volume.

3.3.3 Nick Translation

Fluorescently label 1 μg of PCR products using the Vysis Nick Translation (NT) kit following the manufacturer's instructions. Calculate the volume needed for each probe according to their concentration.

Ex1: For the **human *XIST*** probe, as the PCR product concentration obtained after DNA purification was 0.158 μg/μL, 6.3 μL of PCR product is needed for a 50 μL of NT reaction.

Ex2: For the **human *TLR7*** probe, as the PCR product concentration obtained was 0.207 μg/μL, 4.8 μL of PCR product is needed for a 50 μL of NT reaction.

Incubate the NT reaction for 1 h30 at 15 °C in the dark, then stop the reaction for 10 min at 70 °C. Store the probes at −20 °C until their use. Probes are ready for the hybridization step. Note that other DNA templates, such as plasmids, fosmids, or BACs, can be used for the NT reaction (*see* Ref. [17] for details). In this case, the reaction should be incubated overnight.

3.3.4 Probe Precipitation, Resuspension, and Denaturation

Note that as the fluorophores are light-sensitive, the tubes should be kept in the dark as much as possible.

3.3.4.1. In a 1.5 mL Eppendorf tube, precipitate the probes (quantities and volumes are given for one spot of cells on a slide). Mix ~100 ng of each probe (i.e., ~5 μL of each Nick Translation solution) and 20 μg of salmon sperm DNA. Add 0.1 volume of 3 M Sodium Acetate (NaAc, pH 5.2) and mix well by vortexing. Add 0.6 volume of Isopropanol and mix well by vortexing.

Ex: 5 μL of *TLR7* probe +5 μL of *XIST* probe +2 μL of salmon sperm DNA = 12 μL. Add 1.2 μL of NaAc and vortex. Add 8 μL of Isopropanol and vortex.

Do a quick Spin down and incubate the probes at −80° for 20 min to facilitate the precipitation.

3.3.4.2. Quickly defreeze the tube in your hands, and centrifuge for 30 min at 15,000 g at 4 °C. Discard the supernatant, add 250 μL of 70% ethanol, and centrifuge for 5 min at 13000 rpm at 4 °C. Discard the supernatant and air dry the probes at room temperature for few minutes in the dark. Resuspend the pellet with 5 μL of Formamide per spot of cells and denature the probes for 7 min at 75 °C in the dark. Add 5 μL of the hybridization buffer per spot and proceed straight to the overnight hybridization in Subheading 3.4. Note that if the slides are not ready, you can keep the probes on ice in the dark until hybridization.

3.4 Overnight Hybridization

If the slides are in 2× SSC (from Subheading 3.2.3), take them out of their coplin jar and gently wipe off the liquid around the spots of cells with a clean kimwipe paper so that only a minimum amount of liquid remains on the cells.

If the slides are in 100% Ethanol (from Subheading 3.2.4), briefly air dry them.

3.4.1. Add 10 μL of the probe mix (Subheading 3.3.4) onto each spot of cells on the slides and cover with a glass coverslip. Use a small round one if only 1 spot, a larger square or rectangle one if there are 2 or 3 spots on the slide.

3.4.2. Incubate for overnight hybridization at 42 °C in a humid chamber in the dark. Use a solution of 50% formamide/2X SSC to humidify the chamber. Note that the time of hybridization should be no longer than 16 h to avoid background. Note that the temperature may be troubleshooted from 37 °C to 42 °C depending on the probes and cells: the higher the temperature is, the more stringent the hybridization will be meaning less background but lower signal.

3.5 Washing and Mounting

Note that as the fluorophores are light-sensitive, the washes should be performed in the dark and light should be avoided as much as possible.

3.5.1. The day after, incubate the slides (with the coverslips still on) in a coplin jar containing pre-warmed 50% formamide/ 2X SSC solution for 5 min at 45 °C in the dark.

IMPORTANT: note that to avoid damaging the cells, don't try to remove the round coverslips straight. Instead, the coverslips should slide off by themselves during the incubation. If they don't, slowly remove the slides from the coplin jar, while observing how the coverslips slide down and fall to the bottom of the coplin jar.

3.5.2. Do two more washes in 50% formamide /2X SSC solution and three washes in 2X SSC at 45 °C for 5 min in the dark.

Note that we perform the washes at 45 °C and with 2X SSC, but the temperature (from 37 °C to 45 °C), concentration of SSC (from 0.2X to 2X), as well as the length of incubation may be troubleshooted: the higher the temperature is and/or the lower the concentration of SSC is, the more stringent the washings will be, meaning less background but lower signal.

3.5.3. During the last washing step, prepare 1:10,000 DAPI (2 μg/mL finale concentration) in sterile 1× PBS, and incubate the slides for 10 min at room temperature, in the dark.

3.5.4. Wash the slides three more times in 2X SSC buffer for 5 min at room temperature in the dark.

3.5.5. Leave the slides drying in the dark for ~15 min, then mount the slides with 10 μL of DAKO mounting medium (Agilent) per spot of cells, and cover them with a glass coverslip (use a square coverslip if only one spot, or a larger rectangle one if several spots on the slide).

3.5.6. Store the slides at 4 °C until image acquisition. It is recommended to leave the slides at least 2-4 h before acquisition. We use a Leica TCS SP8 or Zeiss LSM710 microscope, using a 63× oil immersion objective.

3.6 Image Analysis (Fig. 3)

The aim is first to create vectorial surfaces of cell nuclei (e.g., create an artificial solid object) in order to separate each nucleus individually. The machine learning-based object classification is then used to train the classifier for recognizing nuclei with the four different *XIST* cloud types. The final workflow allows for an automated labeling-based cell nuclei classification of the different *XIST* cloud types (e.g., type I = big and compact domain localized with one X, types II = more diffuse pinpoints of *XIST* RNA encompassing a nuclear area of the size of one X, III = dispersed pinpoints extending beyond the X chromosome territory, type IV = no *XIST* RNA signal) as described [19]. This workflow can be used with various immune cells to obtain a statistical analysis [15, 19]. The same approach can be used to classify nuclei with zero, one, or two signals of *TLR7* primary transcripts.

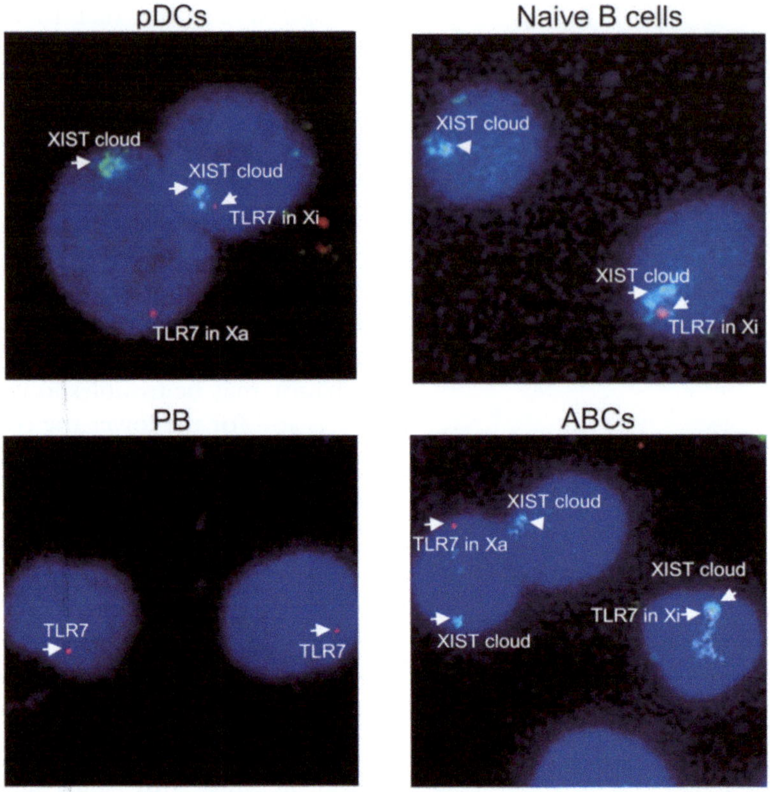

Fig. 3 Representative images of acquired RNA FISH slides processed with the Fiji software of pDCs, Naïve B cells, plasmablasts (PB) or ABCs. 3D confocal Z-stacks were converted into 2D stacks. The primary transcripts of TLR7 (red dots) are observed on the inactive X (Xi) chromosome (green cloud of XIST RNA) or Xa chromosome (no XIST cloud). No XIST cloud was observed in PB

3.6.1 File Conversion and 3D to 2D Projection

Confocal images are first converted to IMARIS file with the "Imaris File Converter." Then, in order to facilitate further analysis process, 3D confocal Z-stacks are converted into 2D stacks, with a Maximum Intensity Projection (MIP).

1. Open the IMARIS file converter.
2. Drag and drop or open the confocal images (Leica ".lif" files in our case) and press convert images. Images are converted in ". ims" files and are ready to be open in the IMARIS program.
3. Open the IMARIS program 10.1.1 version and the converted images.
4. Use the "Project to 2D" XTension to do a 3D to 2D projection. First select the projection plane XY and then select the projection mode MIP (max intensities over the Z-projection line divided by the number of voxels).
5. Then save the 2D image as a ".ims" file.

3.6.2 Channel Name and Color Reassignment

Before starting, if needed, it may be useful to reassign colors and name of the channels according to the configuration of the fluorochromes selected for the RNA FISH procedure. For that:

1. Click the "display adjustment" task bar on "Edit".
2. Then click "Reset display adjustments".
3. A new window with the channel settings will open, then click in each channel to change the display settings, Example:
   (a) Channel 1 (DAPI-Blue).
   (b) Channel 2 (TLR7-ATTO-550-red).
   (c) Channel 3 (XIST-ATTO647N-magenta).

The colors for each channel can be changed by clicking on base color, choose the desired color.

3.6.3 Channel Filtering and Denoising

Some filters must be applied to reduce the image background and facilitate the detection and classification of the types of *XIST* clouds, without modifying their perception.

1. Background Subtraction.

   The background subtraction applies a Gaussian filter to define the background at each voxel and then performs a Baseline Subtraction of this variable background. The filter width is the only parameter of the background subtraction method.

   (a) Select the Menu Image Processing→Thresholding→-Background Subtraction. The Background Subtraction box is displayed.
   (b) Select the channels by ticking the check-box on the left of each channel.
   (c) Enter the requested filter width in the parameter field. Use 1 μm for *TLR7* or any gene of interest, for which small signals (pinpoints) are detected; for larger *XIST* RNA clouds, use 3 μm. Click "Apply" and then "OK".

2. Median filter.

   The median replaces the intensity of each voxel with the statistical median of the intensities of neighboring voxels. There are several choices for the neighboring voxels to be included in the median calculation. The median filter is useful when image data are corrupted by "salt and pepper" noise. It is the same for all channels.

   (a) Select the Menu Image Processing→Median Filter. The Median Filter box is displayed.
   (b) Select the channels to be filtered by ticking the check-box.

   (c) Check the requested Filter Size (3 × 3 × 1, same for all channels) and click on "Apply". The effect of the change can be seen in the image.

3. Save the resulting filtered image as a ".ims" file.

3.6.4 Segmentation of Nuclei and Object Classification Training with Machine Learning

In order to separate each nucleus individually, vectorial surfaces of nuclei are first generated. It creates an artificial solid object in order to visualize the range of interest of the surface nuclei. The machine learning-based object classification feature is then used to train the classifier for recognizing nuclei with the four different *XIST* cloud types. The final workflow allows for an automated labeling-based cell nuclei classification of the different *XIST* cloud types.

1. **STEP 1: Creation of new surfaces (e.g., nuclei)**
   (a) Click the button "Surface" to create a Surface object out of the Contour object. The Surface object is added in the Surpass Tree and displayed in the viewing area.
   (b) Select the option "Classify surfaces" for the machine learning. Note that the "Object-object statistics" button is only needed to measure distance between objects or spots (but not needed in general).
   (c) Select "Start creation with slicer view".
   (d) Then go to next step with the blue arrow.
2. **STEP 2: Smoothing of the nucleus**
   (a) In "Source channel", select the Channel 1 (DAPI) or the one that corresponds to the nucleus. Select the "smooth" option in order to smooth the structure of the nucleus and to use less computational resources.
   (b) Select "Absolute intensity".
   (c) Go to next step with the blue arrow.
3. **STEP 3: Definition of the threshold**
   (a) This step is based on fluorescence intensity of the nuclei: the intensity threshold delimits the outside and the inside of the nucleus, so it delimits the contour. For that, adjust the threshold (positive signal is into the yellow part of the graph) to separate nuclei from background.
   (b) Sometime, nuclei are touching each other. It is possible to separate them individually, by selecting the "enable split touching object" option, with a snake algorithm. This needs to define a seed points diameter (7 for mice immune cells (B cells) and 2.5 for human immune cells (naïve B cells, PB, ABCs, and pDCs)).
   (c) Select the option "morphological split".
   (d) Go to next step with the blue arrow.

4. **STEP 4: Filter based on the intensity of each nucleus**
   (a) Adjust the yellow graph to select or unselect desirable or undesirable cells (nuclei with background, apoptotic nuclei, etc.).
   (b) Change the color of the surface to transparent to be able to appreciate the spots in the nuclei.
   (c) Go to next step with the blue arrow.
5. **STEP 5: Second filter based on the surface of each nucleus.**
   (a) Adjust the surface threshold keeping surfaces of the nuclei above the threshold corresponding to yellow graph to select or unselect desirable or undesirable nuclei (incomplete nuclei on the border of the image, fused nuclei...).
   (b) Add an extra filter selection "distance to image border x/y", in order to remove the nuclei at the edge of the image.
   (c) Go to next step with the blue arrow.
6. **STEP 6: classification of nuclei by *XIST* cloud types.**

General comments before starting nuclei classification with Machine Learning (ML)-based classification do not need a feature selection; it includes all features; and based on your data and labels, it chooses the best ones for the user.

Notes on review and adjust features

1. In "Features" you can explore the relevant features the classifier uses for the current prediction.
2. In "Settings" you can select or de-select features used for classification. Usually this is not needed but can be used to bias the classifier towards some aspects of your data.
   (a) Create the number of classifications needed. Here, five classifications are needed, four for each *XIST* cloud type, and one for "unclassified" nuclei.
   (b) Select the corresponding cells to be classified according to the criteria of the size and density of the *XIST* cloud (from type I = big and dense cloud, to type IV = no XIST signal). For each classification, select at least 10–50 nuclei.
   (c) Click "Train and Predict" to train the classifier using the current labels.
   (d) Check the classification result and refine the classifier by adding more labels or reassigning objects to different classes.
   (e) Retrain the classifier by re-running "Training and Predict".

(f) Once all the classifications have been performed, click on the green arrow to finish the training and to obtain the numbers of cells according to the classifications.

(g) In the "Statistics" tab of the surfaces, check the *XIST* types and explore the respective class statistics.

(h) Classification can also be included in a semi-automatic processing. For that, click on the menu "creation" "Store Parameters for Batch" and save as a "favorite creation parameter" after successful classification, or for continuing the ML training on another image.

4 Notes

1. *TLR7* is an Interferon Simulated Gene (ISG) which expression can be induced upon culture of B cells or pDCs for 2 h with human IFN-β (2 ng/mL) before processing the cells for RNA FISH. This step boosts *TLR7* primary transcript expression from both X chromosomes, as previously described [4, 15]. This step can be omitted if the scientific question is to assess the status of *TLR7* allelic expression on ex vivo purified immune cells from patients versus controls as recently described for pDCs from women with SSc [15].
2. To gain further insights into *TLR7* allelic expression relative to the Xi territory, we designed an in vitro system to induce *XIST* relocalization to the Xi in female human pDCs. For this, female pDCs were cultured for 48 h in the presence of IL-3. We observed a major upregulation of *XIST* clouds Type I-II in 48 h-stimulated pDCs [15]. Because *TLR7* is an interferon stimulated gene (ISG) and to assess the accessibility of the *TLR7* loci to the transcriptional machinery cells were stimulated with IFN-β 2 h before RNA FISH analysis. GM-CSF, like IL-3, promotes pDC survival in vitro [20], and we also found strong *XIST* relocalization to the Xi and *TLR7* expression by RNA FISH in pDCs cultured for 48 h with GM-CSF (5 ng/mL; Peprotech).

Acknowledgments

We wish to thank members of JCG's lab for their critical reading of the manuscript. We thank the Imaging Core Facilities of INFINITY for their help with Imaris analyses. The research carried out in the laboratory of JC Guéry and J Chaumeil benefits from the financial support of the Agence Nationale de la Recherche (ANR-20-CE15-0014-01), and the Guéry's lab is further supported by ANR-23-CE15-0002-01, FOREUM Foundation

(2020-041), Fondation pour la Recherche Médicale (EQU202403018065), Novartis DREAMER (2023), and Société Française de Rhumatologie (2025). B.F.-L. was supported by a post-doctoral fellowship from the Inspire Program from Région Occitanie/Pyrénées-Méditerranée (#1901175) and by ANR-23-CE15-0002-01.

References

1. Dossin F, Heard E (2022) The molecular and nuclear dynamics of X-chromosome inactivation. Cold Spring Harb Perspect Biol 14: a040196
2. Carrel L, Willard HF (2005) X-inactivation profile reveals extensive variability in X-linked gene expression in females. Nature 434:400–404
3. Tukiainen T et al (2017) Landscape of X chromosome inactivation across human tissues. Nature 550:244–248
4. Souyris M et al (2018) TLR7 escapes X chromosome inactivation in immune cells. Sci Immunol 3:eaap8855
5. Hagen SH et al (2020) Heterogeneous escape from X chromosome inactivation results in sex differences in type I IFN responses at the single human pDC level. Cell Rep 33:108485
6. Abbas F et al (2022) HIV-1 infection enhances innate function and TLR7 expression in female plasmacytoid dendritic cells. Life Sci Alliance 5: e202201452
7. Youness A et al (2023) TLR8 escapes X chromosome inactivation in human monocytes and CD4(+) T cells. Biol Sex Differ 14:60
8. Deane JA et al (2007) Control of toll-like receptor 7 expression is essential to restrict autoimmunity and dendritic cell proliferation. Immunity 27:801–810
9. Guiducci C et al (2013) RNA recognition by human TLR8 can lead to autoimmune inflammation. J Exp Med 210:2903–2919
10. Souyris M et al (2019) Female predisposition to TLR7-driven autoimmunity: gene dosage and the escape from X chromosome inactivation. Semin Immunopathol 41:153–164
11. Jiwrajka N, Anguera MC (2022) The X in seX-biased immunity and autoimmune rheumatic disease. J Exp Med 219:e20211487
12. Forsyth KS et al (2024) The conneXion between sex and immune responses. Nat Rev Immunol 24:487
13. Yu B et al (2021) B cell-specific XIST complex enforces X-inactivation and restrains atypical B cells. Cell 184:1790–1803. e17
14. Huret C et al (2024) Altered X-chromosome inactivation predisposes to autoimmunity. Sci Adv 10:eadn6537
15. Du Y et al (2025) Altered X-chromosome inactivation of the TLR7/8 locus and heterogeneity of pDCs in systemic sclerosis. J Exp Med 222:e20231809
16. Syrett CM, Anguera MC (2019) When the balance is broken: X-linked gene dosage from two X chromosomes and female-biased autoimmunity. J Leukoc Biol 106:919–932
17. Salataj E et al (2023) Single-cell detection of primary transcripts, their genomic loci and nuclear factors by 3D immuno-RNA/DNA FISH in T cells. Front Immunol 14:1156077
18. Zumaquero E et al (2019) IFNgamma induces epigenetic programming of human T-bet(hi) B cells and promotes TLR7/8 and IL-21 induced differentiation. elife 8:e41641
19. Wang J et al (2016) Unusual maintenance of X chromosome inactivation predisposes female lymphocytes for increased expression from the inactive X. Proc Natl Acad Sci USA 113: E2029–E2038
20. Ghirelli C et al (2010) Systematic cytokine receptor profiling reveals GM-CSF as a novel TLR-independent activator of human plasmacytoid predendritic cells. Blood 115:5037–5040

Chapter 14

Single-Cell Spatial Transcriptomic of Sjögren's Disease Salivary Glands

Charlotte Gandubert, Simon Höllerer, Lisa Sériau, Soizic Garaud, Christelle le Dantec, Valérie Devauchelle-Pensec, Sophie Hillion, Divi Cornec, and Marie Frutoso

Abstract

Sjögren's Disease (SjD) presents a challenge as a chronic, progressive, and systemic autoimmune disorder affecting approximately 0.05% of the population. SjD primarily targets exocrine glands, leading to severe dryness of the mouth and eyes, but can impact multiple organs and increase lymphoma risk. Despite advances in B cell-targeted therapies, no cure exists, and treatment remains symptom-focused, highlighting gaps in our understanding of SjD inflammatory environment. One roadblock to study the heterogeneity of the landscape in affected salivary glands in humans is the limited amount of tissue that is typically available from biopsies of minor salivary glands (MSGs). Spatial transcriptomics has emerged as a cutting-edge approach to address this challenge by preserving tissue architecture and enabling precise mapping of immune and structural cell interactions. This chapter explores how the Visium HD protocol from 10X Genomics can be leveraged to dissect the cellular and molecular landscape of formalin-fixed paraffin-embedded human MSGs, uncovering insights into disease mechanisms and therapeutic opportunities.

Key words Sjögren's Disease, Human FFPE tissue, Minor salivary glands, Spatial transcriptomics, Bioinformatic analysis, Visium HD

1 Introduction

Sjögren's Disease (SjD) is a chronic, progressive, and systemic autoimmune disease affecting around 0.05% of the population [1]. While SjD typically originates in exocrine glands, the disease can affect the function of almost any organs in the body [2] and increase the risk of lymphomas [3]. The immune dysregulation in SjD can arise from various factors, including genetic predisposition and environmental triggers, like viruses [4]. These factors lead to the activation of salivary gland epithelial cells, resulting in the release of auto-antigens and production of cytokines and

Rami Bechara (ed.), *Autoimmunity: Methods and Protocols*, Methods in Molecular Biology, vol. 2990,
https://doi.org/10.1007/978-1-0716-4997-8_14,

chemokines (IFNα/β, IL-6, IL-7, IL-12, IL-18, IL-22, CXCL13, CCL17, CCL19, CCL21, CCL22) [4, 5]. In response to these signals, antigen presenting cells (APC) accumulate, process, and present viral and self-antigens, ultimately activating autoreactive T and B cells. Subsequently, activated B and T cells differentiate, respectively, in autoantibody-producing plasma cells and in autoreactive T cells secreting cytotoxic granules, further disrupting the epithelium. This aberrant activation leads to an inflammatory cascade, involving the recruitment and activation of other immune cells through mediators like IFNγ or BAFF [4]. Very recently, promising drugs targeting B cell activation have demonstrated efficacy in small phase 2 trials by reducing disease activity, with larger clinical trials currently ongoing [6]. However, the underlying mechanisms inside the tissue leading to B cell overactivation are still misunderstood. A more detailed and deeper understanding of the cellular and molecular network within the inflamed tissues in SjD is needed to enhance our knowledge of the disease and help to develop new therapeutic targets in SjD. A major challenge in studying the microenvironment in affected tissues lies in the limited tissue availability from minor salivary gland (MSG) biopsies, making it difficult to capture the heterogeneity of immune cell infiltration. Traditional transcriptomic approaches, such as bulk RNA sequencing, mask cellular diversity, while single-cell RNA sequencing, despite its resolution, lacks the crucial spatial context. Spatial transcriptomics has emerged as a groundbreaking solution to bridge this gap by preserving tissue architecture while simultaneously enabling high-resolution gene expression profiling [7–9]. Among the latest innovations, the Visium HD spatial gene expression protocol from 10x Genomics offers unprecedented resolution to dissect the immune and stromal cell interactions within formalin-fixed paraffin-embedded (FFPE) MSG biopsies [10, 11]. In this chapter, we explore how Visium HD can provide novel insights into SjD pathogenesis, offering a powerful tool to decode the intricate immune landscape driving this disease. This chapter describes the Visium HD methodology tailored for human FFPE biopsies from 5 μm tissue sections to single-cell analysis. Ultimately, the insights gained from spatial transcriptomic profiling in SjD may have broader implications for understanding tissue-specific immune dysregulation across autoimmune disorders.

2 Materials

Components and kits from 10X Genomics are indicated in parenthesis. All 3rd party reagents/equipment used are the ones suggested in the Protocol Planner (CG000698) from 10X Genomics. Indicated mix are prepared for two tissue sections + 10%.

Required for most steps: 15 mL and 50 mL conical tubes, 1 L beakers, pipette tips, Milli-Q water, Lint-free laboratory wipes, 1.5 mL DNA LoBind tubes, 0.2 mL PCR tubes, P1000 P200 and P20 pipettes and filtered pipette tips, 80% ethanol (freshly prepared dilution of molecular grade ethanol 99.6%), nuclease-free water, ice, Visium slide seals (cut in half in lengthwise) (Component: 2000283, Kit: 1000669), thermocycler and thermocycler adapters (Component 3000823, Accessory Kit 1000499), water bath, or thermomixer.

2.1 Step 1 (See Note 1)

1. 100% Xylene.
2. 100%, 96% and 70% EtOH.
3. Milli-Q water.
4. Gill II Hematoxylin, Sigma, GHS232.
5. Bluing buffer, Dako, CS70230-2.
6. Eosin Y solution, Alcoholic, Merck HT110116.
7. 85% glycerol (mounting medium).
8. Coverslips.
9. Digital slide scanner.

2.2 Step 2

1. 0.1 N HCl.
2. TE buffer pH 8.0.
3. 1X PBS, pH 7.4.
4. Diluted Perm Enzyme B: Add reagents in the order listed. Mix thoroughly and maintain at room temperature.

- 998 μL 1X PBS.
- 2 μL Perm Enzyme B (Component 3000553, Kit: 1000668: Remove from −20 °C shortly before use. Maintain at room temperature. Pipette mix and centrifuge briefly. Do not vortex).

5. Decrosslinking Mix: Add reagents in the order listed. Pipette mix thoroughly. Centrifuge briefly. Maintain at room temperature.

- 203.50 μL Decrosslinking Buffer B (Component 2001094, Kit 1000668: Thaw decrosslinking buffer B at room temperature and then incubate at 37 °C in water bath for 30 min. After 30 min, cool to room temperature for 5 min. Vortex 30 s and centrifuge briefly).
- 13.75 μL 8 M Urea.
- 2.75 μL Diluted Perm Enzyme B.

6. Visium Tissue Slide cassette S3 (Component: 3001826, 3001828, 3001927, 3001825, Kit: 1000669).
7. Visium slide seals (cut in half in lengthwise) (Component 2000284, Kit 1000669).
8. 10X PBS, pH 7.4.
9. Pre-Hybridization Mix:

- 295.3 μL Nuclease-free water.
- 33.0 μL 10X PBS pH 7.4.
- 1.7 μL 10% Tween-20.

10. Probe Hybridization Mix: Add reagents in the order listed. Pipette mix and centrifuge briefly. Keep at room temperature.

- 22.0 μL Nuclease-free water.
- 154.0 μL FFPE Hybridization Buffer (Component 2000423, Kit 1000466 or 1000667: Thaw at room temperature. If precipitation persists, heat at 37 °C until dissolved. Avoid vortexing to prevent bubble formation. Pipette mix 10x. Maintain at room temperature).
- 22.0 μL Human WT probes v2—RHS (Component 2000657, Kit 1000466: Thaw at room temperature. Vortex and centrifuge briefly. Maintain at room temperature).
- 22.0 μL Human WT probes v2—LHS (Component 2000658, Kit 1000466: Thaw at room temperature. Vortex and centrifuge briefly. Maintain at room temperature).

11. FFPE post-hybridization wash buffer (Component 2000424, Kit 1000466 or 1000667). Thaw at room temperature. If precipitate persists, heat at 37 °C until dissolved. Vortex briefly. Flick tube until liquid settles at the bottom of the tube.
12. 20X SSC buffer.
13. 2X SSC buffer: Add reagents in the order listed. Maintain at room temperature.

- 315 μL 20X SSC.
- 2,835 μL Nuclease-free water.

14. Probe Ligation Mix: Add reagents in the order listed. Pipette mix 10x and centrifuge briefly. Do not vortex. Maintain on ice.

- 52.8 μL Nuclease-free water.
- 66.0 μL 2X Probe ligation buffer (Component 2000445, Kit 1000466 or 1000667: Thaw at room temperature until no precipitate remains. Vortex and centrifuge briefly).

- 13.2 μL Probe ligation enzyme (Component 2000425, Kit 1000466 or 1000667: Centrifuge briefly. Maintain on ice).

15. Post-ligation wash buffer (Component 2000419, Kit 1000466 or 1000667: Thaw at room temperature. If precipitate persists, heat at 37 °C until dissolved. Vortex).

2.3 Step 3

1. 20X SSC buffer.
2. Visium HD slide (Kit: 2000970).
3. Visium Cassette (Component 3001830/3001831, Kit 1000669).
4. Visium HD slide seals (Component 2000283, Kit 1000669).
5. 0.1X SSC Buffer: Add reagents in the order listed. Pipette mix. Maintain at room temperature.

- 59.7mL Nuclease-free water.
- 0.3mL 20X SSC.

6. 1X PBS: Vortex and maintain at room temperature.

- 5940 μL Nuclease-free water.
- 660 μL 10X PBS.

7. 10% Eosin: Add reagents in the order listed. Vortex and centrifuge briefly. Eosin should be prepared fresh for each CytAssist instrument run.

- 33 μL Alcoholic Eosin.
- 297 μL 1X PBS.

8. Slide equilibration mix: Add reagents in the order listed. Pipette mix and centrifuge briefly. Maintain on ice.

- 96.8 μL Nuclease-free water.
- 110.0 μL 2X RNase Buffer (Component 2000411, Kit 1000668: Thaw at room temperature. Pipette mix slowly and thoroughly. Do not vortex).
- 13.2 μL RNase Enzyme (Component 3000605, Kit 1000668: Pipette mix, centrifuge briefly. Maintain on ice until ready to use).

9. Probe Release mix: Add reagents in the order listed. Pipette mix thoroughly until solution is homogenous. Avoid generating bubbles. Maintain on ice.

- 20.0 μL 2X RNase Buffer (Component 2000411, Kit 1000668: Thaw at room temperature. Pipette mix slowly and thoroughly. Do not vortex).

- 17.5 μL RNase Enzyme (Component 3000605, Kit 1000668: Pipette mix, centrifuge briefly. Maintain on ice until ready to use).

10. CytAssist with software Firmware version 2.0.1 or higher.
11. Perm enzyme B (Component 3000553, Kit 1000668): Pipette mix and centrifuge briefly. Maintain at room temperature until ready to use. Perm Enzyme B is added to the Probe Release Mix immediately before running the CytAssist instrument.
12. Probe Extension mix: Add reagents in the order listed. Pipette mix. Centrifuge briefly. Maintain on ice.

- 323.4 μL Extension buffer (Component 2000409, Kit 1000668: Thaw at room temperature, vortex, centrifuge briefly).
- 6.6 μL Extension enzyme (Component 2000389, Kit 1000668: Pipette mix, centrifuge briefly. Maintain on ice until ready to use).

2.4 Step 4

1. KOH Mix: Add reagents in the order listed. Vortex and centrifuge briefly. Maintain at room temperature (Prepare shortly before use).

- 108.9 μL Nuclease-free water.
- 1.1 μL 8 M KOH.

2. Tris 1 M, pH 8.0 (Tris–HCl).
3. Alcoholic eosin.
4. Qiagen buffer EB.
5. 10X PBS, pH 7.4.
6. TS primer mix B (Component 2000537, Kit 1000668): Thaw at room temperature, vortex, and centrifuge briefly.
7. Pre-Amplification mix: Add reagents in the order listed. Pipette mix and centrifuge briefly. Maintain on ice.

- 42.9 μL Nuclease-free water.
- 55.0 μL Amp Mix B (Component 2000567, Kit 1000668: Vortex, centrifuge briefly, maintain on ice).
- 5.5 μL TS Primer Mix B (Component 2000537, Kit 1000668: Thaw at room temperature, vortex, and centrifuge briefly).

8. Beckman Coulter SPRIselect reagent.
9. Qiagen Buffer EB.
10. 80% ethanol (Prepare fresh—1 mL per reaction is needed).
11. 10x Magnetic separator (Component 230003, Kit 1000499).

2.5 Step 5

1. Agilent Bioanalyzer High Sensitivity DNA kit.
2. KAPA SYBR Fast Universal.
3. Real time qPCR system.
4. MicroAmp Fast 96-well reaction plate.
5. PCR plate seals.
6. TS Primer Mix B (Component 2000537, Kit 1000668): Thaw at room temperature, vortex, and centrifuge briefly.
7. qPCR mix (for two duplicates per sample and one negative control): Add reagents in the order listed. Vortex and centrifuge briefly. Maintain on ice.

- 27.5 μL KAPA SYBR FAST qPCR Master Mix (minimize light exposure).
- 5.5 μL Diluted TS primer mix B.
- 16.5 μL Nuclease-free water.

8. Dual Index Plate TS Set A (Component 3000511, Kit 1000251: Thaw at room temperature, vortex, and centrifuge briefly).
9. Qiagen buffer EB.
10. 10x Magnetic separator (Component 230003, Kit 1000499).
11. 80% ethanol (Prepare fresh).
12. Beckman Coulter SPRIselect reagent.
13. Amplification master mix:

- 99 μL Nuclease-free water.
- 55 μL Amp Mix B (Component 2000567, Kit 1000668: Vortex, centrifuge briefly, maintain on ice).

2.6 Step 6

1. Bioanalyzer.
2. Kit Agilent High Sensitivity DNA.
3. Qiagen buffer EB.

2.7 Step 7

1. LoupeBrower v8.2.1 (10X Genomics).
2. NovaSeq X Plus.
3. FastQ file (Raw NGS data).
4. SpaceRanger v3.1.2 (10X Genomics).
5. R version 4.4.2 with Seurat 5.2.0.

3 Methods

Entire protocol should take approximately two and a half days to be completed (*see* **Note 2**).

3.1 Step 1 (About 3.40 h)

3.1.1 Deparaffinization (3 h)

1. Retrieve slides with tissue sections after overnight drying (*see* **Note 3**).
2. Place slides on a thermocycler with adapter without closing the thermocycler lid for 2 h at 60 °C.
3. During the two-hour incubation, prepare (*see* **Note 4**).
   (a) 2 50 mL conical with 30 mL 100% xylene (X1 and X2)
   (b) 2 50 mL conical with 30 mL 100% ethanol (E1 and E2)
   (c) 2 50 mL conical falcons with 30 mL 96% ethanol (E3 and E4)
   (d) 1 50 mL conical with 30 mL 70% ethanol
   (e) 1 50 mL conical with 30 mL Milli-Q water
4. Remove slides from the thermocycler and allow to cool down to room temperature for 5 min.
5. Gently immerse slides 2x in 100% xylene X1 and immerse for 10 min.
6. Gently immerse slides 2x in 100% xylene X2 and immerse for 10 min.
7. Gently immerse slides 2x in 100% ethanol E1 and immerse for 3 min.
8. Gently immerse slides 2x in 100% ethanol E2 and immerse for 3 min.
9. Gently immerse slides 2x in 96% ethanol E3 and immerse for 3 min.
10. Gently immerse slides 2x in 96% ethanol E4 and immerse for 3 min.
11. Gently immerse slides 2x in 70% ethanol and immerse for 3 min.
12. Gently immerse slides 2x in Milli-Q water and immerse for 20 s.
13. Proceed immediately to staining & coverslip mounting. Do not let the slides dry.

3.1.2 H&E Staining (15 min)

1. Prepare 8 beakers with 350 mL Milli-Q water.
2. Place slide on a flat, clean, non-absorbent work surface. Some residual droplets may remain.
3. Add 1 mL Hematoxylin per slide to uniformly cover all tissue sections. If staining two tissue slides at a time, ensure that

Hematoxylin is discarded quickly after incubation to prevent overstaining.

4. Incubate 1 min at room temperature.
5. Discard reagent by holding slides at an angle with the bottom edge in contact with a lint-free laboratory wipe.
6. Immerse slides 5x in Milli-Q water beaker 1.
7. Immerse slides 15x in Milli-Q water beaker 2.
8. Immerse slides 15x in Milli-Q water beaker 3.
9. Wipe excess liquid from the back of the slides without touching the tissue section.
10. Place slide on a flat, clean, non-absorbent work surface. Some droplets may remain.
11. Add 1 mL Bluing Buffer per slide to uniformly cover all tissue sections.
12. Incubate 1 min at room temperature.
13. Discard reagent by holding slides at an angle with the bottom edge in contact with a lint-free laboratory wipe.
14. Immerse slides 15x in Milli-Q water beaker 4.
15. Wipe excess liquid from the back of the slides with a lint-free laboratory wipe without touching the tissue section. Place on a flat, clean, and non-absorbent work surface. Some droplets may remain.
16. Gently immerse slides in a 50-mL conical falcon containing 30 mL of undiluted alcoholic eosin solution.
17. Incubate 2 min at room temperature.
18. Discard reagent by holding slides at an angle with the bottom edge in contact with a lint-free laboratory wipe.
19. Immerse slides for 30 s in Milli-Q water beaker 5.
20. Immerse slides 10x in Milli-Q water beaker 6.
21. Wipe excess liquid from the back of the slides with a lint-free laboratory wipe without touching the tissue section. Do not air-dry the slides.

3.1.3 Coverslip Mounting (5 min)

1. Place slides on a flat, clean, non-absorbent work surface. Some residual droplets may remain.
2. Using a wide-bore pipette tip, add 100–150 μL mounting medium to cover all tissue sections on slides uniformly (*see* **Note 5**).
3. Slowly apply coverslip at an angle on one end of the slides (*see* **Note 6**). Allow mounting medium to spread and settle. Do not use Cytoseal or nail polish for securing the coverslip.

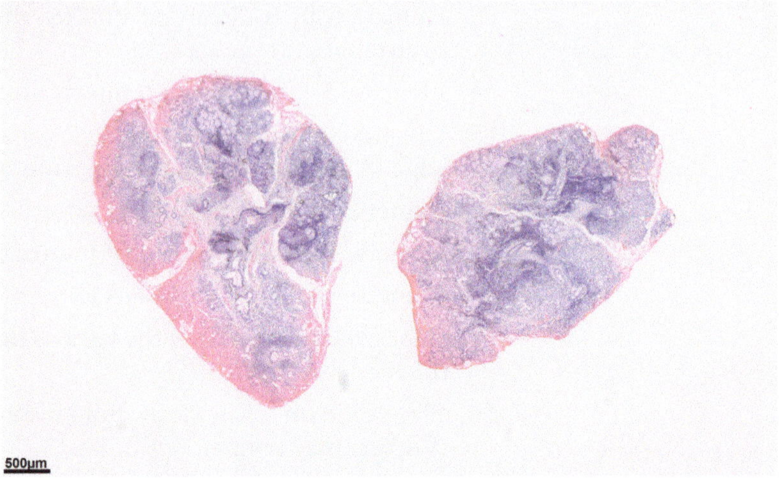

Fig. 1 H&E image of human minor salivary gland tissue section. Image was acquired on the Hamamatsu NanoZoomer at 20X

4. Remove any large excess of mounting medium by carefully wicking away from the edge of the coverslip using a lint-free laboratory wipe. Do not move coverslip and disturb tissue.
5. Immediately proceed with imaging or store slides laying flat in a slide holder. Coverslip will dry out over time. Do not let the attached coverslip dry out (*see* **Note 7**).

3.1.4 Imaging (10 min)

1. Image tissue section of interest at desired magnification using brightfield imaging settings as shown in Fig. 1 (*see* **Note 8**).
2. Proceed immediately to coverslip removal.

3.1.5 Coverslip Removal (10 min)

1. Immerse slides sideways in Milli-Q water beaker 7 with coverslipped surface fully sideways to prevent the coverslip from dragging across the tissue.
2. Hold slides in water until the coverslip separates away from the slide. To avoid tissue section damage or detachment, do not move the slide up and down, shake forcibly, or manually move the coverslip.
3. Gently immerse slides 30x in Milli-Q water beaker 8 to ensure all mounting medium is removed.
4. Wipe the back of the slide with a lint-free laboratory wipe.
5. Place on a flat, clean, non-absorbent work surface and air dry for 5 min.
6. Incubate slide on low profile thermocycler adapter with thermocycler lid open for 3 min at 37 °C. Ensure tissue is completely dry. If necessary, wipe droplets around tissue with a lint-free laboratory wipe.

STOP: Proceed immediately to destaining or store slides laying flat in a slide mailer or slide holder at 4 °C in the dark with desiccant for up to 2 weeks. Ensure slides do not touch one another and that desiccant does not come in contact with the tissue slide.

3.2 Step 2 (About 21.10 h)

3.2.1 Destaining (30 min)

1. While incubating decrosslinking buffer B at 37 °C, proceed to **step 2**.
2. Place a low-profile thermocycler adapter in the thermocycler and start the below program.

Step	Temperature	Time
Pre-equilibrate	42 °C	Hold
Destaining	42 °C	15 min
Hold	22 °C	Hold

(Lid temperature: 42 °C, Reaction volume 100 μL)

3. Place slide in a tissue slide cassette, centering the tissue or area of interest.
4. Add 150 μL 0.1 N HCl along the side of the wells to uniformly cover tissue sections, without introducing bubbles. Tap cassette gently to ensure uniform coverage.
5. Remove HCl from wells (*see* **Note 9**).
6. Add 100 μL 0.1 N HCl along the side of the wells to uniformly cover tissue sections, without introducing bubbles. Tap cassette gently to ensure uniform coverage.
7. Apply pre-cut slide seal on cassette, place on the low-profile thermocycler adapter at 42 °C.
8. Close thermocycler lid. Skip pre-equilibrate step to initiate destaining.
9. Remove cassettes from the adapter and place on a flat, clean work surface. Some color remaining in the tissue after thermocycler incubation is normal.
10. Peel back slide seal, and using a pipette, remove all the HCl from well corners.
11. Add 150 μL TE buffer along the side of the wells.
12. Incubate for 5 min.
13. Remove TE buffer from the wells.
14. Repeat steps 10–12 two times for a total of three washes.
15. Add 100 μL 1X PBS along the side of the wells.
16. Re-apply slide seal.

3.2.2 Decrosslinking (40 min)

1. Ensure decrosslinking buffer B has been thawed correctly.
2. Place a low-profile thermocycler adapter in the thermocycler and start the below program.

Step	Temperature	Time
Hold	22 °C	Hold
Decrosslinking	80 °C	30 min
Re-equilibrate	22 °C	10 min
Hold	22 °C	Hold

(Lid temperature: 80 °C, Reaction volume 100 μL)

3. Prepare diluted perm enzyme B.
4. Prepare decrosslinking mix.
5. Remove slide seal and remove 1X PBS from the wells.
6. Add 100 μL decrosslinking mix along the side of the wells.
7. Apply a new pre-cut slide seal on the cassette and place cassette on the thermocycler adapter.
8. Close and tighten thermocycler lid and skip pre-equilibrate step to initiate decrosslinking.
9. Proceed immediately to library construction with the Visium HD spatial gene expression reagent kits user guide (CG000685).

3.2.3 Probe Hybridization (18.20 h)

1. Prepare pre-hybridization mix shortly before use.
2. Retrieve tissue slide cassettes containing H&E-stained sections.
3. Peel back Visium slide seals.
4. Using a pipette, remove all buffer from each well at well corners.
5. Add 150 μL pre-hybridization mix along the side of each well to uniformly cover tissue sections, without introducing bubbles.
6. Re-apply Visium slide seal on each tissue slide cassette.
7. Incubate for 15 min at room temperature.
8. Place a low-profile thermocycler adapter in the thermocycler and start the below program.

Step	Temperature	Time
Pre-equilibrate	50 °C	Hold
Hybridization	50 °C	18 h
Post Hybridization Wash	50 °C	Hold

(Lid temperature: 50 °C, Reaction volume 100 μL)

9. Prepare probe hybridization mix shortly before use.
10. Remove all pre-hybridization mix from each well.
11. Add 100 μL room temperature probe hybridization mix to each well.
12. Apply a new pre-cut Visium slide seal to each tissue slide cassette and place on the thermocycler adapter. On the thermocycler. Close the thermocycler lid.
13. Skip pre-equilibrate step to initiate hybridization.

3.2.4 Post-hybridization Wash (20 min)

1. Aliquot 495 μL FFPE post-hybridization wash buffer per 6.5 mm sample and pre-heat to 50 °C in a water bath or thermomixer. Maintain pre-heated wash buffer at 50 °C throughout washes (*see* **Note 10**).
2. Prepare 2X SSC buffer. Add reagents in the order listed. Maintain at room temperature. Do not discard excess buffer, as it will be used for subsequent washes in the protocol.
3. Remove Visium tissue slide cassettes from thermocycler adapter and place on a flat, clean work surface.
4. Peel back Visium slide seals, and using a pipette, remove all probe hybridization mix from each well.
5. Immediately add 150 μL pre-heated FFPE post-hybridization wash buffer to each well. Avoid well drying or cooling to room temperature. Removal and addition of buffers should be done quickly.
6. Re-apply Visium slide seal on each tissue slide cassette and place on thermocycler adapter on pre-heated thermocycler. Close thermocycler lid. Ensure that tissue slide cassette sits snugly on the thermocycler adapter for all wash steps.
7. Skip the hybridization step on thermocycler and initiate post-hybridization wash.
8. Incubate Visium tissue slide cassettes in thermocycler at 50 °C for 5 min.
9. Remove tissue slide cassettes from thermocycler adapter and place on a flat, clean work surface. Exercise caution, as slide is hot.
10. Peel back Visium slide seals, and using a pipette, remove all FFPE post-hybridization wash buffer from each well.
11. Immediately add 150 μL pre-heated FFPE post-hybridization wash buffer to wells. Avoid well drying or cooling to room temperature. Removal and addition of buffers should be done quickly.

12. Re-apply Visium slide seal on each tissue slide cassette and place on thermocycler adapter on pre-heated thermocycler. Close thermocycler lid.
13. Incubate in thermocycler at 50 °C for 5 min.
14. Repeat **steps 9–13** for a total of three washes.
15. Peel back Visium slide seal and remove FFPE post-hybridization wash buffer from each well.
16. Add 150 μL 2X SSC to each well and re-apply Visium slide seal.
17. Let the tissue slide cassette cool to room temperature (~3 min) before proceeding to the next step.

3.2.5 Probe Ligation (1.0 h)

1. Place a low-profile thermocycler adapter onto a thermocycler and start the below program.

Step	Temperature	Time
Pre-equilibrate	37 °C	Hold
Ligation	37 °C	1 h
Hold	4 °C	Hold

(Lid temperature: 37 °C, Reaction volume 100 μL)

2. Prepare probe ligation mix shortly before use.
3. Before the end of the ligation, pre-heat 110 μL/sample of post-ligation wash buffer to 57 °C. Proceed to **step 4** after putting on thermomixer or water bath (*see* **Note 11**).
4. Remove Visium slide seals from each tissue slide cassette and remove 2X SSC buffer from each well.
5. Add 60 μL probe ligation mix directly to tissue sections in each well, without introducing bubbles. Tap tissue slide cassettes gently to ensure uniform coverage.
6. Apply a new pre-cut Visium slide seal on each tissue slide cassette and place on the thermocycler adapter on the pre-heated thermocycler. Close the thermocycler lid.
7. Skip pre-equilibrate to initiate ligation.

If proceeding with Visium HD slide preparation, remove Visium HD slide mailer from −80 °C. Keep slide mailer upright, capped, and thaw at room temperature for 30–60 min during ligation step.

3.2.6 Post-ligation Wash (20 min) (See Note 12)

1. Use room temperature post-ligation wash buffer at first wash **step (5)** and pre-heated post-ligation wash buffer for the second wash **step (10)**.

2. Remove tissue slide cassettes from the thermocycler adapter and place on a flat, clean work surface.
3. Start on the thermocycler.

Step	Temperature	Time
Incubate	57 °C	Hold

(Lid temperature: 57 °C, Reaction volume 100 μL)

4. Remove Visium slide seal from each tissue slide cassette, and using a pipette, remove all probe ligation mix from each well.
5. Immediately add 100 μL room temperature post-ligation wash buffer. Removal and addition of buffers should be done quickly.
6. Apply a new pre-cut Visium slide seal on each tissue slide cassette and place on the thermocycler adapter on pre-heated thermocycler. Close thermocycler lid. Ensure that tissue slide cassette sits snugly on the low-profile thermocycler adapter for all wash steps.
7. Incubate at 57 °C for 5 min.
8. Remove tissue slide cassette from thermocycler adapter and place on a flat, clean work surface.
9. Peel back Visium slide seals, and using a pipette, remove all post-ligation wash buffer from each well.
10. Add 100 μL pre-heated post-ligation wash buffer to each well and re-apply Visium slide seal.
11. Place tissue slide cassette on the thermocycler adapter on pre-heated thermocycler. Close thermocycler lid.
12. Incubate at 57 °C for 5 min.
13. Remove tissue slide cassette from the thermocycler adapter and place on a flat, clean work surface.
14. Peel back Visium slide seals, and using a pipette, remove all post ligation wash buffer from each well.
15. Add 150 μL 2X SSC buffer prepared at **step 2** of probe ligation (*see* Subsection 3.2.4).
16. Remove all 2X SSC buffer from each well.
17. Add 150 μL 2X SSC buffer to each well.
18. Re-apply Visium slide seals on each tissue slide cassette.
19. Allow tissue slide cassettes to come to room temperature for 5 min (*see* **Note 13**).

STOP: Store at 4 °C for up to 24 h or proceed to next step.

3.3 Step 3 (About 2.40 h)

3.3.1 Visium HD Slide Preparation (20 min)

1. If Visium HD slide has not been thawed during post-ligation wash, remove Visium HD slide mailer from −80 °C. Remove slide mailer from mylar bag. Do not uncap slide mailer. Keep slide mailer upright and thaw at room temperature for 30–60 min. Do not touch spacer during slide washes.
2. Prepare 0.1X SSC buffer. Pipette mix and maintain at room temperature.
3. Open slide mailer.
4. Add 7 mL 0.1X SSC buffer slowly along the side of the mailer on one side of the slide.
5. Add 7 mL 0.1X SSC buffer slowly along the side of the mailer on the other side of the slide, so that both slides of the slide are immersed in 0.1X SSC buffer.
6. Incubate at room temperature for 1 min. Do not close the mailer.
7. Remove 0.1X SSC buffer by securing slide in place with one finger and pouring SSC buffer from mailer.
8. Add 7 mL 0.1X SSC buffer slowly along the side of the mailer on one side of the slide.
9. Add 7 mL 0.1X SSC buffer slowly along the side of the mailer on the other side of the slide, so that both sides of the slide are immersed in 0.1X SSC buffer.
10. Incubate at room temperature for 5 min.
11. Remove 0.1X SSC buffer by securing slide in place with one finger and pouring SSC buffer from mailer.
12. Repeat **steps 8–11** twice more for a total of three 5 min washes. During the last wash, save SSC buffer in case Visium HD slide needs additional wash.
13. Ensure back (side without spacers) of Visium HD slide is dry. If necessary, remove excess moisture from the back of the slide with a lint-free laboratory wipe.
14. Gently touch the long edge of the Visium HD slide on a lint-free laboratory wipe 3–5× to remove excess SSC buffer.
15. Ensure Visium HD slide is free of particulate matter. If necessary, wash Visium HD slide by immersing in slide mailer with 0.1X SSC buffer.

Record slide serial number before placing Visium HD slide into cassette.

16. Place Visium HD slide in a new 6.5 mm Visium cassette.
17. Add 100 μL 0.1X SSC to each well in the cassette.
18. Apply a new pre-cut Visium slide seal on the Visium cassette.

19. Leave 0.1X SSC buffer in the Visium cassette at room temperature until Visium HD slide equilibration. Do not exceed 2 h before proceeding with a CytAssist run. Keep Visium cassette and Visium HD slide free from debris.

3.3.2 Probe Release (1.10 h)

1. Ensure that the Visium CystAssist is powered on, running firmware version 2.0.1 or higher, and is clean and ready to perform an experimental run.
2. If processing more than two tissue slides, keep remaining tissue slides at 4 °C with 2X SSC buffer.
3. Press blue New Run button on the touchscreen to initiate run.
4. Enter the following assay configuration information: Visium slide serial number, custom run name, temperature and time (37 °C for 30 min is recommended for most applications).
5. Enter the following sample information: sample names and locations on the instrument (A1 for right side, D1 for left side).
6. Prepare 1X PBS. Maintain at room temperature.
7. Prepare 10% eosin, vortex, and centrifuge briefly. Eosin should be prepared fresh for each CytAssist instrument run.
8. Prepare Slide equilibration mix.
9. Retrieve Visium cassette with Visium HD slide.
10. Remove Visium slide seal, and using a P200 pipette, remove 0.1X SSC from the lower right corner of each well in the cassette without scratching the fiducial frame or hydrogel. Use a P20 pipette to remove any remaining liquid.
11. Add 100 μL slide equilibration mix to each well in the cassette. Gently tap to ensure uniform coverage of the capture area.
12. Apply a new uncut Visium slide seal on the Visium cassette and incubate at room temperature for 10 min.
13. Prepare probe release mix during Visium HD slide equilibration. Do not prepare probe release mix for more than two tissue slides at a time.
14. Remove Visium slide seal, and using a P200 pipette, remove slide equilibration mix from the right corner of each well in the cassette without scratching the fiducial frame or hydrogel. Use a P20 pipette to remove any remaining liquid.

Failure to remove slide equilibration mix completely may result in reduced assay performance.

15. Remove top half of Visium cassette, leaving Visium HD slide resting in bottom half of Visium cassette. Rest top half of the Visium cassette such that the gaskets face away from work surface.

16. Remove Visium HD slide from Visium cassette. Avoid touching active surface. If necessary, remove excess moisture from the back of the slide with a lint-free laboratory wipe. The time between completing **step 16** and starting **step 20** should not exceed 5 min.

Load Visium HD slide against the grooves of Visium slide stage, using one hand to load the slide and the other to close the Visium slide lock.

17. Allow Visium HD slide to dry on the Visium slide stage for 10 min. Inspect entire well. If liquid remains on the slide or anywhere in the well, continue drying and proceed immediately when no liquid remains within the well (*see* **Note 14**). Ensure area around the Visium CytAssist instrument is free from debris. While the Visium HD slide is drying, prepare tissue slides as described in **steps 18–27**.
18. Retrieve tissue slide cassettes.
19. Using a pipette, remove all 2X SSC buffer from each well of the tissue slide cassettes.
20. Remove tissue slides from tissue slide cassettes.
21. Add 100 μL 10% eosin to uniformly cover each tissue section per slide.
22. Incubate 1 min at room temperature.
23. Remove 10% eosin by holding slide at an angle over a liquid waste container.
24. While holding the slide over the liquid waste container, rinse with 1 mL 1X PBS. Do not pipette directly onto tissue.
25. Repeat **step 24** two more times for a total of three washes.
26. Gently waft/flick slide back and forth to remove excess PBS. Remove any excess PBS with a lint-free laboratory wipe in areas outside of the tissue, without touching the tissue sections.
27. Wipe back of tissue slides with a lint-free laboratory wipe and load into Visium CytAssist. Ensure tissue sections are completely dry prior to instrument run. If necessary, repeat slide wafting.

Complete tissue slide alignment before Visium HD slide is done drying.

Before proceeding to next step, ensure that Visium HD slide is completely dry. See **step 17**.

28. Pipette mix Perm enzyme B and centrifuge briefly. Remove probe release mix from ice. Add 2.5 μL perm enzyme B to 37.5 μL of Probe release mix (prepared at **step 13**). Pipette mix 15x with pipette set to 30 μL. Avoid generating bubbles. Centrifuge for 5 s.

The time between adding Perm enzyme B to probe release mix and starting the Visium CytAssist instrument run should be less than 5 min.

29. Slowly aspirate 17 μL of probe release mix and inspect the pipette tip. Ensure that no bubbles were drawn up into the pipette tip.
30. Slowly dispense 17 μL of probe release mix into the center of each spacer well on the Visium HD slide, using a fresh pipette tip for each dispense. Do not push plunger beyond first stop to avoid generating bubbles.
31. Close the lid and press next. The home screen will now display a play symbol and run information along the bottom of the screen.
32. Press the play button to start the run. 37 °C for 30 min is recommended for most applications.

Midrun progress bar will show the time remaining in the run.
Yellow before a run begins indicates a user-recoverable error. If an error occurs, follow on-screen prompts.

33. Place a low-profile thermocycler adapter in the thermocycler and start program below.

Step	Temperature	Time
Pre-equilibrate	53 °C	Hold
Probe extension 1	53 °C	30 min
Cool down	4 °C	3 min
Hold	4 °C	Hold
Probe extension 2	53 °C	30 min
Cool down	4 °C	3 min
Hold	4 °C	Hold

(Lid temperature: 53 °C, Reaction volume 100 μL)

34. Prepare probe extension mix.
35. At the end of a run, the button will display "Done" and a "Run info" tab at the bottom of the screen. Do not allow sample to sit in the Visium CytAssist after run completion. Immediately move to next step (*see* **Note 15**). Click "Done" button and open lid. Do not power off the instrument at this time, as it needs to process support data. It is normal for the Visium slide to move after opening instrument.
36. Immediately remove Visium HD slide. It is normal if tissue remains on tissue slides after running completion.

37. While holding Visium HD slide over liquid waste container, rinse each capture area with 1 mL buffer EB three times. Do not pipette directly onto capture areas, which are surrounded by the fiducial frames.
38. Place Visium HD slide in the Visium cassette from earlier in this step.

    Some moisture remaining on the Visium HD slide is normal.

39. Proceed immediately to probe extension. CytAssist instrument should only be cleaned and data transferred at a safe stopping point.

3.3.3 Probe Extension (1.10 h)

1. Probe extension occurs over two rounds, probe extension 1 and probe extension 2. Add 75 μL probe extension mix to each well. Gently tap Visium cassettes to ensure uniform coverage of capture area. Keep remaining probe extension mix on ice.
2. Apply a new uncut Visium slide seal on visium cassette and place on the thermocycler adapter on pre-heated thermocycler. Close thermocycler lid.
3. Skip pre-equilibrate step to initiate probe extension 1.
4. After the first cool down, open thermocycler lid and remove cassette from thermocycler adapter.
5. Peel back Visium slide seal and remove probe extension mix from each well.
6. Add 75 μL probe extension mix to each well. Gently tap Visium cassettes to ensure uniform coverage of capture area.
7. Re-apply Visium slide seal and place on the thermocycler adapter. Close thermocycler lid.
8. Skip hold step to initiate probe extension 2.
9. Clean CytAssist instrument.

STOP: Sample may remain at 4 °C in the thermocycler for up to 24h.

3.4 Step 4 (About 1.25 h)

3.4.1 Probe Elution (15 min)

1. Prepare 0.08 M KOH mix shortly before use.
2. Remove Visium cassette from the thermocycler adapter and place on a flat, clean work surface after probe extension is complete.
3. Remove Visium slide seal, and using a pipette, remove all probe extension mix from wells.
4. Add 150 μL buffer EB to each well and remove.
5. Add 50 μL 0.08 M KOH mix to each well. Gently tap Visium cassettes to ensure uniform coverage of capture area.

6. Incubate at room temperature for 10 min.
7. Add 3 μL 1 M Tris–HCl pH 8.0 to a microtube strip for each sample and transfer all solution for each sample. Vortex, centrifuge briefly and place on ice. Do not leave behind any solution in the wells.

3.4.2 Pre-amplification Cleanup—SPRIselect (1.10 h)

1. Prepare pre-amplification mix.
2. Add 47 μL pre-amplification mix to each tube, pipette mix, and centrifuge briefly.
3. Incubate in a thermocycler with program below.

Step	Temperature	Time
1	98 °C	3 min
2	98 °C	15 s
3	63 °C	20 s
4	72 °C	30 s
5	Go to **step 2**, repeat 9X for a total of 10 cycles	
6	72 °C	1 min
7	4 °C	Hold

(Lid temperature: 105 °C, Reaction volume 100 μL)

4. Vortex to resuspend the SPRIselect reagent. Add 120 μL SPRIselect reagent (1.2X) to each pre-amplification reaction and pipette mix 15x (pipette set to 175 μL).
5. Incubate 5 min at room temperature.
6. Place on the magnet High until the solution clears.
7. Remove the supernatant.
8. Add 300 μL 80% ethanol. Wait 30 s.
9. Remove the ethanol.
10. Add 200 μL 80% ethanol. Wait 30 s.
11. Remove the ethanol.
12. Centrifuge briefly and place on the magnet low.
13. Remove any remaining ethanol without disturbing the beads. Air dry for 2 min. Do not exceed 2 min as this will decrease elution efficiency.
14. Remove from magnet, add 105 μL buffer EB and pipette mix 15x (pipette set to 100 μL).
15. Incubate 2 min at room temperature.

16. Place tube strips on magnet High until solution clears.
17. Transfer 100 μL sample to a new tube strip.

STOP: Store at 4 °C for up to 72 h or −20 °C for up to 4 weeks or proceed to next step.

3.5 Step 5 (About 3.25 h)

3.5.1 Cycle Number Determination—qPCR (3 h)

1. Dilute TS primer mix B 1:10 in nuclease-free water for a final volume of 10 μL.
2. Prepare qPCR mix.

A passive reference dye, such as ROX, may be required. Consult the Roche KAPA SYBR FAST qPCR Kit website or the appropriate qPCR system manufacturer for guidance.

3. Add 9 μL qPCR mix to two wells per sample in a qPCR plate (a well for negative control may be included).
4. Dilute 2 μL sample from pre-amplification cleanup—SPRIselect in 8 μL nuclease-free water. Pipette mix and centrifuge briefly.
5. Transfer 1 μL diluted sample from pre-amplification cleanup—SPRIselect to each qPCR plate well containing qPCR mix. For the negative control, add 1 μL nuclease-free water to the corresponding well. Pipette Mix.
6. Apply seal and centrifuge briefly. Record which sample is in which well of the qPCR plate.
7. Start program below on the qPCR system.

Step	Temperature	Time
1	98 °C	3 min
2	98 °C	5 s
3	63 °C	30 s
4	Go to step 2, 24X (total of 25 cycles)	

(Lid temperature: 105 °C, Reaction volume 10 μL)

Record the Cq value for each sample: set the y-axis to a linear scale and plot RFU on the y-axis if not using a reference dye or ΔRN if using a reference dye. The threshold for determining the Cq value should be set along the exponential phase of the amplification plot, at 25% of the peak fluorescence value. If the Cq value is >20 cycles, it can be a troubleshooting.

Round Cq values up to the nearest whole number and add one cycle (examples below). Samples within ± 1 cycle numbers can be combined in a single SI-PCR run at the higher cycle number.

Do not combine samples if the cycler number difference is greater than 1 to avoid library overamplification. See example for cycle numbers below.

Cycle number for single libraries		
Cq value from qPCR	**+1**	**Total cycles**
7.2	+1	9
8.5	+1	10
13.7	+1	15

Cycle number for combined libraries			
Cq value from qPCR	**+1**	**Total cycles**	**Batched cycles**
7.2	+1	9	10
9.5	+1	11	
13.7	+1	15	16
14.6	+1	16	

3.5.2 Sample Index PCR (40 min)

1. Choose the appropriate sample index sets to ensure that no sample indices overlap in a multiplexed sequencing run. Record the 10x sample index name used.
2. Prepare amplification master mix and pipette mix shortly before use.
3. Add 70 μL amplification master mix to a tube in an 8-tube strip for each sample.
4. Add 25 μL of each sample from pre-amplification to a separate tube previously aliquoted with amplification master mix.
5. Add 5 μL of an individual Dual Index TS Set A to each tube and record well ID used. Pipette mix and centrifuge briefly.
6. Incubate in a thermocycler with the program below.

Step	Temperature	Time
1	98 °C	3 min
2	98 °C	15 s
3	63 °C	20 s
4	72 °C	30 s
5	Go to **step 2**, use the Cq value +1 as the total number of cycles	
6	72 °C	1 min
7	4 °C	Hold

(Lid temperature: 105 °C, Reaction volume 100 μL)

STOP: Any remaining pre-amplification material can be stored for up to 72 h or at -20 °C for up to 4 weeks for generating additional libraries.

3.5.3 Post-sample Index PCR Cleanup—SPRIselect (10 min)

1. Vortex to resuspend SPRIselect reagent. Add 85 μL SPRIselect reagent (0.85X) to each sample. Pipette mix 15x (pipette set to 175 μL).
2. Incubate 5 min at room temperature.
3. Place on the magnet High until solution clears.
4. Remove supernatant.
5. Add 200 μL 80% ethanol to pellet. Wait 30 s and remove ethanol. Repeat for a total of two washes.
6. Centrifuge briefly. Place on the magnet low.
7. Remove any remaining ethanol. Air dry for 2 min. DO NOT exceed 2 min as this will decrease elution efficiency.
8. Remove from the magnet. Add 27 μL buffer EB. Pipette mix.
9. Incubate 2 min at room temperature.
10. Place on the magnet low until solution clears.
11. Transfer 25 μL sample to a new tube strip.

STOP: Store at −20 °C for long-term storage.

3.6 Step 6 (About 30 min)

1. Dilute sample (1:50 dilution, i.e., 1 μL sample in 49 μL of buffer EB) until it is at an appropriate concentration for the Bioanalyzer.
2. Run 1 μL of sample on an Agilent Bioanalyzer High Sensitivity Chip. If peak is too small or flat, retry with a lower dilution.
3. Determine the average fragment size from the Bioanalyzer trace. The expected average fragment is 250 bp. This will be used as the insert size for library quantification.

3.7 Step 7

3.7.1 Sequencing

1. Use Loupe Browser to calculate sequencing depth by multiplying the percentage of the capture area covered by tissue by 275 million to get the minimum total number of read pairs. In Fig. 2, the sample covers 27.9% of total surface. The total read pairs for this sample is 74 M (0.27 * 275 M) (*see* **Note 16**).
2. Sequence your samples on compatible sequencers (NextSeq 2000, NovaSeq 6000, NovaSeq X Plus) as indicated: Paired-end, dual indexed sequencing, Read 1: 43 cycles, I7 index: 10 cycles, I5 index: 10 cycles, Read 2S: 50 cycles.

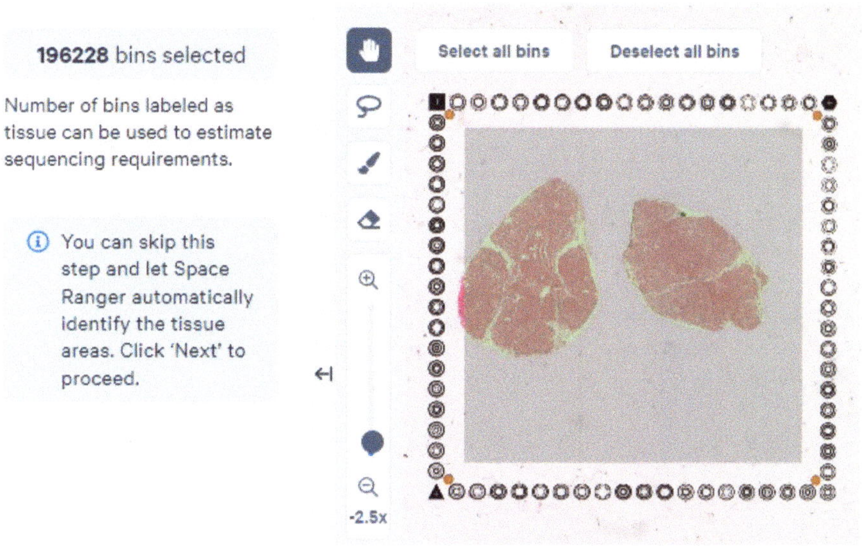

Fig. 2 Coverage area using LoupeBrowser. Tissue is highlighted in green and covers 196,228 bins out of 702,244 total bins (27.9% of coverage)

3.7.2 Alignment and Analysis

1. Generate JSON files for fiducial identification, (manual) tissue selection, and H&E image alignment using Loupe Browser (v8.2.1).
2. Process raw NGS data with SpaceRanger v3.1.2 (10x Genomics), providing the JSON file as loupe-alignment. Use the refdata-gex-GRCh38-2020-A (10x Genomics) as the transcriptome reference and the Human Transcriptome Probe Set v2.0 GRCh38-2020-A (10x Genomics) as the probe set.
3. Conduct downstream analyses on the processed data in the binned_outputs folder using RStudio and custom R scripts. For Fig. 3, analyses were performed in RStudio Pro 2024.12.0 with R v4.4.2, utilizing Seurat v5.2.1 with data aggregated into 8 μm bins.
4. Assess data quality based on standard quality control metrics, including nFeatures/nCounts per spot and mitochondrial gene percentages).
5. Normalize data using log-normalization and identify variable features. Custom marker lists can be incorporated at this step.
6. Downsample ("sketch") the dataset to 10,000 cells using LeverageScore, followed by dimensionality reduction via principal component analysis (PCA).
7. Integrate data across multiple samples. For Fig. 3, four samples were integrated using Harmony.

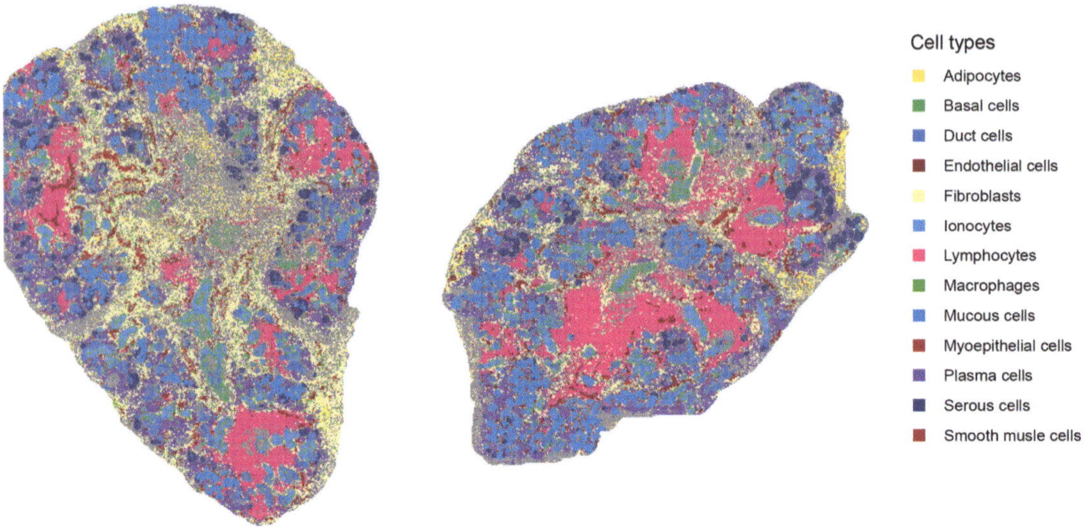

Fig. 3 Spatial projection of the dataset. Projection of the identified cell types after QC, integration and clustering. (Integration of 4 different samples)

8. Perform clustering. For Fig. 3, the following parameters were used: resolution = 2.0, k.param = 10, and dims = 1:50. Cell types were manually annotated based on known marker genes.
9. Finally, the "project" identified cell types from the sketched dataset onto the full dataset and visualize cells in their spatial locations.

4 Notes

1. Protocol has been split in different steps according to possible stopping points.
2. Four tissue sections can be processed at the same time if two thermocyclers and two thermocycler adapters are available (two tissue sections fit on one Visium slide). Before starting the experiment, take the time to save the different programs on the thermocycler and the qPCR system.
3. This section entails having cut a 5 μm section in the FFPE block of interest ahead of starting. After sectioning, tissue slides should be incubated at 42 °C for 3 h followed by overnight drying in a 50 mL conical with desiccant sachets. Tissue slides can be kept up to 6 months at room temperature with dessicant sachets before processing.
4. 30 mL should be sufficient to fully cover tissue sections.

5. Smaller or bigger tissue sections require less or more mounting medium respectively.
6. H&E images are used for later analysis, avoiding introducing bubbles.
7. Slides can be stored at 4 °C for up to 24 h.
8. H&E staining is crucial for determining the tissue position on the Visium CytAssist instrument and for later analyses.
9. During reagent removal steps, ensure all liquid is removed from wells using a P200 pipette first followed by a P20 pipette.
10. Aliquoting 3 Eppendorf with 330 μL FFPE post-hybridization wash buffer enables to take one Eppendorf per wash without removing the others from the 50 °C water bath.
11. Will be used for post-ligation wash step.
12. Visium HD slide preparation can be carried out in parallel to the post-ligation wash.
13. Up to 10 min at room temperature and store at 4 °C if longer.
14. This step can take more than 10 min depending on ambient moisture of the room but should not take more than 20 min.
15. Green indicates a successfully completed run, while red and yellow indicate a failed run/error or an incomplete run, respectively. However, a failed run may also result from an error in retrieving all run data and does necessarily mean unsuccessful transfer.
16. Capture area zone is slightly smaller than visual square. This difference should be considered when positioning tissue on the CytAssist.

Acknowledgments

This work was supported by the European Union (MSCA Postdoctoral Fellowships 2022 No 1011078471) and the LabEx IGO program (n° ANR-11-LABX-0016) funded by the "Investment into the Future" French Government program, managed by the National Research Agency (ANR). Views and opinions expressed are however those of the authors only and do not necessarily reflect those of the European Union or Research Executive Agency. Neither the European Union nor the granting authority can be held responsible for them.

Cohort The study was performed in accordance with the Declaration of Helsinki and was approved by ethical committees. The Diagnostic Suspicion of Primitive Syndrome Sjogren's (DIApSS) study is registered on ClinicalTrials.gov (identifier: NCT03681964).

References

1. Cornec D, Chiche L (2015) Is primary Sjögren's syndrome an orphan disease? A critical appraisal of prevalence studies in Europe: Table 1. Ann Rheum Dis 74:e25. https://doi.org/10.1136/annrheumdis-2014-206860
2. Sène D, Oaklander AL, Lefaucheur J, Lal S (2018) Primary Sjögren's syndrome. New Engl J Med
3. Brito-Zerón P, Baldini C, Bootsma H et al (2016) Sjögren syndrome. Nat Rev Dis Primers 2:16047. https://doi.org/10.1038/nrdp.2016.47
4. Nocturne G, Mariette X (2013) Advances in understanding the pathogenesis of primary Sjögren's syndrome. Nat Rev Rheumatol 9: 544–556. https://doi.org/10.1038/nrrheum.2013.110
5. Ciccia F, Guggino G, Rizzo A et al (2012) Potential involvement of IL-22 and IL-22-producing cells in the inflamed salivary glands of patients with Sjögren's syndrome. Ann Rheum Dis 71:295–301. https://doi.org/10.1136/ard.2011.154013
6. Bowman SJ, Fox R, Dörner T et al (2022) Safety and efficacy of subcutaneous ianalumab (VAY736) in patients with primary Sjögren's syndrome: a randomised, double-blind, placebo-controlled, phase 2b dose-finding trial. The Lancet 399:161–171. https://doi.org/10.1016/S0140-6736(21)02251-0
7. Williams CG, Lee HJ, Asatsuma T et al (2022) An introduction to spatial transcriptomics for biomedical research. Genome Med 14:68. https://doi.org/10.1186/s13073-022-01075-1
8. Meylan M, Petitprez F, Becht E et al (2022) Tertiary lymphoid structures generate and propagate anti-tumor antibody-producing plasma cells in renal cell cancer. Immunity 55: 527–541.e5. https://doi.org/10.1016/j.immuni.2022.02.001
9. Lyubetskaya A, Rabe B, Fisher A et al (2022) Assessment of spatial transcriptomics for oncology discovery. Cell Rep Methods 2:100340. https://doi.org/10.1016/j.crmeth.2022.100340
10. Warner B, Pranzatelli T, Perez P et al (2023) GZMK+CD8+ T cells target a specific acinar cell type in Sjögren's disease
11. Nayar S, Turner JD, Asam S et al (2025) Molecular and spatial analysis of tertiary lymphoid structures in Sjogren's syndrome. Nat Commun 16:5. https://doi.org/10.1038/s41467-024-54686-0

Chapter 15

Establishment of Adult Stem Cell-Derived Organoid Cultures: General Principles from the Mouse Gut

Kadi Lõhmussaar

Abstract

Adult stem cell (ASC)-derived organoid cultures have become an important model system for studying mammalian epithelial tissues in health and disease, including autoimmune conditions. Somatic stem cells are isolated and embedded into basement membrane-like matrix to allow for the in vitro outgrowth of miniature versions of functional epithelium in the presence of supportive growth factor niche. Organoids derived from tissues affected by autoimmune diseases, such as the intestine in inflammatory bowel disease or salivary glands in Sjögren's disease, may provide an invaluable platform for exploring disease mechanisms, immune-epithelial interactions, and therapeutic interventions. Here, we describe the basic establishment and maintenance principles of mouse intestinal organoid cultures, suitable for many downstream applications, including for disease modeling, drug discovery, and regenerative medicine.

Key words Stem cells, Intestine, 3D organoids, In vitro models

1 Introduction

3D organoid technology, first described by Clevers and colleagues in 2009, has transformed the field of stem cell and tissue engineering biology [1]. Derived from adult stem cells, organoids are miniature, self-organizing, three-dimensional structures that replicate key functional and structural characteristics of native tissues. These organoids can be generated from a variety of epithelial tissues, including the intestine, liver, pancreas, and lung [1–4]. Crucially, the core principles for establishing organoid cultures remain the same: adult stem cells are cultured in an extracellular matrix, such as Matrigel, and supplemented with tissue-specific growth factors that support epithelial differentiation and self-organization.

The ability to culture genetically stable ASC-derived organoids for long-term has been instrumental in advancing our understanding of tissue development, homeostasis, and disease biology. Organoid systems can be used to recapitulate complex tissue

Rami Bechara (ed.), *Autoimmunity: Methods and Protocols*, Methods in Molecular Biology, vol. 2990,
https://doi.org/10.1007/978-1-0716-4997-8_15,

architecture, cell-cell interactions in vitro, which were previously challenging to study using traditional two-dimensional cell cultures. This innovative technology has enabled the study of patient-derived organoids for personalized medicine, allowing for disease modeling and drug discovery [5–9]. Additionally, it has also paved the way for new research into epithelial infection biology across a range of pathogens [10–14].

Since its inception, advances in protocols and the identification of optimized culture conditions have expanded the range of tissues that can be modeled with organoids, providing new insights into developmental biology and opening the door to potential therapeutic applications. Here we describe a methodology for derivation, expansion, and maintenance of the long-term ASC-derived organoid cultures using mouse small intestine as an example.

2 Materials

2.1 Tissue Processing and Organoid Establishment

Equipment

1. CO2 incubator with stable temperature (37 °C) and CO2 level (5%).
2. Water bath (at 37 °C).
3. Pipette boy.
4. Pipettes (P200 and P1000).
5. Aspirator.
6. Centrifuge.
7. Swirling platform.
8. Dissection tools (forceps, surgical scissors).
9. Hemocytometer.
10. Cell counter.

Consumables and Reagents

1. Bucket with ice.
2. 70% ethanol spray.
3. Cotton swabs.
4. Thick coverslip.
5. 100 μm mesh strainers.
6. 20 mL syringe and fitted pipet tips (trim approximately 0.5 cm from the larger end to fit onto a syringe).
7. 50 mL Falcon tubes.
8. 10 cm Petri dishes.
9. T175 flask (optional).
10. Pipette tips (200 μL and 1000 μL volume).

11. Serological pipettes (e.g., 10 mL, 25 mL, 50 mL volume).
12. Glass pipettes with a narrow head (for aspiration).
13. Culture plates (e.g., 12-, 24- or 48-well format).
14. Matrigel.
15. Rho kinase inhibitor Y-27632.

Media and Buffers

1. 1×PBS.
2. Wash buffer: 1× PBS, 0.1% BSA.
3. Dissociation buffer: 1× PBS, 10 mM EDTA.
4. Base medium: Advanced DMEM/F12, 1× Glutamax, 10 mM HEPES buffer, 50 U/mL Penicillin-Streptomycin.
5. Growth medium: Advanced DMEM/F12, 1× Glutamax, 10 mM HEPES buffer, 50 U/mL Penicillin-Streptomycin, 10% R-spondin 1 conditioned medium, 100 ng/mL recombinant human Noggin, 50 ng/mL recombinant human EGF.

2.2 Organoid Passaging

Equipment

1. CO2 incubator with stable temperature (37 °C) and CO2 level (5%).
2. Water bath (at 37 °C).
3. Pipette boy.
4. Pipettes (P200 and P1000).
5. Aspirator.
6. Centrifuge.

Consumables and Reagents

1. Bucket with ice.
2. 70% ethanol spray.
3. 15 mL Falcon tubes.
4. Pipette tips (200 μL and 1000 μL volume).
5. Serological pipettes (e.g., 10 mL, 25 mL, 50 mL volume).
6. Culture plates (e.g., 12-, 24- or 48-well format).
7. Glass pipettes with narrow head (for aspiration).
8. Matrigel.
9. Optional: Cell Recovery Solution.

Media and Buffers

1. Wash buffer: 1× PBS, 0.1% BSA.
2. Base medium: Advanced DMEM/F12, 1× Glutamax, 10 mM HEPES buffer, 50 U/mL Penicillin-Streptomycin.

3. Growth medium: Advanced DMEM/F12, 1× Glutamax, 10 mM HEPES buffer, 50 U/mL Penicillin-Streptomycin, 10% R-spondin 1 conditioned medium, 100 ng/mL recombinant human Noggin, 50 ng/mL recombinant human EGF.

2.3 Organoid Cryofreezing

Equipment

1. Pipette boy.
2. Pipettes (P1000).
3. Aspirator.
4. Centrifuge.
5. Freezing container (e.g., Mr. Frosty).
6. –80C freezer (or liquid nitrogen tank).

Consumables and Reagents

1. Bucket with ice.
2. 70% ethanol spray.
3. 15 mL Falcon tubes.
4. Cryovials (e.g., 1.5 mL volume).
5. Pipette tips (1000 μL volume).
6. Serological pipettes (e.g., 10 mL, 25 mL, 50 mL volume).
7. Glass pipettes with narrow head (for aspiration).
8. Optional: Cell Recovery Solution.

Media and Buffers

1. Wash buffer: 1× PBS, 0.1% BSA.
2. Base medium: Advanced DMEM/F12, 1× Glutamax, 10 mM HEPES buffer, 50 U/mL Penicillin-Streptomycin.
3. 2× Freezing Medium (2×FM): 40% of base medium, 40% of FBS and 20% of DMSO.

2.4 Organoid Thawing

Equipment

1. CO_2 incubator with stable temperature (37 °C) and CO_2 level (5%).
2. Water bath (at 37 °C).
3. Pipette boy.
4. Pipettes (P200 and P1000).
5. Aspirator.
6. Centrifuge.

Consumables and Reagents

1. Bucket with ice.
2. 70% ethanol spray.

3. 15 mL Falcon tubes.
4. Pipette tips (200 μL and 1000 μL volume).
5. Serological pipettes (e.g., 10 mL, 25 mL, 50 mL volume).
6. Culture plates (e.g., 12-, 24- or 48-well format).
7. Glass pipettes with narrow head (for aspiration).
8. Matrigel.
9. Rho kinase inhibitor Y-27632.

Media and Buffers

1. Wash buffer: 1× PBS, 0.1% BSA.
2. Base medium: Advanced DMEM/F12, 1× Glutamax, 10 mM HEPES buffer, 50 U/mL Penicillin-Streptomycin.
3. Growth medium: Advanced DMEM/F12, 1× Glutamax, 10 mM HEPES buffer, 50 U/mL Penicillin-Streptomycin, 10% R-spondin 1 conditioned medium, 100 ng/mL recombinant human Noggin, and 50 ng/mL recombinant human EGF.

3 Methods

3.1 Tissue Processing and Organoid Establishment

1. Prepare the Setup:
   (a) Spray the work surfaces, flow hood, and pipettes with 70% ethanol.
   (b) Place a 500 mL bottle of 1× PBS on ice for cooling down.
   (c) Prepare 50 mL dissociation buffer.
   (d) Warm up the growth medium at 37 °C water bath (15 min before use).
   (e) Pour 5 mL of ice-cold 1× PBS into a 10 cm Petri dish and keep it on ice.
   (f) Optional: Pre-warm the suitable format culture plates (e.g., 12-well, 24-well, 48-well) at 37 °C incubator for at least 60 min (or overnight) prior to use (*see* **Note 1**).
2. Dissect the desired section of the small intestine tube (proximal, middle, distal, or entire tube) from a euthanized mouse using surgical scissors. Place the tissue on the ice-cold PBS in the 10 cm Petri dish.
3. Attach a trimmed pipette tip to a 20 mL syringe. Slowly flush the intestinal tube with 10 mL of ice-cold 1× PBS to remove fecal matter.
4. Place the tissue on a clean surface (e.g., the top of a T175 flask, flat tray). Using scissors, open the intestinal tube longitudinally

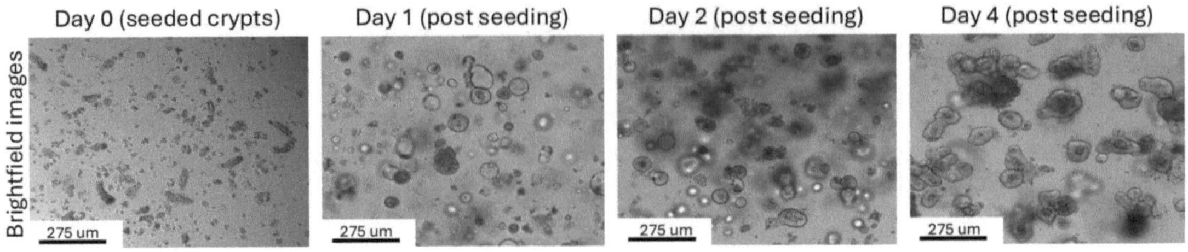

Fig. 1 Overview images of mouse small intestinal organoids in culture after derivation

to create a flat rectangular sheet with the luminal side facing upward.

5. Remove excessive mucus by gently rolling over with the soft moist cotton swab.
6. Use a thick coverslip to mechanically scrape away the villi (*see* **Note 2**). Discard the villi.
7. Transfer the villi-depleted tissue into a new 10 cm dish containing 5 mL ice-cold 1× PBS. Cut the tissue into smaller pieces, approximately 0.5 cm in size, using two scalpels.
8. Use forceps to move the tissue fragments into a 50 mL Falcon tube filled with 50 mL of dissociation buffer.
9. Digest the samples by gently rocking the tube horizontally on a swirling platform on ice (or in a cold room) for 30 min.
10. Stand the tube upright on ice and allow all tissue fragments to settle to the bottom by gravity (~1 min).
11. Carefully aspirate the dissociation buffer without disturbing the sedimented tissue and add 10 mL of ice-cold 1× PBS to the pellet.
12. Shake the tube vigorously by hand 50 times, do not vortex (*see* **Note 3**).
13. Allow the tissue fragments to settle by gravity for ~1 min.
14. Collect the crypt-enriched supernatant to a new tube.
15. Add 10 mL ice-cold 1× PBS to the pellet.
16. Repeat **steps 11–13** in total of two more times or until the suspension becomes clear and pool the supernatants from each round into the same tube.
17. Filter the pooled supernatant fraction through a 100 μm mesh strainer into a new 50 mL Falcon tube and fill the tube with PBS+0.1% BSA to a final volume of 50 mL (*see* **Note 4**).
18. Centrifuge the crypt-enriched fractions at 200× *g* for 5 min at room temperature (RT) (*see* **Note 5**).
19. Aspirate the supernatant and add 500 μL base medium to the tube.

20. Gently tap the bottom of the tube to resuspend the crypts in the remaining solution.
21. Use a hemocytometer to count the crypts under a microscope.
22. Resuspend the crypts at a density of 200–500 crypts per dome in 30% of base medium before adding 70% Matrigel. Mix well (*see* **Note 6**).
23. Plate the crypt-matrix suspension as 30 ul domes on the pre-warmed 12- or 24-well plate (adjust the volume to 20 μL in the 48-well format).
24. Incubate the plate in the in 37 °C incubator for ca 20–30 min. It is optional to flip the plate and incubate the domes in the hanging-droplet fashion (*see* **Note 7**).
25. Once solidified, add appropriate volume of pre-warmed growth medium supplemented with 10 μM Y-27632, a specific Rho kinase inhibitor, to improve cell survival.
26. Organoid outgrowth and expansion can be observed within days (*see* Fig. 1).
27. Refresh the medium every 2 days until splitting.

Graphical Overview of the Organoid Establishment Protocol

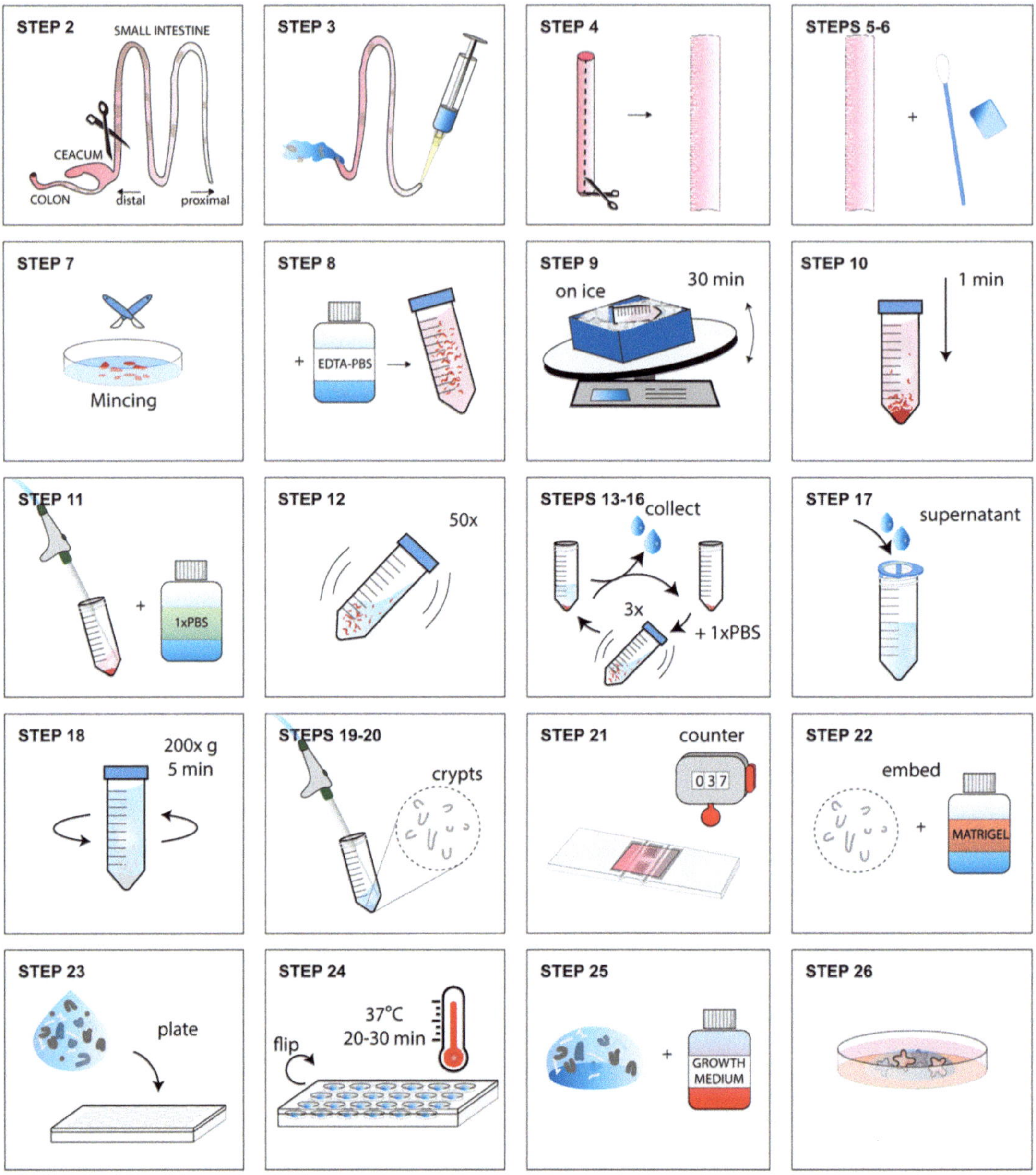

3.2 Passaging the Organoids

1. Prepare the setup:
   (a) 5–6 days old mouse small intestinal organoid culture.
   (b) Spray the work surfaces, flow hood and pipettes with 70% ethanol.
   (c) Prepare wash buffer and place it on ice.

(d) Warm up the growth medium at 37 °C water bath (15 min before use).

(e) Optional: Pre-warm the suitable format culture plates (e.g., 12-well, 24-well, 48-well) at 37 °C incubator for at least 60 min (or overnight) prior to use (*see* **Note 1**).

2. Label and prepare a 15 mL tube with ca 5 mL of wash buffer.
3. Aspirate the medium from the plates and add ca 1 mL ice-cold wash buffer to the wells.
4. Suspend the Matrigel-embedded organoids in the cold wash buffer several times and transfer the suspensions into the previously prepared 15 mL tube (**step 2**) containing 5 mL wash buffer.
5. Pellet the organoids via centrifugation at 500 g for 5 min at RT (*see* **Note 5**).
6. Optional: use Cell Recovery Solution to dissolve excess of Matrigel (*see* **Note 8**).
7. Aspirate the supernatant and place the tube on ice.
8. Add ca 200 ul cold wash buffer to the pellet and vigorously suspend the cell mix with P200 pipette ca 60× up and down.
9. Add up to 1 mL additional wash buffer and vigorously suspend the cell mix with P1000 pipette ca 60× up and down.
10. Pellet the fragments via centrifugation at 500 g for 5 min at RT.
11. Aspirate the supernatant and place the tube on ice.
12. Calculate the passaging ratio based on the organoid pellet size/ number of cells. In the case of mouse small intestine, the ratio is normally 1:4.
13. Keep the cells on ice and resuspend the pellet in 30% of base medium before the addition of 70% of Matrigel. Mix well (*see* **Note 6**).
14. Plate the matrix-embedded organoids in 30 μL droplets on a pre-warmed 24-well format (adjust to 20 μL drops in the 48-well format).
15. Incubate the plate in the in 37 °C incubator for ca 20–30 min. It is optional to flip the plate and incubate the domes in the hanging-droplet fashion (*see* **Note** 7).
16. Add appropriate volume of pre-warmed growth medium (e.g., 1 mL/well in 12-well format, 500 μL/well in 24-well format, 300 μL/well in 48-well format).
17. Refresh the medium every 2 days until splitting.

3.3 Cryofreezing Organoids

It is the best practice to freeze organoids when they are in highly proliferative state. This state depends on the particular organoid line characteristics. In the case of mouse small intestinal organoids: ca 2–3 days post mechanical splitting. It is also possible to freeze single-cell suspension; however, this procedure is not described in the current protocol.

1. Label 1.5 mL cryovials with necessary details (e.g., line name, passage number, cell number/volume, medium, date).
2. Add 500 μL of 2×FM to each tube.
3. Label and prepare 15 mL Falcon tubes with ca 5 mL of ice-cold wash buffer.
4. Collect the organoids in ice-cold wash buffer and add the cell mix to the 5 mL wash buffer in the previously prepared tubes (**step 3**). Collect max 12 × 30 μL droplets per cryovial.
5. Pellet the organoids via centrifugation at 500 g for 5 min at RT (*see* **Note 5**).
6. Optional: use Cell Recovery Solution to dissolve excess of Matrigel (*see* **Note 8**).
7. Aspirate supernatant and resuspend the pellet in 500 μL of wash buffer.
8. Combine the 500 μL cell suspension (from **step 7**) with the 2× Freezing medium (dilution ratio 1:1) in the ready-prepared cryovial (from **step 2**). Mix well by suspending ca 10–20 times up and down.
9. Seal the cryovials with lids and place in an appropriate freezing container.
10. Transfer the freezing container to −80 °C freezer for overnight.
11. Next day, transfer the cryovials to liquid nitrogen tank for long-term storage.

3.4 Thawing Organoids

1. Prepare the Setup:
   (a) Spray the work surfaces, flow hood and pipettes with 70% ethanol.
   (b) Place a bottle of wash buffer on ice for cooling down.
   (c) Warm up the growth medium at 37 °C water bath (15 min before use).
   (d) Optional: Pre-warm the suitable format culture plates (e.g., 12-well, 24-well, 48-well, etc.) at 37 °C incubator for at least 60 min (or overnight) prior to use (*see* **Note 1**).
2. Label and prepare a 15 mL tube with 5 mL of wash buffer containing 10 mM Rho kinase inhibitor Y-27632.

3. Use dry ice to transfer the cryovial(s) from the cryobank into the tissue culture facility.
4. Hold the cryovial from the cap and quickly shake the vial in the 37 °C water bath back and forth ca 2 min to thaw up the contents of the tube (*see* **Note 9**).
5. Once thawed, spray the cryovial with 70% ethanol to clean from water and bring it under the flow hood for the next steps.
6. Screw off the cap from the cryovial and slowly (drop-wise) add ca 1 mL of wash buffer with 10 μM Y-27632 added and mix by gently tapping the bottom of the tube with your finger during the process (*see* **Note 10**).
7. Transfer the diluted cell suspension from the cryovial into the previously prepared 15 mL tube, combining it with the remaining 4 mL of the wash buffer.
8. Pellet the cells via centrifugation at 400 g for 5 min at RT.
9. Aspirate the supernatant and place the tube on ice.
10. Keep the cells on ice and, according to the organoid pellet size or cell number (ideally recorded on the cryovial), resuspend the pellet in 30% of base medium before the addition of 70% of Matrigel. Mix well (*see* **Note 6**).
11. Plate the matrix-embedded organoids in 30 μL droplets on a pre-warmed 24-well format (adjust to 20 μL drops in the 48-well format).
12. Incubate the plate in the in 37 °C incubator for ca 20–30 min. It is optional to flip the plate and incubate the domes in the hanging-droplet fashion (*see* **Note 7**).
13. Add appropriate volume of pre-warmed growth medium supplemented with 10 μM Y-27632 (e.g., 1 mL/well in 12-well format, 500 μL/well in 24-well format, 300 μL/well in 48-well format).
14. Refresh the medium every 2 days until passaging.

4 Notes

1. Pre-warming the culture plates will enable Matrigel to solidify uniformly and also helps to prevent the flatness/smearing of the Matrigel droplets/domes when plating the organoids.
2. Removing the villi will help to enrich for the crypt fraction. Try to operate fast at this step to not let the tissue dry up.
3. During this process, many crypts and smaller cell fragments will be released to the supernatant. Thus, expect the supernatant to become a bit cloudy after shaking due to the cellular contribution.

4. Make sure to wet the 100 μm mesh beforehand to prevent it getting clogged with air bubbles/cells during filtering the supernatant.
5. All centrifugations in the protocol are performed at room temperature. It is optional to perform the centrifugation steps in a pre-cooled centrifuge set to 4C for better Matrigel solubility and cell survival.
6. Matrigel is a temperature-sensitive extracellular matrix. When kept at low temperatures (around 4 °C), Matrigel remains in a liquid state, making it easier to handle and mix with organoids. However, at warmer temperatures, like room temperature or above, Matrigel rapidly solidifies, which is necessary to support 3D cell growth but can be challenging during the plating process. Therefore, Matrigel should be added on ice and as a last step before plating the cells.
7. Incubating the plates in the hanging-droplet fashion will reduce cell sinking and adhering to the bottom of the plate for better circulation of nutrients.
8. In case a thick viscous phase is observed on top of the pellet, Matrigel volume was likely too high for it to get dissolved during the brief washing step. In this case, you can carefully remove the viscous phase via pipetting. However, there is a risk of losing cells that are still stuck in the matrix. Alternatively, resuspend the Matrigel pellet in ca 5–10 mL of ice-cold Cell Recovery Solution and incubate the samples on ice for 20–30 min, occasionally inverting the tube to dissolve the matrix before continuing with the protocol.
9. The thawing procedure should be as fast as possible. Water can access the tube all around and results in faster and more homogenous thawing compared to the bead paths. Beads will have a lot of air trapped between them that will introduce differences in heat conduction and slow down the thawing.
10. During thawing, a sudden change in the extracellular osmolarity can induce osmotic stress, which reduces cell viability after thawing. Therefore, slow introduction of the wash buffer is recommended.

References

1. Sato T, Vries RG, Snippert HJ, van de Wetering M, Barker N, Stange DE, van Es JH, Abo A, Kujala P, Peters PJ, Clevers H (2009) Single Lgr5 stem cells build crypt-villus structures in vitro without a mesenchymal niche. Nature 459(7244):262–265
2. Huch M, Dorrell C, Boj SF, van Es JH, Li VS, van de Wetering M, Sato T, Hamer K, Sasaki N, Finegold MJ, Haft A, Vries RG, Grompe M, Clevers H (2013) In vitro expansion of single Lgr5+ liver stem cells induced by Wnt-driven regeneration. Nature 494(7436):247–250
3. Huch M, Bonfanti P, Boj SF, Sato T, Loomans CJ, van de Wetering M, Sojoodi M, Li VS, Schuijers J, Gracanin A, Ringnalda F, Begthel H, Hamer K, Mulder J, van Es JH,

de Koning E, Vries RG, Heimberg H, Clevers H (2013) Unlimited in vitro expansion of adult bi-potent pancreas progenitors through the Lgr5/R-spondin axis. EMBO J 32(20): 2708–2721

4. Sachs N, Papaspyropoulos A, Zomer-van Ommen DD, Heo I, Böttinger L, Klay D, Weeber F, Huelsz-Prince G, Iakobachvili N, Amatngalim GD, de Ligt J, van Hoeck A, Proost N, Viveen MC, Lyubimova A, Teeven L, Derakhshan S, Korving J, Begthel H, Dekkers JF, Kumawat K, Ramos E, van Oosterhout MF, Offerhaus GJ, Wiener DJ, Olimpio EP, Dijkstra KK, Smit EF, van der Linden M, Jaksani S, van de Ven M, Jonkers J, Rios AC, Voest EE, van Moorsel CH, van der Ent CK, Cuppen E, van Oudenaarden A, Coenjaerts FE, Meyaard L, Bont LJ, Peters PJ, Tans SJ, van Zon JS, Boj SF, Vries RG, Beekman JM, Clevers H (2019) Long-term expanding human airway organoids for disease modeling. EMBO J 38(4):e100300
5. Clevers H (2016) Modeling development and disease with organoids. Cell 165(7): 1586–1597
6. Matano M, Date S, Shimokawa M, Takano A, Fujii M, Ohta Y, Watanabe T, Kanai T, Sato T (2015) Modeling colorectal cancer using CRISPR-Cas9-mediated engineering of human intestinal organoids. Nat Med 21(3): 256–262
7. Boj SF, Hwang CI, Baker LA, Chio II, Engle DD, Corbo V, Jager M, Ponz-Sarvise M, Tiriac H, Spector MS, Gracanin A, Oni T, Yu KH, van Boxtel R, Huch M, Rivera KD, Wilson JP, Feigin ME, Öhlund D, Handly-Santana A, Ardito-Abraham CM, Ludwig M, Elyada E, Alagesan B, Biffi G, Yordanov GN, Delcuze B, Creighton B, Wright K, Park Y, Morsink FH, Molenaar IQ, Borel Rinkes IH, Cuppen E, Hao Y, Jin Y, Nijman IJ, Iacobuzio-Donahue C, Leach SD, Pappin DJ, Hammell M, Klimstra DS, Basturk O, Hruban RH, Offerhaus GJ, Vries RG, Clevers H, Tuveson DA (2015) Organoid models of human and mouse ductal pancreatic cancer. Cell 160(1–2):324–338
8. Driehuis E, van Hoeck A, Moore K, Kolders S, Francies HE, Gulersonmez MC, Stigter ECA, Burgering B, Geurts V, Gracanin A, Bounova G, Morsink FH, Vries R, Boj S, van Es J, Offerhaus GJA, Kranenburg O, Garnett MJ, Wessels L, Cuppen E, Brosens LAA, Clevers H (2019) Pancreatic cancer organoids recapitulate disease and allow personalized drug screening. PNAS 116(52):26580–26590
9. Kopper O, de Witte CJ, Lõhmussaar K, Valle-Inclan JE, Hami N, Kester L, Balgobind AV, Korving J, Proost N, Begthel H, van Wijk LM, Revilla SA, Theeuwsen R, van de Ven M, van Roosmalen MJ, Ponsioen B, Ho VWH, Neel BG, Bosse T, Gaarenstroom KN, Vrieling H, Vreeswijk MPG, van Diest PJ, Witteveen PO, Jonges T, Bos JL, van Oudenaarden A, Zweemer RP, Snippert HJG, Kloosterman WP, Clevers H (2019) An organoid platform for ovarian cancer captures intra- and interpatient heterogeneity. Nat Med 25(5):838–849
10. Bartfeld S, Bayram T, van de Wetering M, Huch M, Begthel H, Kujala P, Vries R, Peters PJ, Clevers H (2015) In vitro expansion of human gastric epithelial stem cells and their responses to bacterial infection. Gastroenterology 148(1):126–136
11. Heo I, Dutta D, Schaefer DA, Iakobachvili N, Artegiani B, Sachs N, Boonekamp KE, Bowden G, Hendrickx APA, Willems RJL, Peters PJ, Riggs MW, O'Connor R, Clevers H (2018) Modelling cryptosporidium infection in human small intestinal and lung organoids. Nat Microbiol 3(7):814–823
12. Pleguezuelos-Manzano C, Puschhof J, Rosendahl Huber A, van Hoeck A, Wood HM, Nomburg J, Gurjao C, Manders F, Dalmasso G, Stege PB, Paganelli FL, Geurts MH, Beumer J, Mizutani T, Miao Y, van der Linden R, van der Elst S, Genomics England Research Consortium, Garcia KC, Top J, Willems RJL, Giannakis M, Bonnet R, Quirke P, Meyerson M, Cuppen E, van Boxtel R, Clevers H (2020) Mutational signature in colorectal cancer caused by genotoxic pks$^+$ E. coli. Nature 80(7802):269–273
13. Lamers MM, Beumer J, van der Vaart J, Knoops K, Puschhof J, Breugem TI, Ravelli RBG, Paul van Schayck J, Mykytyn AZ, Duimel HQ, van Donselaar E, Riesebosch S, Kuijpers HJH, Schipper D, van de Wetering WJ, de Graaf M, Koopmans M, Cuppen E, Peters PJ, Haagmans BL, Clevers H (2020) SARS-CoV-2 productively infects human gut enterocytes. Science 369(6499):50–54
14. Lõhmussaar K, Oka R, Espejo Valle-Inclan J, Smits MHH, Wardak H, Korving J, Begthel H, Proost N, van de Ven M, Kranenburg OW, Jonges TGN, Zweemer RP, Veersema S, van Boxtel R, Clevers H (2021) Patient-derived organoids model cervical tissue dynamics and viral oncogenesis in cervical cancer. Cell Stem Cell 28(8):1380–1396

Chapter 16

In Vitro Lymphoid Differentiation of Human Hematopoietic Progenitor Cells from Co-cultures in Hypoxia

Françoise Pflumio and Rima Haddad

Abstract

Understanding the identity of cells from which lymphoid differentiation occurs can be addressed using specific cell assays. Ex vivo conditions have been established based on co-culture systems between human hematopoietic stem/progenitor cells or early lymphoid progenitors and stromal cells. These systems provide a functional approach to carefully assess cell potential at both qualitative and quantitative levels. This chapter describes a two-step co-culture system for isolated hematopoietic stem/progenitor cells, using a murine bone marrow stromal cell line (MS-5), a hypoxic environment (3.5% O_2), defined lymphoid medium and cytokine conditions, and limiting dilution settings. This setup enables a robust characterization of the intrinsic lymphoid potential of progenitor cells, closely mimicking physiological condition. Importantly, this approach can be instrumental in investigating how intrinsic or microenvironmental alterations in early hematopoiesis contribute to immune dysregulation and the development of autoimmune diseases.

Key words Umbilical cord blood, Human $CD34^+$ hemato-lymphoid progenitors, MS-5 stromal cells, Co-cultures, Hypoxia, In vitro lymphoid differentiation, Limiting dilution assays

1 Introduction

The differentiation of human hematopoietic stem and progenitor cells into hemato-lymphoid cells is a finely regulated process that is controlled by various environmental and cellular factors [1–3]. The integration of human ossicles in vivo into immunodeficient mice offers promising opportunities for reproducing more physiological hematopoietic niches [4–7]. Recent advances in cellular bioengineering, particularly the development of 3D-engineered human bone marrow analogs using bioreactors, now allow for the in vitro maintenance of hematopoietic stem cells [4, 8, 9]. However, co-culture systems combining hematopoietic progenitor cells with bone marrow-derived stromal cells remain unique and powerful tools to robustly explore lymphoid differentiation in vitro. These systems enable directed lymphopoiesis and allow assessment of

Rami Bechara (ed.), *Autoimmunity: Methods and Protocols*, Methods in Molecular Biology, vol. 2990, https://doi.org/10.1007/978-1-0716-4997-8_16, both the heterogeneity and intrinsic lymphoid potential of early progenitor subsets. Particularly, MS-5 or OP-9 cells can promote B and NK cell differentiation. When engineered to overexpress the Notch Ligand Delta-Like 1 (MS-5/DLL1, OP-9/DLL1), these lines also support T-cell differentiation by mimicking thymic microenvironment [10–23]. Moreover, performing co-cultures under hypoxic conditions has emerged as a promising strategy to enhance lymphoid differentiation, more closely replicating the low-oxygen environment of the bone marrow niche [24, 25]. This article describes the optimization of human hematopoietic stem and progenitor cells differentiation into NK, B, and T lymphoid cells using a co-culture approach of umbilical cord blood-derived specific early hemato-lymphoid progenitors in hypoxia (3.5% O_2) in comparison to the usual normoxia/hyperoxia (21% O_2) condition. This process also covers the evaluation of the quantitative frequency of NK, B, and T lymphoid precursors derived from original subpopulations of human hemato-lymphoid progenitors alongside the assessment of their inherent lymphoid potential. These insights are critical for understanding how early defects in lymphoid differentiation may contribute to immune dysregulation and the emergence of autoimmune diseases.

2 Materials

2.1 Culture Media, Cytokines, and Consumables

1. 1X Phosphate Buffered Saline (PBS).
2. Ficoll®.
3. CD34 MicroBead Kit.
4. EDTA 500 mM.
5. Dimethyl sulfoxide (DMSO).
6. Trypsin.
7. DNAse.
8. Maintenance medium for CD34$^+$ cells: H5100 human long-term culture medium (STEMCELL Technologies™).
9. Collection medium after cell sorting: IMDM medium with 20% FCS, 1% antibiotics, and 1% Glutamine.
10. Maintenance medium for MS5and MS5-Delta-like 1 cell lines: α-MEM with nucleosides, 10% fetal calf serum (FCS), 1% antibiotics (penicillin-streptomycin), 1% Glutamine.
11. NK-cell differentiation medium: RPMI, 5% FCS, 10% human AB serum, 1% antibiotics, 1% Glutamine, supplemented with human recombinant Stem Cell Factor (SCF) (50 ng/mL), Interleukin (IL)-2 (5 ng/mL), IL-15 (1 ng/mL) [14–17, 24].

12. B-cell differentiation medium: IMDM, 5% FCS, 1% antibiotics, 1% Glutamine [16–18, 25].
13. T-cell differentiation medium: α-MEM with nucleosides, 10% FCS, 10% human AB serum, 1% antibiotics, 1% Glutamine, supplemented with human recombinant SCF (50 ng/mL), Flt3-Ligand (20 ng/mL), IL-7 (10 ng/mL), and Insulin (20 nM) [25].
14. 50 mL and 15 mL plastic conical centrifuge tubes.
15. Plastic 96 wells plates.
16. Flow cytometry polystyrene and polypropylene tubes.
17. 5 mL polypropylene collection tube.
18. 1.5 mL collection tubes for cell sorting.
19. 1.5 mL cryotubes.

2.2 Mouse Anti-human Monoclonal Antibodies and Reagents for Flow Cytometry

CD3 PerCP-Cy5.5 (clone SK7), CD4 PE-Cy7 (clone SK3), CD5 PE (clone UCHT2), CD7 APC (clone CD7-6B7), CD8 APC-ef780 (clone Skl), CD10-FITC, -APC, -PerCP-ef710 (clone SN5C), CD14 PE (clone 63D3), CD15 PE (clone HI98), CD19-FITC,-PE, -APC, APC-Cy7 (clone HIB19), CD34-PerCP-Cy5.5 (clone 581), CD38-APC-ef780 (clone HB-7), CD45-FITC (clone HI30), CD45RA-PE (clone HI100), CD56-APC (clone CMSSB), CD62L-PE-Cy7 (clone DREG-56), CD90-BV421 (clone 5E10), compensation bead tubes, Hoechst 33342.

2.3 Tissue, Cell Lines, and Instrumentation

1. Human umbilical cord blood.
2. MS-5 and MS-5/Delta-Like 1 (overexpressing Notch Ligand, MS-5/DLL1) murine stromal cell lines.
3. Hypoxia Chamber: PrOox culture chamber (BioSpherix).
4. Flow cytometry (FACS) cell sorter: BD Influx™ and BD FACSAria™ cell sorters (BD Bioscience).
5. FACS analyzer: BD SORP LSR II and BD FACSCanto (BD Bioscience).

3 Methods

3.1 Umbilical Cord Blood (UCB)-Derived $CD34^+$ Hemato-Lymphoid Progenitor Cells and Primary Co-cultures Under Normoxia (21% O_2) and Hypoxia (3.5% O_2)

1. UCB is distributed into 50 mL conical centrifuge tubes.
2. After centrifugation for 20 min at 970 g and Room Temperature (RT) without brake, recover the buffy coat (10–15 mL) (*see* **Note 1**).
3. Complete with 1X PBS to reach a total of 30 mL of cell suspension.

A

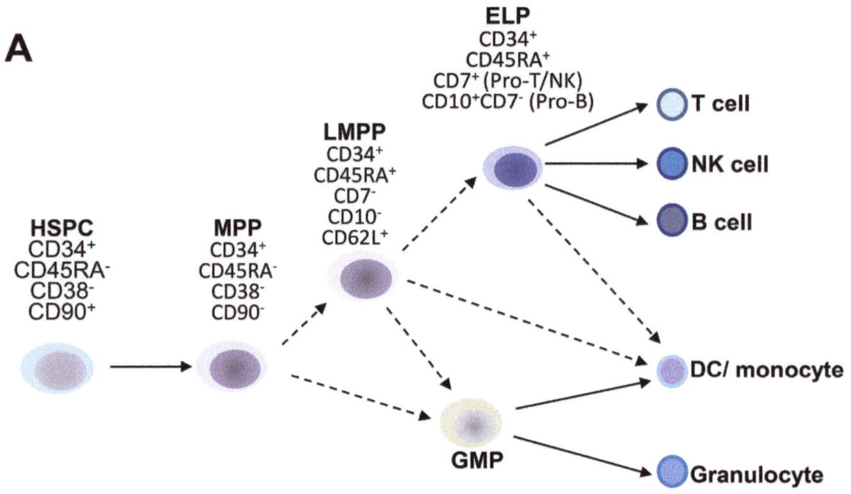

Population	Immunophenotype
HSPC	$CD34^{+}CD38^{lo/-}CD45RA^{-}CD90^{+}$
MPP	$CD34^{+}CD38^{lo/-}CD45RA^{-}CD90^{-}$
LMPP	$CD34^{+}CD45RA^{+}CD7/10/19^{-}CD62L^{hi}$
ELP: Pro-T/NK	$CD34^{+}CD45RA^{+}CD7^{+}$
ELP: Pro-B	$CD34^{+}CD45RA^{+}CD7/19^{-}CD10^{+}$

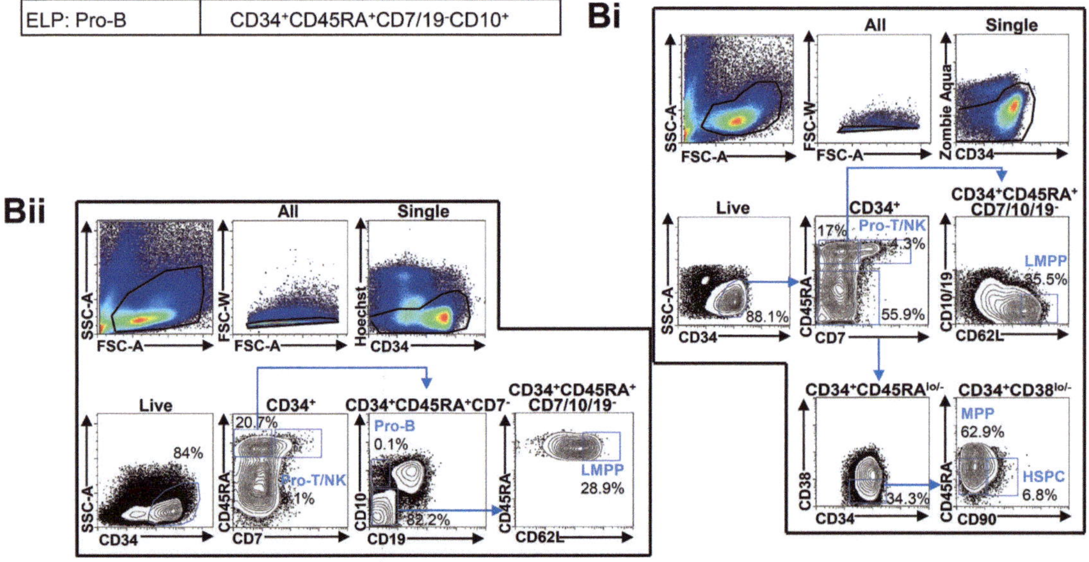

Fig. 1 [25]: Immunophenotypic analyses and cell sorting strategy of hemato-lymphoid progenitors. (**a**) Immunophenotypes of UCB HSPCs, MPPs, LMPPs, and early lymphoid progenitors (ELPs) Pro-T/NK and Pro-B cells [24]. (**b**) Flow cytometry isolation of HSPC, MPP, LMPP, Pro-T/NK, and Pro-B cell populations. Cell sorting strategies (**b**(i)) Purified UCB $CD34^{+}$ cells were sorted into $CD34^{+}CD38^{lo/-}CD45RA^{-}CD90^{+}$ (HSPCs), $CD34^{+}CD38^{lo/-}CD45RA^{-}CD90^{-}$ (MPPs), $CD45RA^{+}CD7^{+}$ (Pro-T/NK), $CD45RA^{+}CD7^{-}CD10^{-}CD19^{-}CD62L^{hi}$ (LMPPs). (**b**(ii)) Purified UCB $CD34^{+}$ cells were sorted into $CD45RA^{+}CD7^{+}$ (Pro-T/NK), $CD45RA^{+}CD7^{-}CD10^{-}CD19^{-}CD62L^{hi}$ (LMPPs), and $CD45RA^{+}CD7^{-}CD19^{-}CD10^{+}$ (Pro-B). HSPCs: hematopoietic stem progenitors cells; MPPs: multipotent progenitors; LMPPs: lymphoid-primed multipotent progenitors

4. Proceed with gradient centrifugation using Ficoll® for 25 min at 650 g, without brake, at RT.
5. Recover the cell ring, perform counting on a counting chamber, and isolate human CD34$^+$ cells using the CD34 MicroBead Kit according to the manufacturer's instructions.
6. CD34$^+$ cells can be frozen at a concentration of 1 to 3 × 10^6 cells per vial in FCS with 10% DMSO.
7. If needed thaw the CD34$^+$ cells and count them using a counting chamber.
8. Resuspend the cells at a concentration of 4 × 10^6 cells/mL in IMDM 20% FCS, 1% antibiotics, and DNase 1X (for 30 min).
9. Filter the cells through a 70 μm mesh.
10. Wash the cells (5 min, 300 g, RT), then resuspend the cells in 1X PBS at a concentration of 10 × 10^6 cells/mL.
11. Prepare antibodies for CD34, CD45RA, CD38, CD90, CD10, CD19, CD7, CD62L. Prepare a control tube (without labeling, ~50 × 10^3 cells) and the bead tubes (Life Technologies, one drop/tube) for single staining to set the compensations (*see* **Note 2**).
12. Stain the cells to be sorted with antibodies diluted 1/50.
13. Prepare collection tubes of 1.5 mL containing 100 μL of IMDM medium with 20% FCS, 1% antibiotics, and 1% Glutamine.
14. Proceed with cell sorting of hemato-lymphoid progenitor cells (HSPCs, MPPs, LMPPs, Pro-T/NK, and Pro-B cells) by flow cytometry as described in Fig. 1 below.

3.2 Primary Co-cultures Under Normoxia (21% O_2) and Hypoxia (3.5% O_2)

1. After detachment of MS-5 cells with trypsin, plate MS-5 cells in 24-well plates at 3 × 10^4 cells/cm^2 24–48 h under 21% O_2, in MS-5 maintenance medium before co-culture initiation with hemato-lymphoid progenitors (HSPCs, MPPs, LMPPs, Pro-T/NK and Pro-B cells). Let the MS-5 cell adhere and grow to ~70% well-coverage (*see* **Note 3**).
2. Remove the MS-5 medium before the co-culture initiation.
3. Plate FACS-sorted HSPC, MPP, LMPP, and Pro-T/NK cells at (5 × 10^3–10^4 cells/cm^2) and Pro-B (300–500 cells/cm^2) over adherent MS-5 cells during 7 days under hypoxia or normoxia in 500 μL H5100 medium/well and in triplicates if the number of stem/progenitor cells is high enough (*see* **Note 4**).

3.3 Processing of Primary Co-cultured-Derived Cells for Lymphoid Cell Differentiation and Assessment of NK, B, and T Cell Potential and Precursor Frequencies at Limiting Dilution from Primary Co-cultured-Derived Cells in Hypoxia and Normoxia

1. Three days prior to NK-cell and B-cell differentiation from primary co-culture-derived cells in hypoxia and normoxia, plate MS-5 cells in 96-well plates at 7.5 × 10^3 cells/cm^2 24–48 h under 21% O_2, in MS-5 maintenance medium before co-culture initiation. Let the MS-5 cell adhere and grow to ~70% well-coverage (*see* **Note 3**).
2. After seven days in contact with MS-5 cells under hypoxia or normoxia, carefully recover all the cells from each well by detaching them from the bottom.
3. Filter the cell suspension through a 70 μm mesh into a 5 mL collection tube (one collection tube per well).
4. Centrifuge the cells for 5 min at 290 g, RT.
5. Resuspend the cell pellet in 300 μL of IMDM medium.
6. Count each cell suspension from each co-culture condition using a counting chamber before testing the harvested cells for their lymphoid potentials (*see* Subheading 3.3).
7. Seed normoxia and hypoxia-derived cells at 300, 100, 30, 10, 3, and 1 cell(s)/well onto MS-5 coated 96-well plates (12 wells/cell dilution) in 200 μL/well of NK-cell or B-cell differentiation medium (*see* **Note 5**).
8. For T-cell differentiation condition, mix in the same conical tube, in T-cell differentiation medium, 7.5 × 10^3 MS-5/DLL1 cells/well of a 96-well plates with normoxia or hypoxia-derived cells for each cell dilution (300, 100, 30, 10, 3, and 1 cell(s)/well of a 96-well plates). Prepare cell suspensions allowing the seeding of 12 wells/cell dilution (*see* **Note 5**). Centrifuge the conical tubes for 5 min at 300 g at RT. Remove the supernatant. Resuspend each dilution tube with T-cell differentiation medium (200 μL × number of wells/dilution). Distribute 200 μL of cell suspension per well of the dilution.
9. Set cultures in normoxia for up to 3 weeks for NK and B conditions and up to 5 weeks for T conditions.
10. Change half medium and add fresh cytokines every 6–7 days.
11. Examine weekly the plates, and score cell-containing wells under the microscope.
12. At the end of the cultures, identify NK cells by flow cytometry analysis of human CD56 cell surface expression, B cells by CD19 cell surface expression and T cells by CD7/CD5/CD4/CD8/CD3 expressions (*see* **Notes 6–8**). Results from the study by Chabi et al. [25] are presented below (Fig. 2).

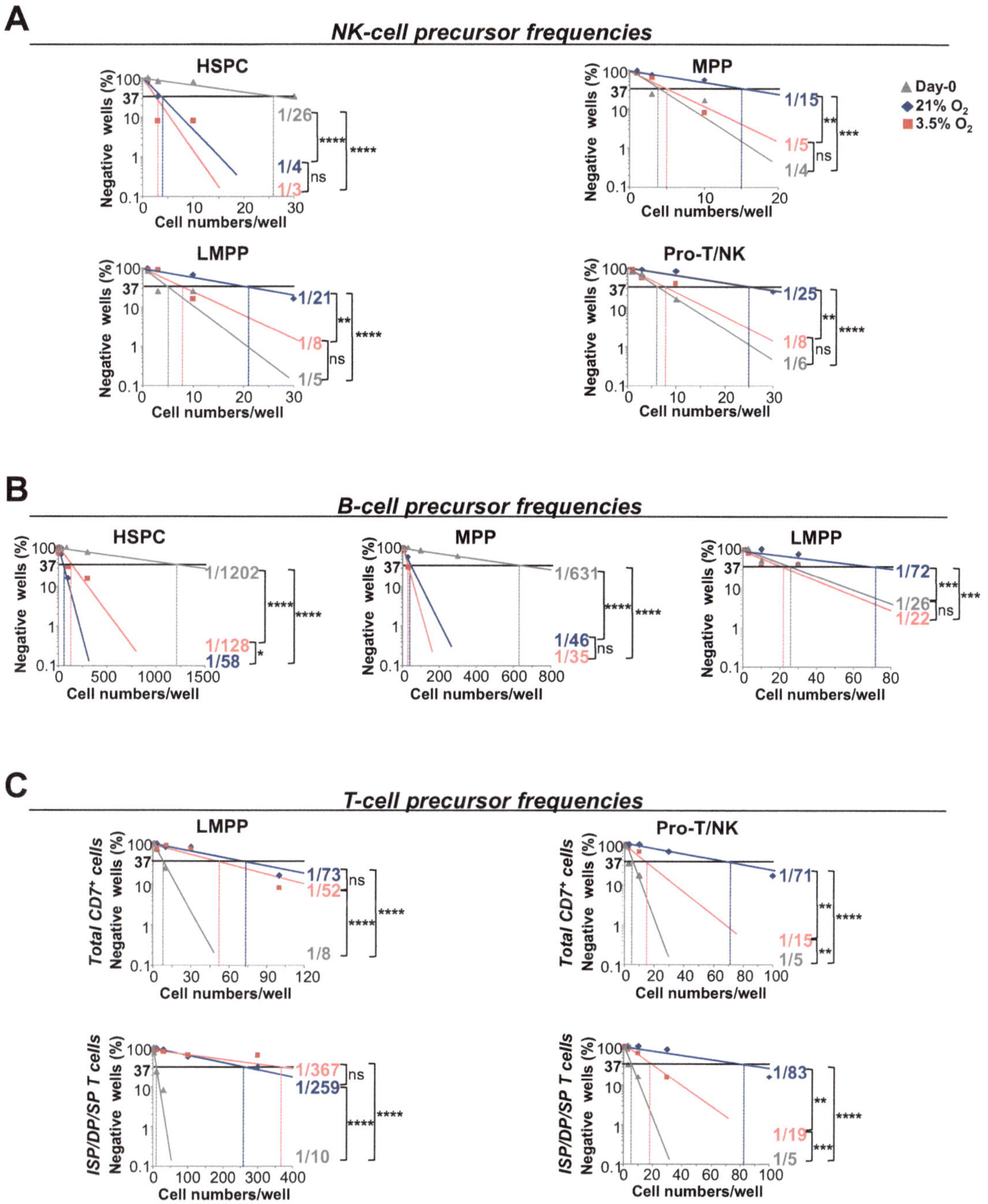

Fig. 2 [25]: NK, B, and T cell precursor frequencies among HPC-derived cells under 21% O_2 and 3.5% O_2. (**a**) Comparative analysis of day-0 HPCs and HPC co-cultures-derived cells from 21% O_2 and 3.5% O_2 conditions in limiting-dilution assays under NK condition cultures (one experiment out of three for HSPCs, MPPs, and Pro-T/NK and 1 experiment out of 2 for LMPPs are shown here). (**b**) Same as in (**a**) but in **b** condition cultures (one experiment out of two for HSPCs, MPPs, and LMPPs is shown here). (**c**) Same as in (**a**) but in T condition cultures (one experiment out of two for LMPPs and Pro-T/NK is shown here). Note that T cell precursor

4 Notes

1. Ensure to carefully collect the interface to avoid contamination with too many red blood cells and platelets.
2. Ensure to have tubes with $CD34^+$ cells to set the FMO (Fluorescence minus one color) in case of rare populations ($\sim 50 \times 10^3$ cells/tube).
3. Prepare one 24-wells plate for normoxia co-culture condition and another 24-wells plate for hypoxia co-culture condition.
4. Following FACS cell sorting of HSPCs, MPPs, LMPPs, Pro-T/NK, and Pro-B cells, the sorted cells should be counted in counting chambers to obtained the most rigorous cell counts.
5. Prepare cells for 13 or 14 wells to accurately distribute the required number of cells per well. In summary, for each cell dilution, the number of MS5-DLL1 cells to be mixed with cells derived from hypoxic or normoxic cultures, in each conical tube, should be adjusted according to the number of dilution wells in the 96-well plate.
6. Day-0 non-manipulated hemato-lymphoid progenitor cells (i.e., not co-cultured during 7 days in normoxia or hypoxia) should be also assessed in NK, B, and T cell differentiation to provide an internal control of lymphoid cell differentiation.
7. For phenotyping, collect the wells in which substantial amount of cells can be identified (≥ 20 cells).
8. The maximum likelihood estimate of NK-, B-, or T-cell precursors is calculated according to the single-hit Poisson model using the L-Calc™ limiting dilution analysis software from STEMCELL Technologies which also provides a confidence interval (95% CI).

Acknowledgments

This work was supported by grants from INSERM, CEA, University Paris-Saclay and University Paris Cité, Ligue Nationale Contre le Cancer (équipe labellisée), Association Laurette Fugain and Agence Nationale pour la Recherche.

Fig. 2 (continued) frequencies among LMPP and Pro-T/NK-derived cells were analyzed for the presence of total $CD7^+$ T cells and the presence of $CD7^+CD4^+CD8^-CD3^-$ intermediate simple positive (ISP) cells and/or $CD7^+CD4^+CD8^+$ double positive (DP) cells and/or $CD7^+CD4^-CD8^+CD3^+$ or $CD7^+CD4^+CD8^-CD3^+$ committed T cells. Wells were scored as positive for NK, B or T cells when % of lymphoid cells was $\geq 1\%$ and represent homogeneous population. $^*p < 0.05$, $^{**}p < 0.01$, $^{***}p < 0.001$, $^{****}p < 0.0001$, ns: no significant

References

1. Wei Q, Frenette PS (2018) Niches for hematopoietic stem cells and their progeny. Immunity 48:632–648
2. Aurrand-Lions M, Mancini SJC (2018) Murine bone marrow niches from hematopoietic stem cells to B cells. Int J Mol Sci 19:2353–2370
3. Pinho S, Frenette PS (2019) Haematopoietic stem cell activity and interactions with the niche. Nat Rev Mol Cell Biol 20:303–320
4. Sommerkamp P, Mercier FE, Wilkinson AC, Bonnet D, Bourgine PE (2021) Engineering human hematopoietic environments through ossicle and bioreactor technologies exploitation. Exp Hematol 94:20–25
5. Dupard SJ, Grigoryan A, Farhat S, Coutu DL, Bourgine PE (2020) Development of humanized ossicles: bridging the hematopoietic gap. Trends Mol Med 26:552–569
6. Reinisch A, Hernandez DC, Schallmoser K, Majeti R (2017) Generation and use of a humanized bone-marrow-ossicle niche for hematopoietic xenotransplantation into mice. Nat Protoc 12:2169–2188
7. Renou L, Sun W, Friedrich C, Galant K, Conrad C, Consalus A et al (2025) Orchestration of human multi-lineage hematopoietic cell development by humanized in vivo bone marrow models. HemaSphere. https://doi.org/10.1002/hem3.70120
8. Bourgine PE, Klein T, Paczulla AM, Shimizu T, Kunz L, Kokkaliaris KD et al (2018) In vitro biomimetic engineering of a human hematopoietic niche with functional properties. Proc Natl Acad Sci USA 115: E5688–E5E95
9. Dupard SJ, Garcia AG, Bourgine PE (2022) Customizable 3D printed perfusion bioreactor for the engineering of stem cell microenvironments. Front Bioeng Biotechnol 10:1081145
10. Amsellem S, Pflumio F, Bardinet D, Izac B, Charneau P, Romeo PH et al (2003) Ex vivo expansion of human hematopoietic stem cells by direct delivery of the HOXB4 homeoprotein. Nat Med 11:1423–1427
11. Robin C, Pflumio F, Vainchenker W, Coulombel L (1999) Identification of lymphomyeloid primitive progenitor cells in fresh human cord blood and in the marrow of nonobese diabetic-severe combined immunodeficient (NOD-SCID) mice transplanted with human CD34(+) cord blood cells. J Exp Med 189: 1601–1610
12. Ravet E, Dubart-Kupperschmitt A, Robin C, Titeux M, Coulombel L, Pflumio F (2002) Successful transduction of human multipotent, lymphoid (T, B, NK) and myeloid, and transplantable CD34+CD38low cord blood cells using a murine oncoretroviral vector. J Hematother Stem Cell Res 11(2):327–336
13. Haddad R, Guardiola P, Izac B, Thibault C, Radich J, Delezoide AL et al (2004) Molecular characterization of early human T/NK and B-lymphoid progenitor cells in umbilical cord blood. Blood 104:3918–3926
14. Haddad R, Guimiot F, Six E, Jourquin F, Setterblad N, Kahn E et al (2006) Dynamics of thymus-colonizing cells during human development. Immunity 24:217–230
15. Haddad R, Caignard A, Visentin G, Vigon I, Fichelson S, Amsellem S (2007) The HOXB4 homeoprotein improves ex vivo generation of functional human NK-cell progenitors. Leukemia 21:1836–1839
16. Haddad R, Pflumio F, Vigon I, Visentin G, Auvray C, Fichelson S et al (2008) The HOXB4 homeoprotein differentially promotes ex vivo expansion of early human lymphoid progenitors. Stem Cells 26:312–322
17. Larbi A, Gombert JM, Auvray C, l'Homme B, Magniez A, Feraud O et al (2012) The HOXB4 homeoprotein promotes the ex vivo enrichment of functional human embryonic stem cell-derived NK cells. PLoS One 7: e39514
18. Larbi A, Mitjavila-Garcia MT, Flamant S, Valogne Y, Clay D, Usunier B et al (2014) Generation of multipotent early lymphoid progenitors from human embryonic stem cells. Stem Cells Dev 23:2983–2995
19. Calvo J, BenYoucef A, Baijer J, Rouyez MC, Pflumio F (2012) Assessment of human multipotent hematopoietic stem/progenitor cell potential using a single in vitro screening system. PLoS One 7:e50495
20. Kohn LA, Hao QL, Sasidharan R, Parekh C, Ge S, Zhu Y et al (2012) Lymphoid priming in human bone marrow begins before expression of CD10 with upregulation of L-selectin. Nat Immunol 13:963–971
21. Alhaj Hussen K, Vu Manh TP, Guimiot F, Nelson E, Chabaane E, Delord M et al (2017) Molecular and functional characterization of lymphoid progenitor subsets reveals a bipartite architecture of human lymphopoiesis. Immunity 47:680–96.e8

22. Alhaj Hussen K, Louis V, Canque B (2024) A new model of human lymphopoiesis across development and aging. Trends Immunol 45: 495–510
23. Karamitros D, Stoilova B, Aboukhalil Z, Hamey F, Reinisch A, Samitsch M et al (2018) Single-cell analysis reveals the continuum of human lympho-myeloid progenitor cells. Nat Immunol 19:85–97
24. Semenza GL (2012) Hypoxia-inducible factors in physiology and medicine. Cell 148:399–408
25. Chabi S, Uzan B, Naguibneva I, Rucci J, Fahy L, Calvo J et al (2019) Hypoxia regulates lymphoid development of human hematopoietic progenitors. Cell Rep 29:2307–20.e6

Chapter 17

Analysis of MicroRNAs Expression Across Cells, Tissues, and Body Fluids

Julie Perrot, Imene Hani, Laetitia Martinerie, and Say Viengchareun

Abstract

MicroRNAs (miRNAs) are key post-transcriptional regulators of gene expression. Accurate quantification of miRNAs is essential for understanding their biological functions. In this chapter, we first describe methods for extracting miRNAs from cells, tissues, and body fluids. We then outline the principles and procedures for reverse transcription and quantification using RT-qPCR and digital droplet PCR (ddPCR).

Key words miRNAs, Extraction, Body fluids, Reverse transcription, Digital droplet PCR

1 Introduction

MicroRNAs (miRNAs) are small, single-stranded, transcribed but untranslated RNAs, typically ranging from 20 to 25 nucleotides in length, which play a crucial role in regulating gene expression at the post-transcriptional level [1]. These post-transcriptional regulators are involved in numerous biological processes, including cell differentiation [2], neurogenesis [3], inflammation and immune response [4], or carcinogenesis [5]. As of 2019, 2300 mature miRNA sequences have been identified in humans, with 1115 of these annotated in the miRBase V22 database [6]. More recently, the miRTarBase database has cataloged 3012 human miRNAs, suggesting an impressive potential for approximately 4,475,477 interactions between miRNAs and their target genes [7]. Quantifying miRNAs is of significant importance for understanding the intricate biological mechanisms in which they participate. Accurate measurement of their expression not only enables the identification of profiles specific to particular pathophysiological conditions but also enhances our understanding of their role in the precise regulation of gene networks. This quantification is especially pivotal as it serves as a prerequisite for the functional study of miRNAs, their application as diagnostic or prognostic biomarkers, and the

Rami Bechara (ed.), *Autoimmunity: Methods and Protocols*, Methods in Molecular Biology, vol. 2990,
https://doi.org/10.1007/978-1-0716-4997-8_17,

advancement of targeted therapeutic strategies. Today, various quantification methods are available, each presenting unique advantages and limitations. Among the most widely utilized techniques, quantitative Polymerase Chain Reaction (qPCR) stands out as a reference method due to its sensitivity, specificity, and relative ease of use. In contrast, droplet digital PCR (ddPCR) allows for absolute quantification without the need for standard curves, offering enhanced precision, especially in detecting low levels of expression. Lastly, high-throughput methods, like miRNA-Sequencing, facilitate a comprehensive analysis of the miRNA transcriptome, delivering an extensive overview of their diversity and expression across different biological and pathological contexts.

2 Materials

2.1 Consumables

2.1.1 For miRNAs Extraction from Cells or Tissues

NucleoSpin miRNA kit from Macherey-Nagel, including NucleoSpin® Filters (violet rings), NucleoSpin® RNA Columns (blue rings), NucleoSpin® Protein Removal Columns (white rings), Collection Tubes (2 mL).

2.1.2 For miRNAs Extraction from Urine

miRNeasy Serum/Plasma Advanced kit from Qiagen, which includes RNAeasy UCP MinElute Columns (each packaged with a 2 mL Collection Tube), collection tubes (1.5 mL), collection tubes (2 mL) stored à +4 °C.

2.1.3 For qPCR

1. MicroAmp™ Optical 96-Well Reaction Plate with BarcodeMicroAmp Optical (Applied Biosystems).
2. MicroAmp™ Optical 8-Cap Strips (Applied Biosystems).

2.1.4 For ddPCR*

3. ddPCR Plates 96-Well, Semi Skirted,25/BX.
4. DG8 Cartridges and Gaskets, QX100/QX200.
5. Pierceable Foil Heat Seal.

2.2 Instrumentation

1. Spectrophotometer.
2. thermocycler or heating block.
3. QX200™ Droplet Generator.
4. PX1™ PCR Plate Sealer, C1000 Touch™, QX200™.
5. QX200™ Droplet Reader.
6. QX Manager2.1™ Software.

(all solutions and reagents are from Macherey-Nagel):

1. Lysis Buffer ML: guanidium thiocyanate 30–60% (store at RT).
2. Ethanol 96–100% (store at RT).

2.3 Solutions and Reagents

2.3.1 For miRNA Extraction from Cells or Tissues

3. Protein Precipitation Buffer: MP (store at RT).
4. Binding Buffer: MX (store at RT).
5. Membrane Desalting Buffer: MDB (store at RT).
6. rDNase: diluted in Reaction Buffer for rDNase (store at +4 °C lyophilized and −20 °C diluted).
7. Wash Buffer MW1: (Guanidinium thiocyanate 1–15% + ethanol 55–75% (store at RT)).
8. Wash Buffer MW2: store at RT.

2.3.2 For miRNA Extraction from Urine

(all solutions and reagents are from Qiagen):

1. Buffer RPL (store at RT).
2. Buffer RPP (store at RT).
3. Buffer RWT (store at RT).
4. Buffer RPE (store at RT).

2.3.3 For Reverse-Transcription

(all solutions and reagents are from Life Technologies):

1. dNTP mix: 100 nM each of dATP, dCTP, dGTP, dTTP (store at −20 °C).
2. Reverse Transcriptase: Multiscribe reverse transcriptase 50 U/µL (store at −20 °C).
3. 10X Reverse transcription buffer (store at −20 °C).
4. RNase Inhibitor (20 U/µL) (store at −20 °C).
5. Tris-EDTA 1X (Store at RT).

2.3.4 For qPCR

1. Forward and Reverse primers: 100 µM, Eurogentec (store at −20 °C).
2. Taqman probes, Life Technologies (store at −20 °C).
3. 2X TaqMan Uni PCR Master Mix, Life Technologies (store at −20 °C).

2.3.5 For ddPCR

1. ddPCR™ Supermix for Probes, BIORAD (No dUTP) 2X (store at −20 °C).
2. Taqman probes, Life Technologies (store at −20 °C).
3. Droplet Generation Oil for Probes, BIORAD (store at RT).

3 Methods

For all manipulations, work in RNase-free environment (laminar flow hood, decontaminated equipment) and always keep RNA on ice.

3.1 MicroRNAs Extraction from Cells or Tissues

(use for animal tissue <30 mg, cells <10^7)

1. Add 300 μL ML buffer directly to all plates and scrape with a spatula and mix the lysate (>5 times) by aspiration-retrieval with a pipette.
2. Place a purple-filtered column (RNA column) in a 2 mL tube and place the mixed solution in the column.
3. Centrifuge for 1 min at 11,000 g and discard purple filter column.
4. Add exactly 150 μL of Ethanol 96–100% to 300 μL of homogeneous lysate (if more than 300 μL, *see* **Note 1**).
5. Vortex for 5 s and incubate for 5 min at room temperature (18–25 °C).
6. Place a blue-filtered column (RNA column) in a 2 mL tube and place the entire solution in the column and centrifuge for 1 min at 14,000 g.
7. Place blue column in new collector tube and add 350 μL of MDB solution and centrifuge for 1 min at 11,000 g and discard filtrate.
8. Add 300 μL of MP buffer to the filtrate containing proteins and miRNAs.
9. Close tube, vortex for 5 s, and centrifuge for 3 min at 11,000 g to precipitate proteins.
10. Place a white filter column in a 2-mL tube and add miRNA-containing aqueous phase to the columns.
11. Centrifuge for 1 min at 11,000 g to completely remove residual proteins.
12. Discard the white filter column. Keep the filtrate.
13. Add 800 μL MX buffer (Dioxane 10–90%) to each tube containing the filtrate.
14. Vortex for 5 s (*see* **Note 2**).
15. Place the previous blue filter column (miRNA binding) in a collector tube and add 600 μL of solution to the filtered miRNA column.
16. Centrifuge for 30 s at 11,000 g, discard the filtrate, and place the filtered miRNA column in the same tube.
17. Repeat if any sample remains.
18. Add 600 μL of buffer MW1 to the blue column containing miRNAs. Centrifuge for 30 s at 11,000 g. Discard the filtrate and place the green column in the same tube.
19. Add 700 μL of MW2 buffer to the blue column containing miRNAs. Centrifuge for 30 s at 11,000 g. Discard filtrate and place blue column in same tube.

20. Add 250 μL of MW2 buffer to the blue column containing miRNAs.
21. Centrifuge for 2 min at 11,000 g to dry the membrane completely (*see* **Note 3**).
22. Place blue column in a new 1.5 mL tube and add 20 μL [normally 30 μL or 50 μL or 100 μL] of RNAse-free water to the column. Volume will depend on desired miRNA concentration.
23. Do not close tubes and Incubate for 1 min at room temperature (18–25 °C).
24. Close tubes and centrifuge for 30 s at 11,000 g.
25. After eluting miRNAs, immediately transfer tubes to an ice tray and store at −20 °C and measure the concentration using Nanodrop and store RNA at −20 °C.

3.2 MicroRNAs Extraction from Body Fluids

1. Thaw frozen samples.
2. Transfer a minimum of 200 μL serum, plasma, or urine into a 2 mL microcentrifuge tube (*see* **Note 4**).
3. Add 60 μL Buffer RPL. Close the tube caps and vortex for >5 s. Leave at room temperature (15–25 °C) for 3 min (*see* **Note 5**).
4. Add 20 μL Buffer RPP. Close the tube caps and mix vigorously by vortexing for >20 s Incubate at room temperature for 3 min (*see* **Note 6**).
5. Centrifuge at 12,000 g for 3 min at room temperature to pellet the precipitate; supernatant should be clear and colorless.
6. Transfer supernatant (~230 μL for 200 μL serum/plasma) to a new microcentrifuge tube. Add 1 volume of isopropanol. Mix well by vortexing.
7. Transfer the entire sample to a RNeasy UCP MinElute column. Close the lid, and centrifuge for 15 s at ≥8000 g. Discard the flow-through.
8. Pipet 700 μL Buffer RWT onto the RNeasy UCP MinElute spin column. Close the lid, and centrifuge for 15 s at ≥8000 g. Discard the flow-through. Reuse the collection tube in the next step.
9. Pipet 500 μL Buffer RPE onto the RNeasy UCP MinElute spin column. Close the lid, and centrifuge for 15 s at ≥8000 g. Discard the flow-through. Reuse the collection tube in the next step.
10. Add 500 μL of 80% ethanol to the RNeasy UCP MinElute spin column. Close the lid gently, and centrifuge for 2 min at ≥8000 g to wash the spin column membrane. Discard the flow-through.

11. Place the RNeasy UCP MinElute spin column in a new 2 mL collection tube (supplied). Open the lid of the spin column and centrifuge at full speed for 5 min to dry the membrane. Discard the flow-through and the collection tube (*see* **Note 7**).
12. Place the RNeasy UCP MinElute spin column in a new 1.5 mL collection tube (supplied). Add 20 μL RNase-free water directly to the center of the spin column membrane and incubate 1 min. Close the lid, and centrifuge for 1 min at full speed to elute the RNA (*see* **Note 8**).

3.3 Nucleic Acid Quantification

Absorbance at 260 nm is measured to quantify miRNA concentration. The quantification reports the nucleic acid concentration along with two absorbance ratios: A260/A280 and A260/A230.

3.4 Reverse Transcription

Protocol adapted from Le Carré et al. using the TaqMan™ MicroRNA Reverse Transcription Kit [8] (*see* **Note 9**).

1. Prepare 400 ng of miRNA sample in a final volume of 6 μL.
2. A customized RT primer pool is prepared by pooling all miRNA-specific stem-loop primers as follows: the primers are combined and diluted in 1X Tris-EDTA (TE) buffer to a final concentration of 0.05X for each primer (a maximum of 12 different stem-loop primers).
3. Prepare 9 μL of RT master mix as follows: 3 μL of primer pool solution, 0.3 μL of 100 mM dNTP mix, 3 μL of enzyme mix, 1.5 μL of 10× RT buffer, and 0.19 μL of RNase inhibitor.
4. Add the RT master mix to each sample to reach a final reaction volume of 15 μL.
5. Perform reverse transcription under the following cycling conditions: 30 min at 16 °C (primer annealing), 30 min at 42 °C (extension), and 5 min at 85 °C (enzyme inactivation). The resulting cDNA is stored at −20 °C.

3.5 Quantitative PCR

Quantitative PCR was performed either as absolute quantification using a standard curve or as relative quantification using the $2^{-\Delta\Delta Ct}$ method.

1. Reverse transcription products are diluted 1:20 in nuclease-free water and stored at −20 °C if not used immediately.
2. For each sample, a 10 μL qPCR mix is prepared as follows: 5 μL of 2× TaqMan™ Universal Master Mix II with UNG and 2 μL of 5× TaqMan™ probe.
3. Add the qPCR mix to the wells first, then add 3 μL of diluted cDNA sample. (Each sample should be analyzed in duplicate).
4. qPCR plates are loaded into the appropriate real-time PCR system and run under the following cycling conditions:

10 min at 95 °C (initial denaturation), followed by 45 cycles of 15 s at 95 °C and 1 min at 60 °C (annealing/extension step; temperature may vary depending on the probe).

3.6 Digital Droplet PCR

1. DNA of interest (approx. 10% of final vol.): QSP 9 μL with H2O (*see* **Note 10**).
2. Homogenize mixture by at least 5 flushes and centrifuge the tube to recover everything.
3. Place the DG8 strip in the appropriate Bio-Rad holder.
4. Deposit DNA mix by transferring 20 μL per well (from the 22 μL of mix previously prepared) into each well in row 2 of the DG8 strip. This is an important step, *see* **Note 11**.
5. Apply Bio-Rad droplet-generating oil.
6. Dispense 70 μL of oil into each well of row 1 of the 8-well strip (*see* **Note 12**).
7. Place the Bio-Rad gasket on top of the strip, and hook the strip onto the holder on either side.
8. Place holder with DG8 strip in Bio-Rad device.
9. Allow approx. 2 min 30 for emulsion preparation.
10. Retrieve the DG8 strip from its holder, removing the single-use seal.
11. The emulsion is recovered from the 3rd well of the DG8 strip, looking slightly milky (the droplets are floating, the oil underneath).
12. Transfer the 40 μL of emulsion (row n°3) from the DG8 strip to the 96-well plate for PCR (it is a crucial step, *see* **Note 13**).
13. Place the green Bio-Rad 96-well plate with emulsions on the silver support/block, followed by the sealing film.
14. Place in PX1 PCR plate sealer (program 180 °C for 5 s).
15. Once finished, remove the block from the unit, recover the 96-well plate, and run the PCR quickly: emulsion stable for 1h30-2h maximum after preparation.
16. Place plate in Bio-Rad S1000 thermal cycler satellite device and program cycle on Bio-Rad Core C1000 thermal cycler. Cycling follows the following conditions: enzymatic activation and emulsion stabilization 10 min at 95 °C, 40 cycles of 95 °C for 30 s/55 °C for 1 min with ramp of 2.5 °C/s (slowly). Fluorescence fixation 98 °C for 10 min.
17. Recover the 96-well plate after PCR and place the plate on the black hole holder of the QX200 and on top of the silver hole cover, lowering the black clips on either side of the plate.

18. Place the system in the QX200 to start reading, checking that the side clips are closed horizontally on both sides of the plate (*see* **Note 14**).
19. Count approximately in reading time: 2 min 30 per well, i.e., 16 min for 8 samples, or 2h30 for a 96-well plate. Each emulsion droplet is analyzed with fluorescence readings at Fam and Hex/Vic wavelengths.

 The result is validated:

If at least 8000 droplets are read and of good quality, if <8000 test not validated, usually 15 to 18,000 droplets read software alarm if <10,000 droplets.

4 Notes

1. Volume must be adjusted according to the lysate if it is greater than 300 μL.
2. Do not centrifuge the mixture even if a precipitate appears.
3. If the eluate touches the column after the 3rd wash, discard the eluate and centrifuge again.
4. It is possible to process up to 600 μL of serum/plasma sample. In this case, adapt the volume of RPL and RPP:
   - For 200 μL of serum or plasma, use 60 μL of RPL and 20 μL RPP.
   - For 400 μL of serum or plasma, use 120 μL of RPL and 40 μL RPP.
   - For 600 μL of serum or plasma, use 180 μL of RPL and 60 μL RPP.
5. If any spike-in controls are used, they should be added to the lysate at this point.
6. Thorough mixing is important for subsequent phase separation.
7. To avoid damage to their lids, place the spin columns into the centrifuge with at least one empty position between columns. Orient the lids so that they point in a direction opposite to the rotation of the rotor (e.g., if the rotor rotates clockwise, orient the lids counterclockwise). It is important to dry the spin column membrane, since residual ethanol may interfere with downstream reactions. Centrifugation with the lids open ensures that no ethanol is carried out during RNA elution.
8. As little as 10 μL RNase-free water can be used for elution if a higher RNA concentration is required, but the yield might be reduced. Do not elute with less than 10 μL RNase-free water, as

the spin-column membrane will not be sufficiently hydrated. The dead volume of the RNeasy UCP MinElute spin column is 2 μL, so elution with 20 μL RNase-free water results in an 18 μL eluate.

9. The TaqMan® technology provided by Applied Biosystems (ABI) uses a target-specific stem-loop RT primer that extends the 3′ end of the targeted miRNA to produce a cDNA template, which can then be amplified and quantified by real-time qPCR.
10. A negative control per gene studied (probe/primer) for baseline fluorescence is needed.
11. Place the cone at the edge of the well (not in the center, as there is a risk of bubbles at the channel entrance) and carefully deposit the 20 μL from the stop (1 pipette stop), inclining the cone at a 15° angle.
12. Oil to be deposited in the strip just a few seconds before installation in the device (do not anticipate) and after sample deposit: do not reverse the order of deposit between sample and oil.
13. Delicate step to maintain droplet integrity: Position the cone in the center of the well (level with the capillary, the pellet), and tilt at an angle of 15° to the bottom of the well.
14. Check the oil level in the bottle and in the waste container on the left-hand side of the unit. If necessary, change the oil bottle (with a "prime" to eliminate bubbles in the system tubing, stable for 2 years). Once a week, run a "system flush" to change the oil in the tubing.

References

1. Bartel DP (2004) MicroRNAs: genomics, biogenesis, mechanism, and function. Cell 116: 281–297. https://doi.org/10.1016/s0092-8674(04)00045-5
2. Gebert LFR, MacRae IJ (2019) Regulation of microRNA function in animals. Nat Rev Mol Cell Biol 20:21–37. https://doi.org/10.1038/s41580-018-0045-7
3. Esteves M, Serra-Almeida C, Saraiva C, Bernardino L (2020) New insights into the regulatory roles of microRNAs in adult neurogenesis. Curr Opin Pharmacol 50:38–45. https://doi.org/10.1016/j.coph.2019.11.003
4. Marques-Rocha JL, Samblas M, Milagro FI, Bressan J, Martínez JA, Marti A (2015) Noncoding RNAs, cytokines, and inflammation-related diseases. FASEB J 29:3595–3611. https://doi.org/10.1096/fj.14-260323
5. Lin S, Gregory RI (2015) MicroRNA biogenesis pathways in cancer. Nat Rev Cancer 15:321–333. https://doi.org/10.1038/nrc3932
6. Alles J, Fehlmann T, Fischer U, Backes C, Galata V, Minet M, Hart M, Abu-Halima M, Grässer FA, Lenhof H-P, Keller A, Meese E (2019) An estimate of the total number of true human miRNAs. Nucleic Acids Res 47:3353–3364. https://doi.org/10.1093/nar/gkz097
7. Huang H-Y, Lin Y-C-D, Li J, Huang K-Y, Shrestha S, Hong H-C, Tang Y, Chen Y-G, Jin C-N, Yu Y, Xu J-T, Li Y-M, Cai X-X, Zhou Z-Y, Chen X-H, Pei Y-Y, Hu L, Su J-J, Cui S-D, Wang F, Xie Y-Y, Ding S-Y, Luo M-F, Chou C-H, Chang N-W, Chen K-W, Cheng Y-H,

Wan X-H, Hsu W-L, Lee T-Y, Wei F-X, Huang H-D (2020) miRTarBase 2020: updates to the experimentally validated microRNA-target interaction database. Nucleic Acids Res 48: D148–D154. https://doi.org/10.1093/nar/gkz896

8. Le Carré J, Lamon S, Léger B (2014) Validation of a multiplex reverse transcription and pre-amplification method using TaqMan(®) MicroRNA assays. Front Genet 5:413. https://doi.org/10.3389/fgene.2014.00413

Chapter 18

Polysome Profiling: A Method for Global and Targeted Analysis of mRNA Translation

Lucilla Fabbri, Ana Luisa Dian, Céline M. Labbé, and Stéphan Vagner

Abstract

Among the regulatory mechanisms that control gene expression, mRNA translation plays a fundamental role in determining the functional output of the transcriptome in cells. This is particularly relevant in immune cells, where rapid and tightly regulated translation is essential for maintaining immune homeostasis. Analyzing the translational status of transcripts, as well as the mechanisms underlying their regulation, can provide valuable insights into the key factors influencing physiological and pathological phenotypes. Here, we describe our routinely used protocol for polysome profiling, a method that enables the analysis of mRNA translation at both the individual transcript and genome-wide levels.

Key words Polysome profiling, Post-transcriptional gene regulation, mRNA translation

1 Introduction

Regulation of gene expression is a complex process that encompasses mRNA transcription and the subsequent post-transcriptional steps, including the synthesis of protein. However, it is now well established that cellular mRNA levels poorly correlate with protein abundance and that such discrepancies can, in part, be attributed to the regulation of mRNA translation [1–3]. Indeed, mRNA translation is regulated by an overgrowing number of RNA-binding proteins and non-coding RNAs, including microRNAs, which interact with specific sequences, secondary/tertiary structures, or ribonucleotide modifications (e.g., methylation) present in the mRNA itself [4]. Dysregulation of translational is implicated in a wide range of diseases, including cancer and autoimmune disorders [5–8]. Therefore, understanding the translational status of transcripts is crucial for determining the actual functional output of the transcriptome in cells and for dissecting the mechanisms underlying its dysregulation in disease pathogenesis.

Rami Bechara (ed.), *Autoimmunity: Methods and Protocols*, Methods in Molecular Biology, vol. 2990,
https://doi.org/10.1007/978-1-0716-4997-8_18,

In recent years, research methods for studying mRNA translation have significantly advanced, enabling the investigation of translation dynamics at both the cell population and single-cell levels. These methods facilitate the study of global mRNA translation by assessing ribosome density on transcripts (e.g., polysome profiling [9] and ribosome profiling [10]) or by measuring nascent peptide synthesis through puromycin-based labeling (e.g., SuNSET [11]) or incorporation of azide-labeled amino acid analogues (e.g., FAN-CAT [12]). They can also be used to detect the translation of a specific mRNA by using recently developed proximity ligation (PLA) assays-based techniques (e.g., Puro-PLA [13], RiboMap [14]).

Among the methods used for the analysis of global mRNA translation, polysome profiling remains the most widely used technique despite its labor-intensive nature and the requirement for specialized equipment. This method can be applied to targeted analyses of specific mRNAs as well as genome-wide assessments of global translation. It is based on the assumption that ribosome density (the number of ribosomes associated with an mRNA) strongly correlates with protein synthesis rate. In brief, after ribosome arrest on mRNAs via cycloheximide treatment, the cytoplasmic lysate is isolated and loaded onto a sucrose gradient. MRNAs associated with different numbers of ribosomes are then separated by ultracentrifugation according to their sedimentation velocity. Within the sucrose gradient, actively translated mRNAs, such as, those bound to multiple ribosomes, are separated from poorly translated or untranslated ones. The sucrose density gradient is subsequently fractionated, and RNAs are extracted from each fraction, enabling the analysis of mRNA distribution. The ratio of mRNAs bound to heavy polysomes relative to total cytosolic mRNAs serves as an indirect measure of translational activity.

Traditionally, northern blotting and DNA microarrays have been used to study the translation of individual mRNAs and global translation, respectively. Here, we present more recent alternative approaches that utilize reverse transcription-quantitative polymerase chain reaction (RT-qPCR) or high-throughput RNA sequencing (RNA-seq) to analyze translation at both the individual mRNA and genome-wide levels, respectively (Fig. 1).

2 Materials

2.1 Buffers

1. 15% sucrose buffer: 20 mM Tris-HCl pH 7.5, 100 mM NaCl, 3 mM $MgCl_2$, 15% sucrose (w/v), 1 mM Dithiothreitol (DTT), 20 U/mL RNase-OUT, 100 μg/mL Cycloheximide (CHX) (*see* **Note 1**).

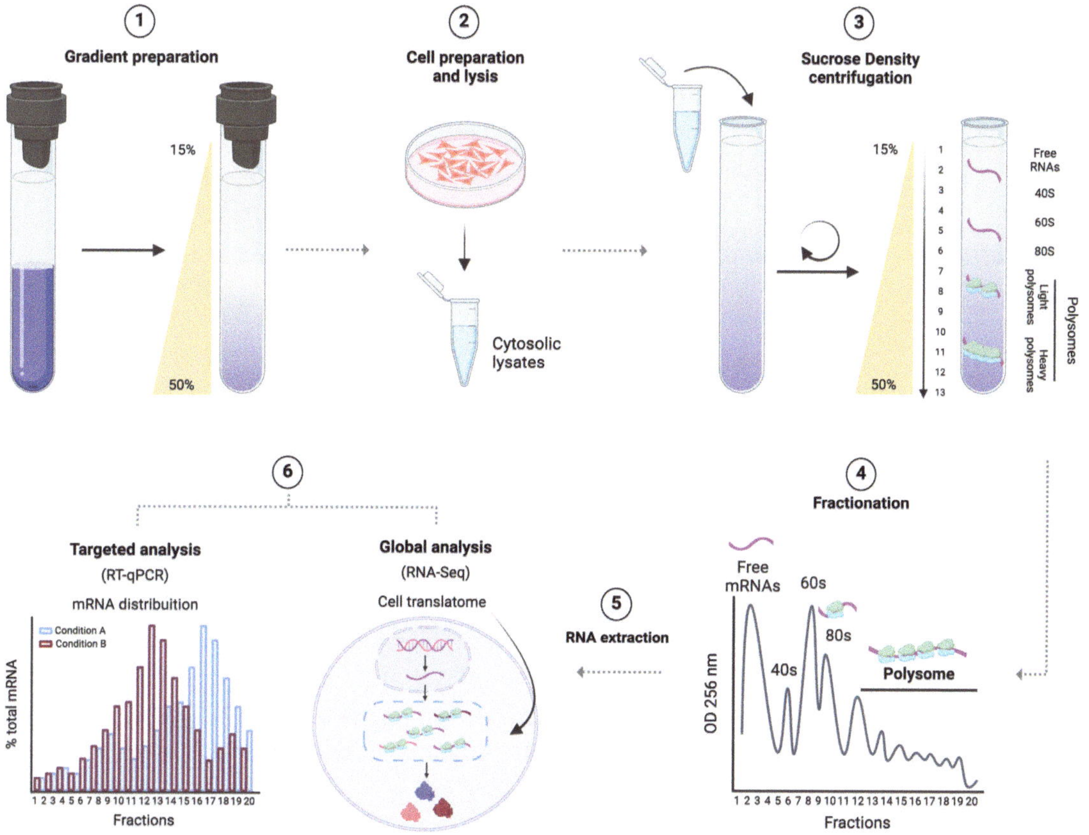

Fig. 1 Schematic representation of polysome profiling workflow. The steps include (1) gradient preparation; (2) cell preparation and lysis; (3) sucrose density centrifugation; (4) gradient fractionation; (5) RNA extraction; and (6) analysis by either RT-qPCR or RNA-sequencing

2. 50% sucrose buffer: 20 mM Tris-HCl pH 7.5, 100 mM NaCl, 3 mM MgCl2, 50% sucrose (w/v), 1 mM DTT, 20 U/mL RNase-OUT, 100 μg/mL CHX (*see* **Note 1**).
3. Lysis Buffer 1: 20 mM Tris-HCl pH 7.5, 100 mM NaCl, 3 mM $MgCl_2$, 1 mM DTT, 100 U/mL RNase-OUT, 100 μg/mL CHX (*see* **Note 2**).
4. Lysis buffer 2: 0.5 M Sucrose, 20 mM Tris-HCl pH 7.5, 100 mM NaCl, 3 mM $MgCl_2$, 1 mM DTT, 2.4% Triton X100, 100 μg/mL CHX (*see* **Note 2**).
5. Reverse Transcription (RT) mix: 1X First Strand buffer, 5 mM DTT, 0.5 mM dNTPs, 40 U RNAseOUT, 200 U SuperScript III.

3 Methods

3.1 Gradient Preparation

The protocol uses the tilted tube rotation process and the Gradient Master instrument developed by Biocomp.

1. After placing the centrifuge tube in the designated marker block provided by the gradient maker, mark the upper half of the tube with the appropriate marker.
2. Gently pour approximately 6 mL of the 15% sucrose buffer until it reaches the marked line (*see* **Note 3**).
3. Fill a 20 mL syringe with a cannula with 50% sucrose buffer and wipe the outside of the cannula with a paper towel (*see* **Note 4**).
4. Insert the cannula at the bottom of the tube, directing it towards the wall and carefully layer the 50% sucrose buffer until it reaches the previously marked line.

   Note that the interface within the two gradients should be sharp.
5. Adjust the volume with the 15% sucrose buffer using the sample layering device provided until it reaches 2–3 mm below the top of the tube.
6. Seal the tube with the cap provided by the gradient maker, making sure that the side with a hole is sealed last (*see* **Note 5**).
7. Ensure to level the plate to which the tube holder is positioned.
8. Place the tube into the tube holder on the gradient maker.
9. Run the sucrose gradient maker by selecting the "10%–50%" program.
10. When finished, place the tubes at 4 °C for up to 30 min before centrifugation.

3.2 Cell Preparation and Lysis

To exemplify the application of polysome profiling, cells have been treated with DMSO (control) or with an inhibitor of the translation initiation factor eIF4A (eIF4Ai). eIF4Ai treatment inhibits the translation of specific mRNAs (Fig. 2).

The protocol described below is suited for experiment on adherent cells but can be adapted for cells cultured in suspension.

1. Based on your experimental conditions, prepare three 15 cm plates of adherent cells at 70% confluency per condition (*see* **Note 6**).
2. Add 100 μg/mL CHX to each cell plate. After gentle shaking, incubate the cells at 37 °C, 5% CO_2 for 15 min (*see* **Note 7**).
3. Transfer the plates on ice, wash cells with ice-cold phosphate buffered saline (PBS) buffer supplemented with 100 μg/mL CHX, and harvest the cells in 4 mL of the same buffer by

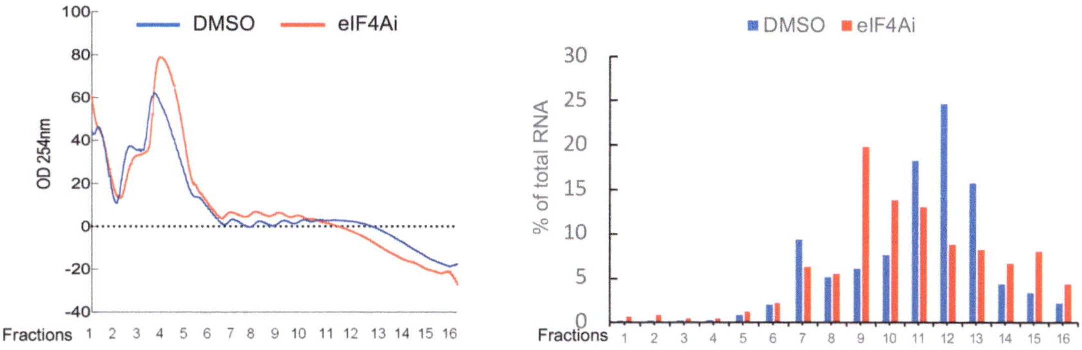

Fig. 2 Polysome profiles of cells treated with DMSO or an eIF4A inhibitor (eIF4Ai) (left). Percentage of transcript in each polysomal fraction obtained by sucrose-gradient ultracentrifugation and quantified by RT-qPCR (right)

scraping. Collect cells of the same condition in a 50 mL falcon tube.

All further steps should be performed at or near 4 °C.

4. Centrifuge the cell suspension at 2.000 rpm for 5 min at 4 °C (*see* **Note 8**). Discard supernatant.
5. Add 400 µL of cold Lysis buffer 1, resuspend, and homogenize the cell suspension by gently pipetting up and down (*see* **Note 9**).
6. Add 400 µL of cold Lysis buffer 2 and gently pipette up and down multiple times to weaken the cell membrane (*see* **Note 9**).
7. Incubate on ice for 30 min, gently pipetting up and down 2–3 times during incubation.
8. Transfer the lysate to a 1.5 mL Eppendorf tube and centrifuge at 12,000× *g* for 5–10 min at 4 °C and collect the cleared lysate supernatant into a new 1.5 mL Eppendorf tube (*see* **Note 10**). ‘.
9. Estimate the volume of cytosolic lysate and adjust a final volume to be close to the original (e.g., 850 µL for 800 µL of lysate) to increase the salt concentration from 100 mM to 150 mM NaCl and from 3 mM to 10 mM $MgCl_2$.
10. Take 10% of the final cytosolic volume as input, while the remaining 90% will undergo polysome fractionation.

    Cytosolic lysates can be flash frozen in liquid nitrogen and stored at −80 °C until use.

3.3 Gradient Centrifugation and Fractionation

Here, we describe a protocol using a Biocomp fractionation system, composed of a Piston Gradient Fractionator and a Triax™ flow cell system for gradient profiling.

1. Carefully load the lysate drop by drop onto the top of the previously prepared sucrose gradient tubes (*see* Subheading 3.1) (*see* **Note 11**).
2. Balance the weight of the tubes prior to centrifugation by carefully adding the 15% sucrose buffer to the top of the tube (*see* **Note 12**).
3. Place the sucrose gradient tube loaded with the samples into the SW41 rotor bucket.
4. Centrifuge at 38,000 rpm, for 2 h at 4 °C in an SW41 rotor. Set the acceleration speed to maximum and deceleration speed to 9 with no brake.
5. Turn on the fractionation system 15 min before the ultracentrifugation ends.
6. Open the Triax™ FlowCell software on the computer.
7. Turn on the Gilson (fraction collector device).
8. Rinse the system by filling with pure, filtered water in the reservoir on the fractionation station and follow the software instruction.
9. In the Triax™ FlowCell software, set the parameters for fractionation (Rotor type: SW41Ti, Wavelength: 260 nm, Sample volume (mL), Gradient sucrose (%), Fractions number, Centrifugation speed: 38 (kRPM), Centrifugation time: 2 h, *T*: 4 °C).
10. Set the number of data points per mm to "10" and the Piston speed to "0.3 mm/s" (*see* **Note 13**).
11. Place 1.5 mL Eppendorf tubes, numbered according to the sequentially collected fractions, onto the fractionation collector.
12. Zero the flow cell using the provided rinse adaptor, injecting it with 5 mL of pure water.
13. Dry the system by following Triax™ FlowCell software instructions.
14. Install the Piston tip in the fractionation system.
15. Once the ultracentrifugation for polysome sedimentation is complete, mount the centrifuged tubes onto the designated tube holder into the fractionation system.
16. Click on "START SCAN" in the software program. This will trigger the piston tip to collect each fraction from top to bottom while simultaneously recording UV absorbance at 260 nm to generate the trace.

3.4 RNA Extraction and RT-qPCR

In this session, we describe the step-by-step protocol for the analysis of the translational status of a given mRNA by RT-qPCR amplification.

1. Transfer 300 μL from each fraction into a new 2 mL Eppendorf tube.
2. Add 1 mL of Trizol LS.
3. Vortex vigorously to allow complete dissociation of the nucleoprotein complexes.
4. Add 300 μL of Chloroform per 1 mL of Trizol LS.
5. Vortex vigorously for 15 s to mix the chloroform thoroughly.
6. Keep the samples at room temperature (20–25 °C) for 5 min.
7. Centrifuge the samples at 12,000× *g* for 20 min at 4 °C (*see* **Note 8**).
8. Transfer the RNA-containing top aqueous phase (around 550 μL) into a clean 1.5 mL Eppendorf tube (*see* **Note 14**).
9. Repeat **steps 5–8** to exclude DNA contamination.
10. Add equal volume of ice-cold isopropanol and 1 μL GlycoBlue™ Coprecipitant (*see* **Note 15**).
11. Mix thoroughly by vortexing.
12. Store the mixture at −20 °C for at least 1 h up or overnight to increase RNA yield.
13. Centrifuge at 12,000× *g* for 30 min at 4 °C (*see* **Note 8**).
14. Discard the supernatant carefully without touching the pellet (*see* **Note 16**).
15. Add 1 mL ice-cold 75% ethanol to wash the RNA pellet.
16. Centrifuge at 12,000× *g* for 5 min at 4 °C.
17. Discard the supernatant carefully and repeat **steps 16–18**.
18. Air dry the pellet for 5 min (*see* **Note 17**).
19. Dissolve the pellet with 10–30 μL RNase-free water.
20. Measure RNA abundance by NanoDrop™ 2000 spectrophotometer (optional).

    The following reverse-transcription steps are performed using SuperScript III reverse transcriptase kit and random primers. Other reverse transcription kits or Oligo(dT), specific for eukaryotic mRNA reverse transcription, can be used.
21. In a PCR multi-well plate, add 1–8 μL of dissolved RNA.
22. Add 1 μL Random primers and RNAse-free water to a final volume of 12 μL.
23. Incubate the mixture at 70 °C for 2 min and then at 4 °C until next use.
24. Prepare RT mix.
25. Add 8 μL of RT mix to the solution from the last step and mix well.
26. Spin quickly.

27. Incubate the samples at 25 °C for 5 min, 50 °C for 45 min, 70 °C for 15 min, and then at 4 °C until next use.
28. Dilute cDNA 1:4 with RNAse-free water and transfer diluted cDNA in new 1.5 mL Eppendorf tubes (*see* **Note 18**).
29. Run the real-time PCR with the appropriate program (1 Cycle: 95 °C for 10 min; 40 Cycles: 95 °C for 15 s and 60 °C for 45 s).
30. Analyze the data using the threshold cycle (Ct) comparative method and quantify as percentage of the total RNA considering the whole fractions stand for 100%.

$$\%\mathrm{RNA} = \frac{2^{-\mathrm{Ct}(\mathrm{fraction}X)}}{2^{-\mathrm{Ct}}\mathrm{fraction1} + 2^{-\mathrm{Ctl}}\mathrm{fraction2} + .. + .2^{-\mathrm{Ct}}\mathrm{fraction}Z)} \times 100$$

3.5 RNA-seq Library Preparation

Alternatively, to analyze the translational status of given mRNAs by RT-qPCR, a global analysis of the content of polysomal fractions can be performed to retrieve the cell translatome, that is, the subset of the transcriptome that is being actively translated. This can be achieved by deep-sequencing analyses of pooled polysomal fractions, as described in the protocol below.

1. Pool heavy polysome fractions by transferring 20 μL from each fraction into a new 2 mL Eppendorf tube for a final volume of 300 μL (*see* **Note 19**).
2. Perform the RNA extraction of total RNA (input samples from Subheading 3.3, **step 13**) and heavy polysome fractions, as described in Subheading 3.4, **steps 1–19**.

   The following library preparation steps are performed using the Illumina TruSeq Stranded mRNA Library preparation kit for the strand specific sequencing of poly-A transcripts, and it is performed according to the manufacturer's instructions [15]. Other library preparation kits, specific for other sequencing platforms, read length and depth, can be used.
3. Prepare RNA sequencing libraries from 0.5 to 1 μg of total RNA or mRNA from polysome fractions (*see* **Note 20**).
4. Measure RNA abundance and purity by NanoDrop™ 2000 spectrophotometer and RNA integrity by Agilent Technologies 2100 Bioanalyzer (RIN > 9).
5. Purify poly-A containing RNA molecules using oligo-dT attached magnetic beads and fragment the mRNA.
6. Synthesize cDNA of the resulting fragments using reverse transcriptase and random primers, followed by dA-tailing (*see* **Note 21**) and ligation of TruSeq indexed adapters (*see* **Note 22**).
7. Perform PCR amplification to generate the barcoded cDNA libraries (*see* **Note 23**).

8. Quantify the prepared libraries using the KAPA Library Quantification Kit.
9. Sequence the libraries using the NovaSeq 6000 instrument.

3.6 Data Analysis

High-throughput sequencing analysis pipelines vary based on the sequencing platform and selected alignment and processing software. Here, we describe a general platform-independent protocol.

1. Demultiplex and convert BCL files to FASTQ files from raw sequencing data. If Illumina platforms are used for library sequencing, this process can be performed using the *bcl2fastq* pipeline (Illumina) [16]. Other tools, such as *Cutadapt* [17] or *Flexbar* [18], can also be used to demultiplex FASTQ files.
2. Trim adapters and check control quality. This step can be performed using the *Trim Galore* tool [19], a Perl wrapper around two tools: *Cutadapt* and *FastQC* [20]. For each condition, apply Cutadapt to remove the indexed adapters that were used for sequencing. FastQC will give general metrics about your reads, such as quality score distribution, the per base sequence content (%T/A/C/G), presence of adapter contaminations, and other overrepresented sequences.
3. Align reads on the human reference genome (e.g., hg38). A commonly used alignment tool is STAR [21], which allows to count the number of reads per gene using gene annotation files (e.g., GENCODE [22] GTF/GFF3 files).
4. Stabilize the variance of the raw count data by applying the variance stabilizing transformation (VST) from *DESeq2* [23]. This allows to perform principal components analyses (PCA) to visualize sample-to-sample distances and assess overall similarity between samples. PCA plots can be generated using the R package *ggplot2* [24] or directly with the plotPCA function supplied by DESeq2.
5. Perform differential translation analysis. To identify genes subjected to translational regulation, we use *Xtail* [25]. This R package is based on the assumption that, if a gene undergoes differential translation, changes in mRNA abundance in the heavy polysome fractions should be decorrelated from total mRNA expression. Xtail takes raw reads counts of mRNA (total RNA-seq) and heavy polysome fractions as input and performs quantification of differential translation for each gene.

4 Notes

1. DTT, RNase inhibitor, and CHX should be added just before use.

2. Lysis buffers can be stored at 4 °C for up to 6 months. DTT, RNase inhibitor, and CHX should be added just before use.
3. Place the pipette at the bottom of the tube and be careful to avoid the formation of bubbles as they may interfere with gradient making.
4. It is recommended to suck and expel a small amount of buffer before filling the syringe to remove any residual air, which can cause bubbles.
5. Ensure to see some liquid inside the cap. If not, a bubble can form in the gradient.
6. Note that for optimal mRNA recovery after fractionation, a minimum of 2×10^7 cells per condition is required.
7. CHX is a translation elongation inhibitor used to prevent ribosomes from dissociating from translating mRNA. Preliminary experiments are required for different cell lines to determine the optimal CHX concentration to generate well-defined polysome peaks.
8. The centrifuge should be pre-cooled to 4 °C before use.
9. Avoid foam or bubbles, as they may cause nuclear disruption.
10. Avoid contamination with precipitants at the bottom of the Eppendorf when transferring the supernatant.
11. Before loading the sample, remove a volume of sucrose buffer equal to the volume of the sample subtracted by 150 μL (e.g., if your sample volume is 500 μL, remove 350 μL of sucrose buffer).
12. A weight difference of up to 0.05 g between opposing buckets is acceptable.
13. This speed is suitable for sucrose gradients up to 50%. If the gradient is higher, reduce the piston speed and set it to 0.2.
14. Avoid contamination with the bottom red organic phase containing protein and the middle, white semi-solid containing DNA.
15. GlycoBlue Coprecipitant enhances nucleic acid precipitation, increasing the visibility of the pellet without interfering with subsequent analysis.
16. Carefully remove all isopropanol remaining as isopropanol can interfere with reverse transcriptases and cDNA synthesis.
17. Do not allow the pellet to dry for more than 5 min, as an overly dry RNA pellet may become difficult to dissolve in water.
18. cDNA can be stored at −20 °C indefinitely.
19. Fractions should be selected according to the trace generated during the fractionation of the samples (Subheading 3.3). For

the study of highly translated mRNAs, only heavy polysome fractions can be selected.

20. Lower amounts can lead to inefficient ligation and low yield.
21. Addition of one adenine (A) nucleotide to the 3′ ends of blunt fragments prevents their ligation to each other during adapter ligation reaction.
22. Multiple indexing adapters ligated to the cDNA fragments ends prepare them for their hybridization onto the flow cell.
23. The PCR is performed using a primer cocktail that anneals to the adapter ends, with the number of cycles kept to a minimum to preserve the library's representation.

Acknowledgments

We are grateful to Sylvain Martineau and Amélie Heneman for the helpful proofreading of this manuscript.

Bibliography

1. Tian Q, Stepaniants SB, Mao M, Weng L, Feetham MC, Doyle MJ et al (2004) Integrated genomic and proteomic analyses of gene expression in Mammalian cells. Mol Cell Proteomics 3(10):960–969. https://doi.org/10.1074/mcp.M400055-MCP200
2. Vogel C, Abreu RS, Ko D, Le SY, Shapiro BA, Burns SC et al (2010) Sequence signatures and mRNA concentration can explain two-thirds of protein abundance variation in a human cell line. Mol Syst Biol 6:400. https://doi.org/10.1038/msb.2010.59
3. Schwanhäusser B, Busse D, Li N, Dittmar G, Schuchhardt J, Wolf J et al (2011) Global quantification of mammalian gene expression control. Nature 473(7347):337–342. https://doi.org/10.1038/nature10098
4. Hershey JW, Sonenberg N, Mathews MB (2012) Principles of translational control: an overview. Cold Spring Harb Perspect Biol 4(12). https://doi.org/10.1101/cshperspect.a011528
5. Fabbri L, Chakraborty A, Robert C, Vagner S (2021) The plasticity of mRNA translation during cancer progression and therapy resistance. Nat Rev Cancer 21(9):558–577. https://doi.org/10.1038/s41568-021-00380-y
6. Bechara R, Vagner S, Mariette X (2023) Post-transcriptional checkpoints in autoimmunity. Nat Rev Rheumatol 19(8):486–502. https://doi.org/10.1038/s41584-023-00980-y
7. Boussemart L, Malka-Mahieu H, Girault I, Allard D, Hemmingsson O, Tomasic G et al (2014) eIF4F is a nexus of resistance to anti-BRAF and anti-MEK cancer therapies. Nature 513(7516):105–109. https://doi.org/10.1038/nature13572
8. Biswas B, Guemiri R, Cadix M, Labbé CM, Chakraborty A, Dutertre M et al (2022) Differential effects on the translation of immune-related alternatively polyadenylated mRNAs in melanoma and T cells by eIF4A inhibition. Cancers (Basel) 14(5). https://doi.org/10.3390/cancers14051177
9. Arava Y, Wang Y, Storey JD, Liu CL, Brown PO, Herschlag D (2003) Genome-wide analysis of mRNA translation profiles in Saccharomyces cerevisiae. Proc Natl Acad Sci USA 100(7):3889–3894. https://doi.org/10.1073/pnas.0635171100
10. Ingolia NT, Ghaemmaghami S, Newman JR, Weissman JS (2009) Genome-wide analysis in vivo of translation with nucleotide resolution using ribosome profiling. Science 324(5924):218–223. https://doi.org/10.1126/science.1168978
11. Schmidt EK, Clavarino G, Ceppi M, Pierre P (2009) SUnSET, a nonradioactive method to monitor protein synthesis. Nat Methods 6(4):275–277. https://doi.org/10.1038/nmeth.1314
12. Kimura Y, Saito H, Osaki T, Ikegami Y, Wakigawa T, Ikeuchi Y et al (2022) Mito-

FUNCAT-FACS reveals cellular heterogeneity in mitochondrial translation. RNA 28(6): 895–904. https://doi.org/10.1261/rna.079097.122

13. Chin A, Lécuyer E (2021) Puromycin labeling coupled with proximity ligation assays to define sites of mRNA translation in Drosophila embryos and human cells. Methods Mol Biol 2381:267–284. https://doi.org/10.1007/978-1-0716-1740-3_15
14. Zeng H, Huang J, Ren J, Wang CK, Tang Z, Zhou H et al (2023) Spatially resolved single-cell translatomics at molecular resolution. Science 380(6652):eadd3067. https://doi.org/10.1126/science.add3067
15. TruSeq Stranded mRNA Reference Guide (n.d.) https://support.illumina.com/downloads/truseq-stranded-mrna-reference-guide-1000000040498.html. Accessed 2 Mar 2025
16. Illumina (2019) bcl2fastq conversion-software. https://support.illumina.com/sequencing/sequencing_software/bcl2fastq-conversion-software.html
17. Martin M (2011) Cutadapt removes adapter sequences from high-throughput sequencing reads. EMBnet J 17:10–12. https://doi.org/10.14806/ej.17.1.200
18. Roehr JT, Dieterich C, Reinert K (2017) Flexbar 3.0—SIMD and multicore parallelization. Bioinformatics 33(18):2941–2942
19. Krueger F (2025) FelixKrueger/TrimGalore. https://github.com/FelixKrueger/TrimGalore. Accessed 2 Mar 2025
20. Babraham Bioinformatics (n.d.) FastQC a quality control tool for high throughput sequence data. http://www.bioinformatics.babraham.ac.uk/projects/fastqc/. Accessed 2 Mar 2025
21. Dobin A, Davis CA, Schlesinger F, Drenkow J, Zaleski C, Jha S et al (2013) STAR: ultrafast universal RNA-seq aligner. Bioinformatics 29(1):15–21. https://doi.org/10.1093/bioinformatics/bts635
22. Frankish A, Carbonell-Sala S, Diekhans M, Jungreis I, Loveland JE, Mudge JM et al (2023) GENCODE: reference annotation for the human and mouse genomes in 2023. Nucleic Acids Res 51(D1):D942–D9d9. https://doi.org/10.1093/nar/gkac1071
23. Love MI, Huber W, Anders S (2014) Moderated estimation of fold change and dispersion for RNA-seq data with DESeq2. Genome Biol 15(12):550. https://doi.org/10.1186/s13059-014-0550-8
24. Wickham H (2009) Getting started with qplot. In: Wickham H (ed) Ggplot2: elegant graphics for data analysis. Springer, New York, pp 9–26. https://doi.org/10.1007/978-0-387-98141-3_2
25. Xiao Z, Zou Q, Liu Y, Yang X (2016) Genome-wide assessment of differential translations with ribosome profiling data. Nat Commun 7: 11194. https://doi.org/10.1038/ncomms11194

Chapter 19

A Versatile Ribo-seq Workflow for Enhanced Analysis of Ribosome Dynamics

Hélène Hérault, Armelle Choquet, and Alexandre David

Abstract

Ribosome profiling (or ribo-sequencing) is a powerful technique that enables high-resolution analysis of active translation by sequencing ribosome-protected mRNA fragments. Developed in 2009 by Nicholas Ingolia and Jonathan Weissman, it provides direct insights into which mRNAs are being translated and at what rate, surpassing traditional transcriptomic and proteomic methods. Ribo-seq allows for precise mapping of ribosome positions, enabling detailed characterization of ribosome dynamics and the identification of alternative translation initiation sites and upstream open reading frames. The standard workflow includes key steps such as ribosome stabilization, nuclease digestion, fragment isolation, and deep sequencing. The protocol described in this chapter incorporates a polysome profiling step prior to RNase treatment, allowing, for instance, the isolation of distinct ribosome species (also known as *k*-somes). Ribo-seq has transformed our understanding of translational regulation and has become an essential tool for omics studies in various fields such as developmental biology, cancer biology, virology, and microbiology.

Key words Ribosome profiling, Ribo-sequencing, Ribosome-protected fragments, Translatome

1 Introduction

Ribo-sequencing (ribo-seq), also known as ribosome profiling, emerged in the late 2000s as a transformative technique to decode the translational landscape of cells with unprecedented resolution. Conceived through a convergence of advances in high-throughput sequencing, molecular biology, and ribosome biochemistry, the method was first introduced by Nicholas Ingolia and Jonathan Weissman in 2009 [1]. Its central goal is to provide a snapshot of active translation by sequencing fragments of mRNA that are protected by ribosomes, thus directly quantifying which mRNAs are being translated and to what extent.

At its core, ribo-seq is based on the principle that ribosomes shield approximately 28–30 nucleotide-long fragments of mRNA from nuclease digestion during active translation. By isolating these ribosome-protected fragments (RPFs) and subjecting them to deep

Rami Bechara (ed.), *Autoimmunity: Methods and Protocols*, Methods in Molecular Biology, vol. 2990,
https://doi.org/10.1007/978-1-0716-4997-8_19,

sequencing, researchers can infer the positions and density of translating ribosomes across the transcriptome. This allows for high-resolution mapping of translation in vivo, surpassing the indirect insights offered by transcriptomic or proteomic methods alone. In particular, ribo-seq enables the identification of alternative mRNA translation start sites, the confirmation of annotated open reading frames (ORFs) as well as upstream ORFs (uORFs) that may be involved in the regulation of translation, the distribution of ribosomes on an mRNA, and the rate at which ribosomes decode codons, therefore providing a strong ground to determine codon usage bias [2]. Over the past 15 years, ribo-seq has enabled the discovery and ongoing annotation of thousands of small ORFs (smORFs) and their microprotein products, driving their functional characterization through clustered regularly interspaced short palindromic repeats (CRISPR) screens and interaction studies [3].

The standard ribo-seq workflow involves several key steps: cell lysis under conditions that preserve ribosome-mRNA complexes; nuclease digestion to trim unprotected RNA regions; isolation of RPFs; purification and conversion of these fragments into a cDNA library; high-throughput sequencing followed by computational analysis. Optimization of parameters such as digestion time, ribosome stabilization, and sequencing depth is critical for accurate interpretation. The ribo-seq protocol described in this chapter (Fig. 1) is based on the method originally developed by Shu-Bing Qian's lab [4] and includes a polysome profiling step prior to RNase treatment, enabling the isolation of specific ribosome species (monosomes, disomes, trisomes, etc.) for targeted applications such as modelling ribosome dynamics [5].

Ribo-sequencing has revolutionized our understanding of gene expression regulation at the translational level. It has been widely applied to uncover translational control mechanisms, detect uORFs, resolve translation initiation sites, and even to discover novel coding regions that were previously missed by traditional annotation. In several fields such as developmental biology [6], disease research [7], and microbiology [8–10], ribo-seq and its derivative methods have proven indispensable for linking mRNA abundance to actual protein synthesis, thus filling a key gap in the central dogma. Ribo-seq is increasingly applied to immunological and autoimmune contexts. Translational profiling of immune cell subsets can reveal disease-specific regulation of cytokines, transcription factors, and stress response pathways.

2 Materials

All solutions must be prepared using ultrapure, DNase-/RNase-free water (referred to as "upH_2O") except stated otherwise. Powders must be weighed without the use of spatula so as not to

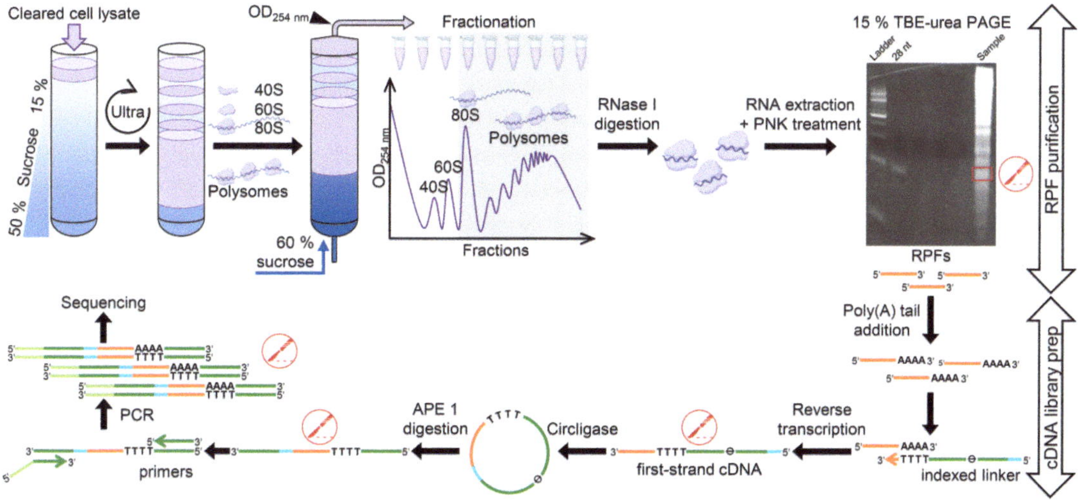

Fig. 1 Ribo-seq experimental protocol overview. After cell harvesting and lysis, the cleared cell lysate is deposited onto sucrose density gradient to separate RNA molecules according to their ribosomal load. Monosomal and polysomal fractions are collected and pooled (indicated by a grey underlay) for RNase I digestion, yielding single ribosomes protecting RNA fragments approximately 28 nucleotides in length. After RNA extraction and treatment with polynucleotide kinase (PNK) to phosphorylate 5′ ends and dephosphorylate 3′ ends, the RNA is separated by denaturing polyacrylamide gel electrophoresis (PAGE) and ribosome-protected fragments (RPFs; shown in orange) are recovered by gel excision (indicated by a red rectangle). The preparation of the cDNA library encompasses several enzymatic steps and gel excisions (indicated by a scalpel symbol), including poly(A) tail addition; reverse transcription using a linker (in dark green) containing a dSpacer abasic site (ϴ) in the middle and a 6-nucleotide index sequence at its 5′ end (in blue); circularization of the first-strand cDNA by circligase; re-linearization using apurinic/apyrimidinic endonuclease 1 (APE 1) targeting the abasic site; amplification by PCR using primers annealing to the linker sequences flanking RPFs, where the forward primer contains a floating tail (adapter sequence; in light green) on its 5′ end to facilitate sequencing; cDNA library outsourcing for high-throughput ribo-sequencing and data analysis

introduce any RNase. Plasticware and commercial solutions must be DNase-/RNase-free. Ideally, reagents used for RNA experiments should be identified and kept separate from those used for other experiments (e.g., when working with DNA or proteins) to avoid cross-contamination. Solutions must be stored at 4 °C unless stated otherwise.

2.1 General

1. DNase-/RNase-free 2 mL, 1.5 mL, and 0.5 mL microcentrifuge tubes treated to reduce nucleic acid binding to the plastic.
2. Refrigerated (4 °C) microcentrifuge accommodating 2 mL and 1.5 mL microcentrifuge tubes.
3. Two heat blocks “A” and “B” accommodating 1.5 mL microcentrifuge tubes.

2.2 Sucrose Density Gradients

1. 40 U/μL RNase-OUT™ (Invitrogen™). Store at −20 °C.
2. cOmplete™ EDTA-free protease inhibitor cocktail (Roche; referred to as "protease inhibitor cocktail," 100×): 2 tablets in 1 mL upH_2O. Store at −20 °C.
3. Fifty percent sucrose solution: 50% w/v sucrose, 50 mM Tris-HCl, 5 mM $MgCl_2$, 25 mM KCl, 1× protease inhibitor cocktail, 10 U/mL RNase-OUT™ (*see* **Note 1**).
4. Fifteen percent sucrose solution: 15% w/v sucrose, 50 mM Tris-HCl, 5 mM $MgCl_2$, 25 mM KCl, 1× protease inhibitor cocktail, 10 U/mL RNase-OUT™ (*see* **Note 1**).
5. Thinwall polypropylene ultracentrifuge tubes for Beckman Coulter SW41 Ti rotor, 13.2 mL, 14 × 89 mm, reference 331372 (Beckman Coulter).

2.3 Cell Harvesting and Lysis

1. Phosphate-buffered saline suitable for cell culture, pH 7.0–7.3.
2. Translation inhibitor stock solution (*see* **Note 2**) such as emetine stock solution (250×): 5 mg/mL emetine dihydrochloride in upH_2O (*see* **Note 3**). Store at −20 °C.
3. Trypsin-emetine solution: 20 μg/mL emetine in 0.05% trypsin-EDTA solution. Prepare extemporaneously.
4. Culture media.
5. Refrigerated centrifuge (4 °C) accommodating 50 mL conical centrifuge tubes.
6. Polysome lysis buffer: 7% w/v sucrose, 50 mM Tris-HCl, 5 mM $MgCl_2$, 25 mM KCl, 1% w/v NP-40, 1× protease inhibitor cocktail, 10 U/mL RNase-OUT™. Prepare extemporaneously.
7. Lysing matrix D tubes containing 1.4 mm zirconium-silicate spheres, 2 mL (MP Biomedicals).
8. FastPrep-24™ bead beater (MP Biomedicals) accommodating 2 mL lysing matrix D tubes and able to perform cell lysis at 4 m/s.

2.4 Fractionation

1. SW41 Ti rotor and associated buckets (Beckman Coulter).
2. Ultracentrifuge accommodating SW41 Ti rotor and able to reach ≈ 150,920 × *g* (average speed), such as Optima LE-80K or Optima-XPN (Beckman Coulter).
3. Sixty percent sucrose solution: 60% w/v sucrose, 0.002% w/v bromophenol blue.
4. Sterile 50 mL syringes.
5. Syringe pump module such as Legato® 110 syringe pump (KD Scientific).

6. Tube piercer module compatible with the fractionation system such as BR-184X tube piercer module (Brandel).
7. UV detector such as UA-6 UV/vis detector and associated type 11 optical unit (Teledyne ISCO).
8. Fraction collector such as Foxy® R1 and associated 1.5 mL microcentrifuge tube rack (Teledyne ISCO).
9. Peak recording system or acquisition software such as Peak-Trak™ (Teledyne ISCO).

2.5 Ribosome-Protected Fragment Purification

1. 100 U/μL RNase I Ambion™ (Invitrogen ™). Store at −20 °C.
2. TRIzol™ LS (Invitrogen™). Store at room temperature.
3. Chloroform. Store at room temperature.
4. Glycogen (*see* **Note 4**). Store at −20 °C.
5. Isopropanol. Store at room temperature.
6. 75% v/v ethanol in upH_2O: one solution stored at room temperature (referred to as "room-temperature") and one solution stored at −20 °C (referred to as "ice-cold").
7. Polynucleotide kinase mix: 3.8× T4 polynucleotide kinase buffer (1.5 μL of 10× buffer), 5 U/μL T4 polynucleotide kinase (2 μL of 10,000 U/mL solution), 2.5 U/μL SUPERase-In™ (0.5 μL of 20 U/μL solution) in a 4 μL reaction volume for each sample. Recommended reagents are 10,000 U/mL T4 polynucleotide kinase and associated buffer from New England Biolabs and 20 U/μL SUPERase-In™ from Invitrogen™. Prepare extemporaneously (*see* **Note 5**).
8. 28 nt standard loading solution: 0.3 μM 28 nt migration standard oligonucleotide added to 1× TBE-urea sample buffer prepared in 1× TBE running buffer. 28 nt-std: 5′-AUG-UAC-ACG-GAG-UCG-ACC-CGC-AAC-GCG-A-3′. Prepare extemporaneously.

2.6 Nucleic Acid Extraction Following Polyacrylamide Gel Electrophoresis

1. XCell *SureLock*™ mini-cell (Invitrogen™) migration apparatus and accessories (i.e., buffer dam if running an odd number of gels, gel knife).
2. Power supply compatible with the migration apparatus.
3. Novex™ 15% denaturing polyacrylamide TBE-urea gels, 10 wells, 1 mm (Invitrogen™).
4. Novex™ 10% denaturing polyacrylamide TBE-urea gels, 10 wells, 1 mm (Invitrogen™).
5. Novex™ 6% polyacrylamide TBE gels, 10 wells, 1 mm (Invitrogen™) (*see* **Note 6**).
6. Novex™ 2× TBE-urea sample buffer (Invitrogen™).
7. Novex™ 5× hi-density TBE sample buffer (Invitrogen™).

8. TBE running buffer (1×): prepared from Novex™ 5× TBE running buffer in autoclaved milli-Q® water. Store at room temperature.
9. 100 or 200 μL sterile, RNase-free, round gel-loading filter tips.
10. TBE-urea empty well loading solution: 1× TBE-urea sample buffer prepared in 1× TBE running buffer. Prepare extemporaneously.
11. High density TBE empty well loading solution: 1× hi-density TBE sample buffer prepared in 1× TBE running buffer. Prepare extemporaneously.
12. TBE-urea ladder loading solution: 2–3 ng/μL adequate ladder such as 10 bp O'RangeRuler DNA ladder (Thermo Scientific™) added to 1× TBE-urea sample buffer prepared in 1× TBE running buffer. Prepare extemporaneously.
13. Hi-density TBE ladder loading solution: 2–3 ng/μL adequate ladder such as 10 bp O'RangeRuler DNA ladder (Thermo Scientific™) added to 1× hi-density TBE sample buffer prepared in 1× TBE running buffer. Prepare extemporaneously.
14. Staining solution: 1× nucleic acid gel stain such as SYBR™ Gold (Invitrogen™) in 1× TBE running buffer. Prepare extemporaneously and store away from light.
15. Orbital shaker.
16. Cling film.
17. Blue light transilluminator.
18. Sterile, single-use scalpels.
19. Sterile, 18 G needle with 11° regular bevel, 38 mm length or longer.
20. RNA elution buffer: 300 mM sodium acetate, 1 mM EDTA, 0.1 U/μL SUPERase-In™ (Invitrogen™) (*see* **Note 7**).
21. DNA elution buffer: 300 mM sodium chloride, 1 mM EDTA. Store at room temperature.
22. Rotating wheel accommodating 2 mL microcentrifuge tubes.
23. Centrifuge tube filters with cellulose acetate membrane, 0.22 μm pore size, such as Costar® Spin-X® centrifuge tube filters (Corning®).

2.7 Nucleic Acid Precipitation

1. Glycogen (*see* **Note 4**). Store at −20 °C.
2. 3 M sodium acetate, pH 5.5. Store at room temperature.
3. Absolute ethanol, ice-cold. Store at −20 °C.
4. 75% v/v ethanol in upH_2O, ice-cold. Store at −20 °C.

2.8 Polyadenylation and Reverse Transcription

1. Polyadenylation mix: 2.9× poly(A) polymerase buffer (0.8 μL of 10× buffer), 3.6 mM ATP (1 μL of 10 mM solution), 2.1 U/μL poly(A) polymerase (0.6 μL of 5000 U/mL solution), 2.9 U/μL SUPERase-In™ (0.4 μL of 20 U/μL solution) in a 2.8 μL reaction volume for each sample. Recommended reagents are 5000 U/mL *E. coli* poly(A) polymerase and associated reagents from New England Biolabs and 20 U/μL SUPERase-In™ from Invitrogen™. Prepare extemporaneously (*see* **Note 5**).
2. 10 mM dNTPs. Store at −20 °C.
3. Reverse transcription mix: 2× first-strand buffer (4 μL of 5× buffer), 20 mM DTT (2 μL of 100 mM solution), 20 U/μL reverse transcriptase (1 μL of 200 U/μL solution), 4 U/μL RNase-OUT™ (1 μL of 40 U/μL solution) brought to a 10 μL reaction volume with upH_2O (2 μL) for each sample. Recommended reagents are SuperScript™ III reverse transcriptase and associated reagents from Invitrogen™. Prepare extemporaneously (*see* **Note 5**).
4. 3 μM indexed linker with 6 nucleotide indexes (*see* **Note 8**). Linkers must be 5′phosphate [PHO] and 3′ wobble (VN, with V corresponding to A, G or C, and N corresponding to A, G, C, or T) and must include a dSpacer abasic [ϴ] site. RT-RIBseqTCTTCT: 5′[PHO]-TCT-TCT-GAT-CGT-CGG-ACT-GTA-GAA-CTC-T[ϴ] C-AAG-CAG-AAG-ACG-GCA-TAC-GAG-ATT-TTT-TTT-TTT-TTT-TTT-TTT-TVN-3′. RT-RIBseqCGCAGA: 5′[PHO]-CGC-AGA-GAT-CGT-CGG-ACT-GTA-GAA-CTC-T[ϴ] C-AAG-CAG-AAG-ACG-GCA-TAC-GAG-ATT-TTT-TTT-TTT-TTT-TTT-TTT-TVN-3′. RT-RIBseqAATGAG: 5′[PHO]-AAT-GAG-GAT-CGT-CGG-ACT-GTA-GAA-CTC-T[ϴ] C-AAG-CAG-AAG-ACG-GCA-TAC-GAG-ATT-TTT-TTT-TTT-TTT-TTT-TTT-TVN-3′. RT-RIBseqTGCTTA: 5′[PHO]-TGC-TTA-GAT-CGT-CGG-ACT-GTA-GAA-CTC-T[ϴ] C-AAG-CAG-AAG-ACG-GCA-TAC-GAG-ATT-TTT-TTT-TTT-TTT-TTT-TTT-TVN-3′. Store at −20 °C.
5. 1 M sodium hydroxide. Store at room temperature.
6. 1 M hydrochloric acid. Store at room temperature.

2.9 Circularization and Re-linearization

1. Circularization mix: 2.4× circligase buffer (2 μL of 10× buffer), 5.9 mM $MnCl_2$ (1 μL of 50 mM solution), 2.4 M betaine (4 μL of 5 M solution), 17.6 U/μL circligase enzyme (1.5 μL of 100 U/μL solution) in an 8.5 μL reaction volume for each sample. Recommended reagents are 100 U/μL CircLigase II ssDNA ligase and associated reagents from LGC Biosearch™ Technologies. Prepare extemporaneously (*see* **Note 5**).

2. Re-linearization mix: 5.7 × apurinic/apyrimidinic endonuclease 1 buffer (1 μL of 10 × buffer), 4.3 U/μL apurinic/apyrimidinic endonuclease 1 (0.75 μL of 10 U/μL solution). Recommended reagents are 10,000 U/mL APE 1 and associated reaction buffer from New England Biolabs. Prepare extemporaneously (*see* **Note 5**).

2.10 PCR Amplification and Sequencing

1. Sanger PCR mix: 1.1× high-fidelity (HF) polymerase buffer (5 μL of 5× buffer), 0.22 mM dNTPs (0.5 μL of 10 mM solution), 0.54 μM forward primer (1.25 μL of 10 μM solution), 0.54 μM reverse primer (1.25 μL of 10 μM solution), 0.04 U/μL high-fidelity polymerase (0.5 μL of 2 U/μL solution) brought to a 23 μL reaction volume with upH_2O (14.5 μL) for each sample. Recommended reagents are 2 U/μL Phusion™ high-fidelity DNA polymerase and associated reagents from Thermo Scientific™. PCR-RIBseq-forward: 5′--AAT-GAT-ACG-GCG-ACC-ACC-GAG-ATC-TAC-ACG-TTC-AGA-GTT-CTA-CAG-TCC-GAC-G-3′. PCR-RIBseq-reverse: 5′-CAA-GCA-GAA-GAC-GGC-ATA-CGA-GAT-3′. Prepare extemporaneously (*see* **Note 5**).
2. 10 μM Sanger-riboseq primer solution in upH_2O. Sanger-riboseq: 5′-AAT-GAT-ACG-GCG-ACC-ACC-3′.
3. Thermocycler accommodating 0.5 mL microcentrifuge tubes.
4. Ribo-seq PCR mix: 1.2× high-fidelity (HF) polymerase buffer (5 μL of 5× buffer), 0.24 mM dNTPs (0.5 μL of 10 mM solution), 0.60 μM forward primer (1.25 μL of 10 μM solution), 0.60 μM reverse primer (1.25 μL of 10 μM solution), 0.05 U/μL high-fidelity polymerase (0.5 μL of 2 U/μL solution) brought to a 21 μL reaction volume with upH_2O (12.5 μL) for each sample. Recommended reagents and primer sequences are the same as used in Sanger PCR mix. Prepare extemporaneously (*see* **Note 5**).

3 Methods

Bench, micropipettes, and pipet controller must be cleaned with 70% ethanol and then with a RNase-inactivating solution to minimize the introduction of RNases while performing the experiment. Disposable gloves must be worn at all times and changed frequently. Samples must be kept on ice and steps must be performed at 4 °C unless stated otherwise. For microcentrifugation steps and especially following nucleic acid precipitation, make sure to orientate the microcentrifuge tube hinges outwards to help locating the pellet.

3.1 Sucrose Density Gradient Preparation

1. Deposit 5 mL of 50% sucrose solution at the bottom of an ultracentrifuge tube. Tilt the tube to maintain a minimal angle between the bench and the side of the tube and gently layer 5 mL of 15% sucrose solution over the 50% sucrose solution. Carefully bring the tube backup in vertical position and seal with parafilm (*see* **Note 9**).
2. Gently lay the tube down in horizontal position on a flat surface. Incubate for 1 h at room temperature before slowly bringing the tube back in vertical position (*see* **Note 10**). Store at 4 °C until needed (*see* **Note 11**).

3.2 Cell Harvesting and Lysis

1. Before starting, refrigerate at 4 °C the centrifuge accommodating 50 mL conical centrifuge tubes, the centrifuge accommodating 2 mL microcentrifuge tubes and the ultracentrifuge, its rotor, and associated buckets. Warm an adequate volume of PBS (to wash cells before harvest; referred to as "warm PBS"), culture media, and trypsin at 37 °C. Place the different conical centrifuge tubes, lysing matrix D tubes and microcentrifuge tubes needed for this section on ice so they are cold when samples are transferred into them.
2. Culture conditions (e.g., cell type, culture media, stress or treatment) must be defined in advance to address the specific scientific questions being asked (*see* **Note 12**). This section will detail the procedure used for adherent cells in exponential growth phase at the time of harvest, ideally 70% confluence. Ensure the cell growth state (i.e., confluence for adherent cells) is similar between the different test conditions (*see* **Note 13**). Moreover, the choice of translation inhibitor used also depends on the goal of the study and the translation step of interest (*see* **Note 2**); this section will detail the procedure when using emetine, an irreversible elongation inhibitor.
3. Remove culture media and wash cells with an adequate volume of warm PBS. Add an adequate volume (e.g., 5 mL for a T-75 culture flask or a 150 mm diameter Petri dish) of trypsin-emetine solution and incubate 5 min at 37 °C. Inactivate the trypsin by adding the same volume of culture media (supplemented with FBS), resuspend the cells and transfer them to a 50 mL conical centrifuge tube on ice (*see* **Note 14**). All subsequent steps must be performed on ice or at 4 °C. Pellet down the cells for 5 min at 400 × *g*, wash once with 10 mL ice-cold PBS, resuspend the cells with 10 mL ice-cold PBS, and count the cells with a method of choice for normalization (*see* **Note 15**).
4. Pellet down the cells for 5 min at 400 × *g*, resuspend them with 1 mL polysome lysis buffer, and transfer them to a lysing matrix D tube. Perform cell lysis using a bead beater set at 4 m/s for

20 s and immediately return tubes on ice. Let the foam settle for 5 min before transferring the suspension to a 2 mL microcentrifuge tube. Pellet down cell debris and nuclei at 16,000 × *g* for 10 min. The supernatant should be clear; if any turbidity is observed, repeat the centrifugation step. Transfer the supernatant to a 1.5 mL microcentrifuge tube and measure RNA concentration using a method of choice. Transfer a volume equivalent to a minimum of 500 ng RNA to a new 1.5 mL microcentrifuge tube and store at −80 °C until needed for RNA extraction: this will be necessary for transcriptomics analysis and be then referred to as "cytosolic RNA" (*see* **Note 16**). The rest of the cleared cell lysate is then used for fractionation on sucrose gradient.

3.3 Polysome Fractionation on Sucrose Density Gradient

1. Gently layer an adequate volume of clear cell lysate on the top of a 15–50% sucrose gradient (*see* **Notes 15** and **17**). Equilibrate the tubes within 10 mg by carefully adding polysome lysis buffer. Place the tubes in the SW41 Ti buckets, load them onto the rotor, and centrifuge at ≈ 150,920 × *g* (average speed; corresponding to 35,000 rpm with a SW41 Ti rotor) for 2.5 h with the maximum acceleration setting and the minimum brake setting (or deactivated brake) depending on the ultracentrifuge used (*see* **Note 18**). Recover the tubes, seal them with parafilm, and store them at 4 °C until ready for fractionation.
2. Prepare the fractionation system: turn the UV light approximately 10 min before starting fractionation for it to be warm enough and select the adequate settings (especially baseline and sensitivity; *see* **Note 19**) on the UV detector. Transfer 50 mL blue 60% sucrose solution to a 50 mL syringe and remove any air bubble trapped within. Place the syringe in the syringe pump and connect it to the tube piercer module. Bring the needle up, set the syringe pump on infuse mode, and inject blue sucrose solution until it comes out of the needle to remove air from the tubing. Bring the needle down and wipe the sucrose solution away. Select an adequate collection setting on the fraction collector (*see* **Note 20**) and start and set up the acquisition software.
3. Connect the ultracentrifuge tube to the tube piercer module and lock it in place. Bring the needle up through the bottom of the tube with the injection hole placed just at the bottom of the tube. Start injecting blue 60% sucrose solution to push the sucrose gradient through the fractionation system; there will be a delay between start of injection, beginning of signal acquisition at 254 nm and collection of the first fraction, corresponding to the tubing dead volume. Collect fractions in 1.5 mL microcentrifuge tubes and transfer them on ice as soon as possible. Store fractions at −80 °C until needed.

4. Once blue 60% sucrose solution starts being eluted, stop the run. Select the withdraw setting on the syringe pump and aspirate the sucrose solution back in the 50 mL syringe, leaving approximately 3–5 mL in the ultracentrifuge tube. Bring the needle down, remove the ultracentrifuge tube and wipe down residual sucrose on connecting parts. Prepare the syringe pump to inject similarly as described in **step 2**. Connect and lock the next ultracentrifuge tube in place and repeat **steps 3** and **4** until all samples are processed. Withdraw the 60% sucrose solution and clean the fractionation system following your usual procedure.

3.4 Ribosome-Protected Fragment Purification

1. Before starting, set heat block A at 37 °C and heat block B at 55–60 °C, and refrigerate the microcentrifuge at 4 °C.
2. Pool 100 μL of each polysomal fraction of interest (classically, starting from the monosomal fraction upwards; *see* **Note 21**) and measure RNA concentration using a method of choice. Transfer a volume corresponding to 9.5 μg RNA (equivalent to 160 uOD_{260nm}) to a new 2 mL microcentrifuge tube and adjust the volume to 300 μL with upH_2O. Add 120 U of RNase I and incubate at 4 °C (ideally in a fridge) for exactly 1 h (*see* **Note 22**).
3. Add 900 μL TRIzol™ LS, homogenize by pipetting up and down, invert a few times, and incubate 10 min at room temperature (*see* **Note 23**). Add 240 μL chloroform, homogenize by pipetting up and down, invert a few times, and incubate 5 min at room temperature. Centrifuge at 13,000 × *g* for 15 min and transfer the clear, aqueous upper phase (containing RNA) using a P200/P100 micropipette to a new 1.5 mL microcentrifuge tube (*see* **Note 24**). To precipitate RNA, add 20 μg glycogen, 600 μL isopropanol, invert a few times, and incubate 10 min at room temperature (*see* **Note 25**). Centrifuge at 13,000 × *g* for 20 min and discard the supernatant while making sure not to dislodge the RNA pellet. Wash the pellet twice by adding 1.2 mL room-temperature 75% ethanol, gently vortexing and centrifuging at 13,000 × *g* for 10 min; carefully discard the supernatant and ensure there is no trace of ethanol left by spinning down samples to remove any drop left on the sides of the tube. Air dry the pellet on ice until see-through (approximately 5–15 min) and resuspend with 11 μL upH_2O. Incubate for 10–15 min at room temperature (or eventually at 55–60 °C if hard to solubilize using block B; after incubation, set block B at 65 °C) to fully dissolve RNA pellet.
4. Add 4 μL polynucleotide kinase mix to each sample and mix by pipetting up and down. Incubate for 2 h at 37 °C (block A; after incubation, set heat block A at 70 °C) to phosphorylate RNA fragments on their 5′ ends and dephosphorylate them on

their 3′ ends. Incubate for 20 min at 65 °C (block B) to inactivate the enzyme. Shortly spin down samples to collect at the bottom of the tubes.

5. Add 15 μL 2× TBE-urea sample buffer to each sample, mix by pipetting up and down, and vortex at low speed. Denature samples and 28 nt standard loading solution (but not TBE-urea ladder loading solution) at 70 °C for 3 min and then incubate on ice for 1 min. Vortex at low speed and spin down to collect at the bottom of the tubes.
6. Prepare the 15% TBE-urea gel: cut open the gel wrapping with clean scissors and take the gel out. Remove the white tape at the bottom of the gel. Carefully remove the comb, ensuring there is no piece of gel blocking any of the wells. Rinse the wells 3 times with 1× TBE running buffer to remove any excess urea and place the gel in the migration apparatus. Fill the tank with 1× TBE running buffer. Only two samples can be loaded on each gel so as not to induce cross-contamination between samples or with the ladder or 28 nt migration standard. Lanes should be loaded as follows: ladder; 28 nt migration standard; empty well; empty well; empty well; sample 1; empty well; empty well; sample 2; empty well. Fill empty wells with 30 μL TBE-urea empty well loading solution, then ladder well with 30 μL TBE-urea ladder loading solution, 28 nt migration standard well with 28 nt standard loading solution, and finally sample wells with 30 μL sample. Run for 95 min at 160 V at room temperature (*see* **Note 26**). After the run is complete, pry open the gel cast and carefully cut the bottom of the gel (just above the hole where the white tape was initially) so as not to tear it. Incubate the gel in staining solution (30 mL in a 150 mm Petri dish for each gel) for 20 min under agitation on an orbital shaker, protected from light.
7. From now on, the 28 nt migration standard should undergo the same procedure as the samples (up to Subheading 3.7, **step** 7 included), providing a migration standard for further electrophoresis steps and a control once the DNA library is generated. For each sample, pierce the bottom of a 0.5 mL microcentrifuge tube with an 18 G, 11° regular bevel, sterile needle and place it inside a 2 mL microcentrifuge tube. Recover the gel and place it on a piece of cling film on the blue light transilluminator, in a dark room, at room temperature. Take a picture before and after gel excision to ensure the band of interest has been entirely recovered (*see* **Note 27**). Cut out the band migrating at the same level as the 28 nt migration standard (Fig. 2) using a sterile scalpel for each sample and transfer it to the pierced 0.5 mL microcentrifuge tube. Disrupt the gel through the hole at the bottom of the tube by centrifuging at 16,000 × *g* for 5 min (*see* **Note 28**). Discard the upper

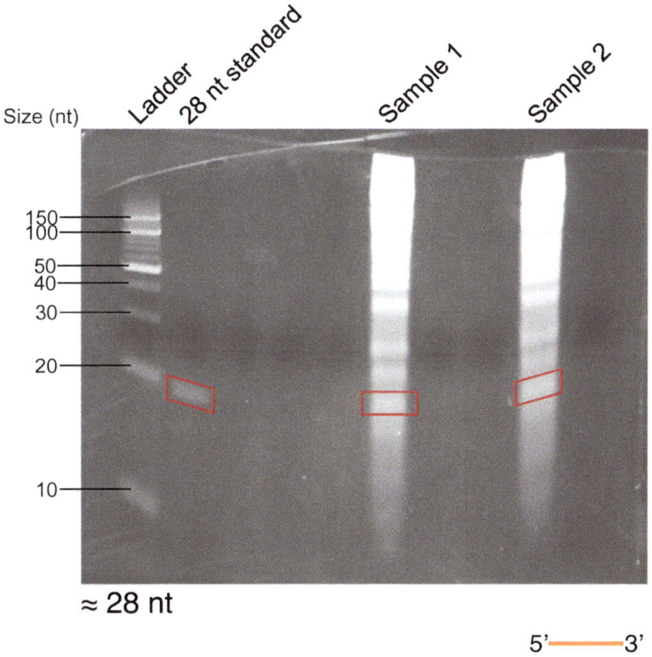

Fig. 2 RPF recovery following denaturing PAGE. RNA fragments were separated on 15% polyacrylamide TBE-urea gels and bands migrating at the same level as the 28 nt size standard (red rectangles), corresponding to the expected RPFs (schematically represented in orange), were excised from the gel prior to the next step

0.5 mL microcentrifuge tube, add 400 μL RNA elution buffer, vortex, and incubate overnight under agitation on a rotating wheel.

8. Spin down samples, widen the end of a 1000 μL tip by cutting it with a clean scalpel, and transfer the gel debris and buffer containing RNA onto a centrifuge tube filter with cellulose acetate membrane and centrifuge at 16,000 × *g* for 5 min. Transfer the eluate to a new 1.5 mL microcentrifuge tube and add 20 μg glycogen, 40 μL 3 M sodium acetate (sodium acetate:sample volume ratio 0.1:1), and 900 μL ice-cold absolute ethanol (absolute ethanol:sample volume ratio 2.25:1). Vortex briefly at low speed and incubate at –20 °C for 1 h for RNA precipitation (*see* **Note 25**). Centrifuge at 16,000 × *g* for 15 min and discard the supernatant carefully. Wash the pellet by adding 900 μL ice-cold 75% ethanol, gently vortexing and centrifuging at 16,000 × *g* for 5 min; carefully discard the supernatant and ensure there is no trace of ethanol left (samples can be spun down to remove any drop left on the sides of the tube). Air dry the pellet on ice until see-through (approximately 5–15 min) and resuspend with 5.2 μL upH_2O.

3.5 Polyadenylation and Reverse-Transcription

1. Before starting, set heat block A at 37 °C and heat block B at 65 °C, and refrigerate the microcentrifuge at 4 °C.
2. Add 2.8 μL polyadenylation mix to each sample and mix by pipetting up and down. Incubate 45 min at 37 °C (block A; after incubation, set block A at 50 °C) and inactivate enzyme at 65 °C for 20 min (block B). Briefly spin down to collect samples at the bottom of the tubes. Add 1 μL of 10 mM dNTPs and 1 μL of 3 μM indexed linker (*see* **Note 8**) and incubate 5 min at 65 °C (block B; after incubation, set block B at 85 °C) followed by 5 min incubation on ice.
3. Add 10 μL reverse transcription mix to each sample and mix by pipetting up and down. Incubate 50 min at 50 °C (block A; adjust block A to 70 °C after incubation) and inactive enzyme for 5 min at 85 °C (block B; adjust block B to 98 °C after incubation). Add 1.8 μL of 1 M sodium hydroxide and briefly vortex and incubate 20 min at 98 °C (block B) to eliminate the leftover RNA. Bring samples back on ice and briefly spin down to collect at the bottom of the tubes. Add 1.8 μL of 1 M hydrochloric acid to neutralize the pH, vortex at low speed, and spin down to collect samples at the bottom of the tubes.
4. Add 20 μL 2× TBE-urea sample buffer to each sample, mix by pipetting up and down, and vortex at low speed. Denature samples (but not TBE-urea ladder loading solution) at 70 °C (block A) for 3 min and then incubate on ice for 1 min. Vortex at low speed and spin down to collect at the bottom of the tubes.
5. Prepare the 10% TBE-urea gel as previously detailed for the 15% TBE-urea gel (*see* Subheading 3.4, **step 6**). Lanes can be loaded as follows: ladder; 28 nt migration standard; empty well; empty well; empty well; sample 1; empty well; empty well; sample 2; empty well. Fill empty wells with 40 μL TBE-urea empty well loading solution, then ladder well with 40 μL TBE-urea ladder loading solution and finally 28 nt migration standard and sample wells with 40 μL of corresponding samples. Run for 105 min at 160 V at room temperature (*see* **Note 26**). After the run is complete, proceed as detailed previously for gel staining (*see* Subheading 3.4, **step 6**).
6. Prepare gel excision of bands of interest as detailed previously (*see* Subheading 3.4, **step 7**). Cut out the band migrating at ≈ 100 nt (Fig. 3) using a sterile scalpel for each sample and transfer it to a pierced 0.5 mL microcentrifuge tube. Disrupt the gel through the hole at the bottom of the tube by centrifuging at 16,000 × *g* for 5 min (*see* **Note 28**). Discard the upper 0.5 mL microcentrifuge tube, add 400 μL DNA elution buffer, and vortex and incubate overnight under agitation on a rotating wheel.

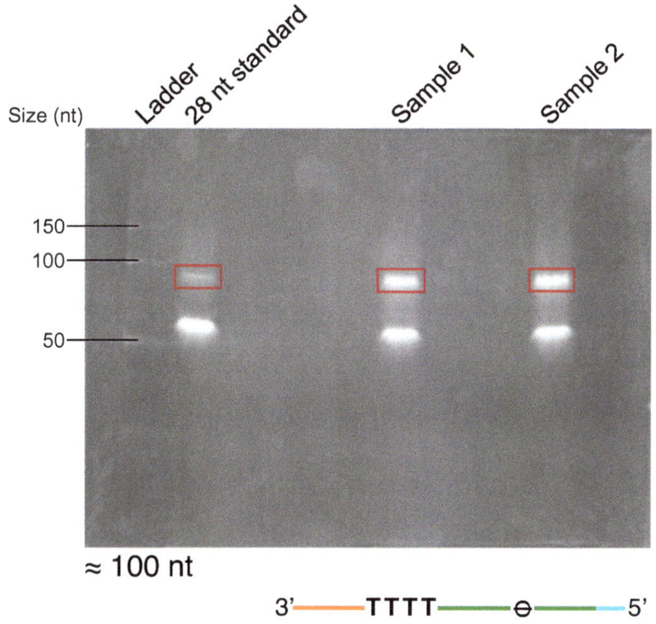

Fig. 3 First-strand cDNA recovery following denaturing PAGE. RPFs (in orange) underwent poly(A) tail addition and reverse transcription using a linker (in dark green) containing a dSpacer abasic site in the middle (ϴ) and a 6-nucleotide index at its 5′ end (in blue). Reaction products were separated on 10% polyacrylamide TBE-urea gels and single-stranded, first-strand cDNAs were extracted from bands migrating at approximately 100 nt (indicated by red rectangles)

7. Proceed to DNA recovery from gel debris and precipitation as detailed previously (*see* Subheading 3.4, **step 8**). Resuspend the resulting DNA pellet with 11.5 μL upH_2O.

3.6 Circularization and Re-linearization

1. Before starting, set heat block A at 60 °C and heat block B at 80 °C, and refrigerate the microcentrifuge at 4 °C.
2. Add 8.5 μL circularization mix to each sample and mix by pipetting up and down (*see* **Note 29**). Incubate at 60 °C for 90 min (*see* **Note 30**) (block A; after incubation, set block A at 37 °C) and inactivate enzyme at 80 °C for 10 min (block B; after incubation, set block B at 70 °C). Spin down to collect samples at the bottom of the tubes.
3. Precipitate samples to remove current buffer before proceeding with re-linearization. To do so, add 380 μL upH_2O, 20 μg glycogen, 40 μL 3 M sodium acetate (≈ 100 mM final concentration), and 900 μL ice-cold absolute ethanol. Vortex briefly at low speed and precipitate DNA at −20 °C for 1 h (*see* **Note 25**). Centrifuge at 16,000 × *g* for 15 min and discard the supernatant carefully. Wash the pellet by adding 900 μL

ice-cold 75% ethanol, gently vortexing, and centrifuging at 16,000 × *g* for 5 min; carefully discard the supernatant and ensure there is no trace of ethanol left (samples can be spun down to remove any drop left on the sides of the tube). Air dry the pellet on ice until see-through (approximately 5–15 min) and resuspend with 8.25 μL upH_2O.

4. Add 1.75 μL re-linearization mix to each sample and mix by pipetting up and down. Incubate at 37 °C for 60 min (block A) (*see* **Note 31**). Spin down to collect samples at the bottom of the tubes.
5. Add 10 μL 2× TBE-urea sample buffer to each sample, mix by pipetting up and down and vortex at low speed. Denature samples (but not TBE-urea ladder loading solution), at 70 °C for 3 min (block B), and then incubate on ice for 1 min. Vortex at low speed and spin down to collect at the bottom of the tubes.
6. Prepare the 10% TBE-urea gel as previously detailed for the 15% TBE-urea gel (*see* Subheading 3.4, **step 6**). Lanes can be loaded as follows: ladder; 28 nt migration standard; empty well or circularization control reaction; empty well; empty well; sample 1; empty well; empty well; sample 2; empty well. Fill empty wells with 20 μL TBE-urea empty well loading solution, then ladder well with 20 μL TBE-urea ladder loading solution and finally 28 nt migration standard and sample wells with 20 μL of corresponding samples. Run for 90 min at 160 V at room temperature (*see* **Note 26**). After the run is complete, proceed as detailed previously for gel staining (*see* Subheading 3.4, **step 6**).
7. Prepare gel excision of bands of interest as detailed previously (*see* Subheading 3.4, **step 7**). Cut out the band migrating at ≈ 100 nt (Fig. 4) using a sterile scalpel for each sample and transfer it to a pierced 0.5 mL microcentrifuge tube (*see* **Note 32**). Disrupt the gel through the hole at the bottom of the tube by centrifuging at 16,000 × *g* for 5 min (*see* **Note 28**). Discard the upper 0.5 mL microcentrifuge tube, add 400 μL DNA elution buffer, and vortex and incubate overnight under agitation on a rotating wheel.
8. Proceed to DNA recovery from gel debris and precipitation as detailed previously (*see* Subheading 3.4, **step 8**). Resuspend DNA pellet with 10 μL upH_2O.

3.7 PCR Amplification and Sequencing

1. This section will be performed twice: a first time to check samples by Sanger sequencing, and if sequencing results are satisfactory, steps will be repeated (with slight modifications) before samples are sent for ribo-seq.

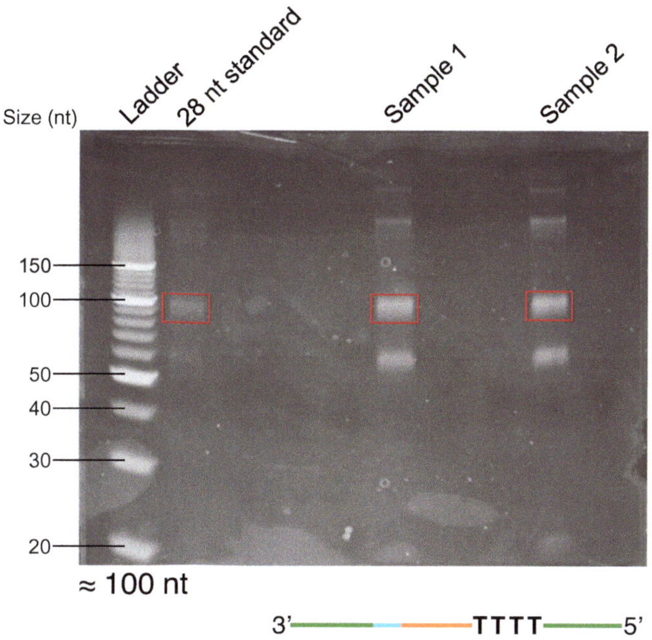

Fig. 4 Re-arranged single-stranded cDNA recovery following denaturing PAGE. First-strand cDNAs were circularized and subsequently re-linearized using APE 1, which targets the dSpacer abasic site in the linker. Linearized products were recovered from bands migrating at approximately 100 nt (indicated by red rectangles) on 10% polyacrylamide TBE-urea gels. Circular forms were still visible in the upper part of the gel lanes, indicating incomplete enzymatic digestion. The circularization/re-linearization step re-arranged the cDNAs so that the RPF (in orange) was flanked by known sequences corresponding to the linker (in dark green) and its 6-nucleotide index (in blue)

2. To prepare samples for Sanger sequencing, transfer 2 μL of each sample to a new 0.5 mL microcentrifuge tube. Store the leftover sample volume at −20 °C until needed. Add 23 μL Sanger PCR mix, mix by pipetting up and down, and spin down to collect samples at the bottom of the tube. Perform PCR for Sanger sequencing as follows: initial denaturation at 98 °C for 30 s followed by 15 cycles of denaturation at 98 °C for 10 s, primer annealing at 60 °C for 20 s and elongation at 72 °C for 10 s.
3. Add 6 μL 5× hi-density TBE sample buffer to each PCR product, mix by pipetting up and down and vortex at low speed. Spin down to collect samples at the bottom of the tubes.
4. Prepare the 6% TBE gel as previously detailed for the 15% TBE-urea gel (*see* Subheading 3.4, **step 6**). Lanes can be loaded as follows: ladder; 28 nt migration standard; empty well; sample 1; empty well; sample 2; empty well; sample 3; empty well; sample 4 (*see* **Note 33**). Fill empty wells with 30 μL hi-density

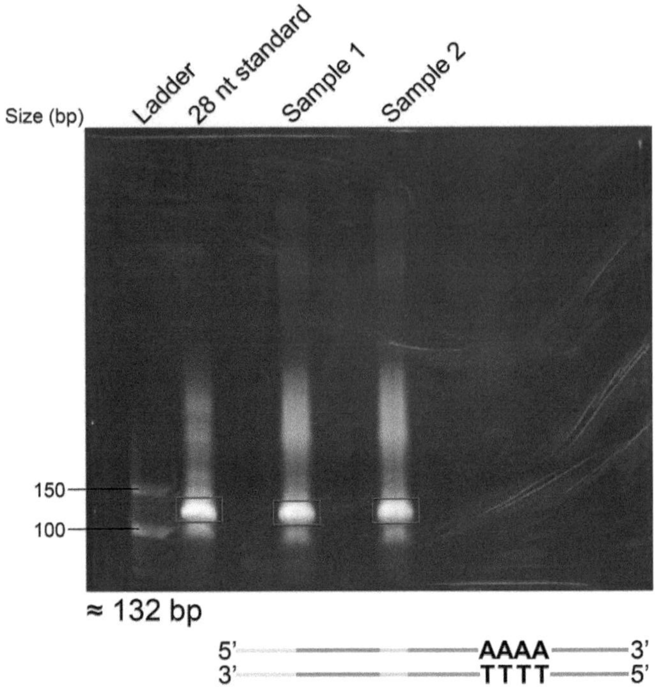

Fig. 5 Double-stranded cDNA library recovery following non-denaturing PAGE. Re-arranged single-stranded cDNAs were amplified by PCR using primers annealing to the flanking, linker sequences (in dark green) and recovered from bands migrating at approximately 132 nt on 6% polyacrylamide TBE gels (indicated by red rectangles). The forward primer included a floating tail at its 5′ end (in light green), extending the final product to facilitate sequencing

TBE empty well loading solution, then ladder well with 30 μL hi-density TBE ladder loading solution and finally 28 nt migration standard and sample wells with 30 μL of corresponding samples. Run for 45 min at 160 V at room temperature (*see* **Notes 6** and **26**). After the run is complete, proceed as detailed previously for gel staining (*see* Subheading 3.4, **step 6**).

5. Prepare gel excision of bands of interest as detailed previously (*see* Subheading 3.4, **step 7**). Cut out the upper band migrating at 132 bp (Fig. 5) using a sterile scalpel for each sample and transfer it to a pierced 0.5 mL microcentrifuge tube. Disrupt the gel through the hole at the bottom of the tube by centrifuging at 16,000 × *g* for 5 min (*see* **Note 28**). Discard the upper 0.5 mL microcentrifuge tube, add 400 μL DNA elution buffer, and vortex and incubate overnight under agitation on a rotating wheel.
6. Proceed to DNA recovery from gel debris and precipitation as detailed previously (*see* Subheading 3.4, **step 8**). Resuspend DNA pellet with 10 μL upH_2O.

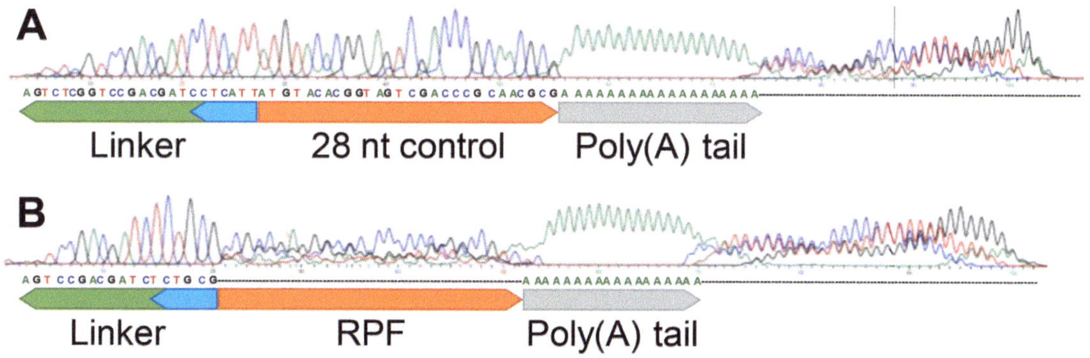

Fig. 6 Sanger sequencing checkpoints after cDNA library amplification. It is recommended to assess the quality of the amplified cDNA libraries by Sanger sequencing before outsourcing for ribo-seq. This should be performed on both the 28 nt migration standard (used as a control; panel **a**) and the biological samples (panel **b**). Arrows indicate the original 5′ (flat end) to 3′ (pointy end) orientation of the fragments. High quality sequences should start with well-defined peaks corresponding to the reverse-complement portion of the linker sequence used for reverse transcription (in dark green) and its 6-nucleotide index (in blue). Here, RT-RIBseqAATGAG and RT-RIBseqCGCAGA linkers were used for the 28 nt control and the sample, respectively. The second half of the sequence must contain a stretch of about 20 "A", corresponding to the poly (A) tail (in grey) added to RPFs. In between (in orange) there should be the 28 nt migration sequence in its original 5′–3′ orientation. Well-defined peaks are expected for the 28 nt control (panel **a**); the samples (panel **b**) typically display mixed signals with overlapping peaks, reflecting the sequence diversity of the cDNA library

7. Determine DNA concentration with a method of choice and send adequate volumes of samples and 10 μM Sanger-riboseq primer solution for Sanger sequencing to the service company of your choice. Several checkpoints should be looked for in the sequences obtained (Fig. 6). The sequences should start with well-separated peaks with low to no overlaps, corresponding to part of the linker reverse complement sequence. In the second half of the sequence, a stretch of approximately 20 "A" peaks should be clearly identified, corresponding to the poly(A) tail added during an earlier step. In between should be a sequence of approximately 20–30 nt corresponding to the purified RPFs; for samples, this signal should look "poor quality" with overlaps between the four different nucleotides, but for the 28 nt control the sequence should correspond to the 5′-3′ sequence indicated previously (*see* Subheading 2.5, **item 8**).
8. After confirmation of the good sample quality by Sanger sequencing, samples can be prepared for ribo-seq. This procedure does not need to be carried out for the 28 nt migration standard anymore. Transfer 4 μL of each sample obtained at the end of the Subheading 3.6, **step 8** (stored at −20 °C) to a new 0.5 mL microcentrifuge tube. Add 21 μL ribo-seq PCR mix, mix by pipetting up and down, and spin down to collect samples at the bottom of the tubes. Perform PCR for ribo-

seq as follows: initial denaturation at 98 °C for 30 s followed by 13 cycles of denaturation at 98 °C for 10 s, primer annealing at 60 °C for 20 s and elongation at 72 °C for 10 s. Then proceed as detailed for Sanger sequencing by repeating **steps 3–6** of this section. Determine DNA concentration with a method of choice and prepare samples for ribo-seq according to the method provided by the service company of your choice (*see* **Note 34**).

3.8 Data Analysis

1. This chapter aims to provide a detailed, step-by-step wet-lab protocol; therefore, bioinformatical treatment of data obtained from ribo-seq will only be outlined in this section. Please refer to relevant references dedicated to this type of analysis for more details.
2. Retrieve the reference nucleotide sequences and gene annotations from NCBI or an equivalent, curated reference database. Only protein-coding transcripts should be retained for downstream analysis.
3. Align ribo-seq reads to spliced mRNA sequences using the CRAC aligner [11], with a *k*-mer size of 25 nucleotides. This longer-than-default *k*-mer length was chosen to better match the minimal size of ribosome-protected fragments (RPFs), enhancing mapping specificity and reducing ambiguous alignments.
4. For each aligned read, extract the 5′ end position and shift by +12 nucleotides to approximate the position of the ribosome A-site on the mRNA. Compute per-gene coverage across codon positions downstream of the start codon, typically beginning at codon +1. This positional information forms the basis for quantifying ribosome occupancy and translation efficiency. Importantly, ribosome positioning (and nucleotide shift) may be influenced by the type of translation inhibitor employed (*see* **Note 2**).
5. The raw data analysis described above does not include a ribosomal RNA depletion (ribo-depletion) step, which is commonly added in some ribo-seq protocols to reduce background noise and improve the signal-to-noise ratio. However, several alternative computational pipelines have been developed and are publicly available, offering various levels of automation and flexibility (*see* **Note 35**). These tools offer complementary features and may be better suited depending on the complexity of the dataset, available metadata, and downstream analyses.

4 Notes

1. At least two sucrose gradients will have to be prepared to be able to equilibrate the rotor during ultracentrifugation. Prepare 5 mL of each sucrose solution per gradient needed, and add 2 mL for dead volume. Sucrose solutions containing RNase-OUT™ and protease inhibitor cocktail must be prepared extemporaneously. Alternatively, they can be prepared without any inhibitor cocktails in larger volumes (e.g., 50 mL) and stored at −20 °C until needed. Then, the RNase-OUT™ and protease inhibitor cocktail can be added later when preparing the sucrose gradients.
2. The choice of translation inhibitor used (if any) should be carefully considered and addressed according to the aim of the study, while taking into account possible experimental constraints. Depending on their mode of actions, translation inhibitors can freeze ribosomes on RNA molecules in different conformations (resulting in different ribosomal footprint sizes [12, 13]), at different steps (initiation or elongation), and in a reversible or irreversible manner. When using a reversible inhibitor, the same concentration should be maintained in all solutions until digestion by RNase I is complete to prevent ribosomes from sliding off mRNA, thus exposing fragments of interest to the effects of the nuclease. Regarding translation initiation inhibitors, harringtonine and lactimidomycin bind to and immobilize ribosomes at the start codon through different mechanisms. Harringtonine binds to free 60S sub-units, but not to those engaged into an 80S ribosome; harringtonine-bound 60S sub-units can still form a complex with 40S sub-units at a start site; however, the resulting 80S ribosome cannot move forward along the mRNA [14]. Harringtonine is a partially reversible inhibitor [15]; it can be added to the culture medium at 2 μg/mL final concentration and incubated for 3 min before cell lysis [16]. Lactimidomycin reversibly binds to the E-site of the large ribosomal sub-unit in already assembled 80S complexes and prevents the translocation reaction. However, due to its large size, lactimidomycin can only bind to ribosomes assembled at the start codon, when the tRNA enters the P-site directly, leaving the E-site empty and accessible [4, 17]. Lactimidomycin can be used at 50 μM final concentration and incubated for 30 min before cell lysis [16]. As for translation elongation inhibitors, cycloheximide is the most widely reported molecule in ribo-seq protocols. Cycloheximide acts similarly to lactimidomycin, but thanks to its smaller size, is able to prevent the translocation reaction in elongating ribosomes as well [17, 18]. Cycloheximide is commonly used at 100 μg/mL final concentration and incubated

for 1–10 min before cell lysis and should be maintained throughout the procedure [16, 19]. Care should be taken when using cycloheximide as continued elongation in its presence has been previously reported [20]. Emetine, used in this protocol, is an irreversible translation elongation inhibitor targeting the ribosomal 40S sub-unit. Emetine prevents the translocation reaction, possibly by impeding the 40S–60S complex movement along mRNAs [18, 21, 22]. Anisomycin inhibits the peptide bond formation by inducing a conformational change upon binding to the peptidyl transferase center of the 60S sub-unit A-site [17, 22], resulting in smaller (20–22 nt) ribosomal footprints [12, 19]. Anisomycin can be added at 100 μg/mL final concentration, incubated for 30 min before cell lysis, and should be maintained in subsequent solutions until RNase I digestion [12].

3. Emetine being highly toxic, the powder must be weighed under a chemical fume hood. Prefer to adjust the volume after weighing the powder to avoid repeating weighing, e.g., weigh approximately 50 mg emetine dihydrochloride and calculate the volume of upH_2O actually needed to prepare a 5 mg/mL solution with the actual amount of powder weighed. Aliquot the solution in smaller volumes (e.g., 500 μL) and store at −20 °C.
4. Avoid using colored glycogen (e.g., GlycoBlue™) as it has been previously observed to interfere with several downstream applications. Prefer using classic, white glycogen for the different precipitation steps.
5. Prepare the mix according to the number of samples with an additional 10–15% to account for dead volume.
6. Separation at this step is critical, as the band of interest (upper band) is really close to a second one just below corresponding to the library adapter background. To maximize separation, samples are ran down to the bottom limit of the gel, which increases the risk of losing samples if the bottom of the gel is torn after recovering it from its cast. We previously used Novex™ 8% polyacrylamide TBE gels and migrated samples for 65 min at 160 V with similar results compared to the procedure described here. This step could be optimized, potentially with the use of higher polyacrylamide percentage gels (e.g., 10%) and increased migration duration to improve band separation while reducing the risk of losing samples.
7. RNA elution buffer can be prepared in large volume and stored at room temperature as long as the SUPERase-In™ has not been added to the solution (referred to as "RNA elution stock solution"). When needed, transfer an adequate volume (i.e., 400 μL per sample with an additional 10–15% to account for

dead volume) of RNA elution stock solution and add the corresponding amount of SUPERase-In™ to reach 0.1 U/μL final concentration.

8. Index sequences will enable Illumina sequencing after the DNA library synthesis at the end of this procedure. The sequences were determined in accordance with the service company performing the Illumina sequencing. Contact your service company of choice to confirm these index sequences will be adequate with their equipment and analysis method. Using different indexed linkers for different samples enables the pooling of up to four samples within the same lane while still being able to decipher the sample of origin of each sequence.
9. For both sucrose solutions, aspirate 6 mL but only deposit 5 mL to avoid introducing bubbles. When layering the 15% sucrose solution onto the 50% sucrose solution, make sure to set the pipet controller at minimum speed and gravity setting if available. The tip of the pipette should stay in contact with the solution already in the tube, and the depositing speed should be kept as slow and steady as possible so as not to mix both sucrose solutions together. After the tube is back in vertical position, the interphase between both sucrose solutions can be observed.
10. Ensure no vibration (e.g., vortex at high speed, bead beater) can impact the flat surface onto which the tubes are incubating as they could disturb the sucrose gradient formation. After the tube is back in vertical position, the interphase between both sucrose solutions should not be visible anymore.
11. Sucrose gradients can be prepared the day before ultracentrifugation if kept at 4 °C.
12. Several culture parameters can influence translational activity and the resulting polysome profile. First, the cell type is critical: some cells naturally exhibit high translational activity, such as certain classical cancer cell lines (e.g., HeLa, SW620). Others, such as neurons, display much lower translation rates. As a result, polysome profiles will be significantly weaker in these cells, and a larger number of cells may be required to obtain sufficient RNA for ribo-seq library preparation. A second parameter is the growth medium: the presence of growth factors can substantially affect translational activity and proliferation. Finally, stress conditions must be considered. Cell proliferation can be significantly reduced by excessive confluence (contact inhibition) or by external stressors that are part of the experimental design. In general, it is best to perform the experiment during the exponential growth phase, at approximately 60–70% confluence for adherent cells. Regardless of the

conditions, conducting a preliminary calibration experiment is highly recommended to evaluate the translation profile and optimize experimental parameters.

13. Translational levels vary during cell growth, with a peak when the cell population is in log phase. If growth states are different between the test conditions (e.g., lag phase or stationary phase), it will not be possible to compare their translational levels as they would not only reflect the differences between treatments but also in global cell population health. This section details the procedure used for adherent cells, where confluence is generally related to growth phase. For cells cultured in suspension, their growth cycle must be studied prior to determine the adequate timepoint at which cells have to be harvested so as to be in their exponential growth phase and/or translationally active.

14. Alternatively, cells can be harvested by scraping instead of trypsinization. Remove culture media, add an adequate volume (e.g., 10 mL for a T-75 culture flask or a 150 mm diameter Petri dish) of fresh culture media containing 20 μg/mL emetine, and incubate for 5 min at 37 °C. Place the culture vessel on ice and keep it on ice until the end of cell harvest. Aspirate the culture media-emetine solution, wash the cells with an adequate (e.g. 5 mL for a T-75 culture flask or a 150 mm diameter Petri dish) volume of ice-cold PBS, scrape the cells, and transfer them into a 50 mL conical centrifuge tube. To maximize cell recovery, ice-cold PBS wash and scraping can be performed 2–3 times. Then, proceed with the initial procedure described above for cell counting and lysis.

15. Normalization between the different conditions can be performed either on total cell or RNA amount. These parameters can be correlated or not depending on the experimental design, and the normalization choice will have to be addressed according to the aim of the study.

16. Transcriptomic data obtained from cytosolic RNA reflects the entire pool of mRNA present in the cytoplasm while, on the other hand, data obtained from ribo-seq (or polysome profiling) corresponds only to molecules bound by one or more ribosomes, therefore actively engaged in translation. Normalization of ribo-seq data against RNA-seq data enables to visualize differences in translation only between the different test conditions. To extract cytosolic RNA from clear cell lysates, bring up the volume to 300 μL with upH_2O, proceed as described in Subheading 3.4, **step 3** and resuspend RNA pellets in an adequate volume of upH_2O. Prepare samples according to the guidelines provided by a service company of choice for RNA-sequencing (library preparation and sequencing).

17. As ultracentrifuge tubes are submitted to important speeds and forces, ensure they are filled with at least 11 mL total volume to prevent them from collapsing. As the sucrose gradients already occupy 10 mL in each ultracentrifuge tube, the adequate sample volume should be brought up to 1 mL with polysome lysis buffer before equilibrating the ultracentrifuge tubes. Prefer layering carefully the adequate volume of polysome lysis buffer first on top of the 15–50% sucrose gradient before gently adding the sample volume.
18. To prevent gradients from being disturbed during the ultracentrifugation run, deceleration should be adjusted to the lowest settings. When using ultracentrifuges such as Optima LE-80K, either "slow" (gradual brake until 500 rpm followed by reduced brake leading to a smooth stop) or "no brake" (free brake) settings can be used. With Optima XPN (or similar) ultracentrifuges, no additional information about deceleration speeds could be retrieved; free brake (corresponding to a deceleration setting of 10) or minimal brake settings (8 or 9) should be preferred. Additional tests should be conducted to find the correct balance between ultracentrifugation run duration (increased when reduced or free brake are used) and preserved polysome separation on sucrose gradients (maximized with smooth to free brake).
19. Depending on the acquisition module used, baseline and sensitivity parameters might have to be manually set up before starting fractionation. Complementarily to the culture condition determination, we advise to run a mock experiment including the fractionation step to ensure the different parameters involved (i.e., cell and/or RNA amount loaded on the sucrose density gradients, combined with baseline and sensitivity parameters set on the acquisition module). These parameters should be kept the same for the different test conditions so as to be able to compare their polysome profiles. Ideally, the resulting polysome profile should show good separation between the different *k*-some peaks; for example, loading too much material can increase the peak width, making it more difficult to decipher between the different peaks (especially relevant for the monosome peak which can overlap with the disome peak). On the other hand, there should be enough RNA loaded on the sucrose density gradients so as to be able to pursue the rest of this protocol. The signal should be high enough to clearly identify the different peaks, but not too high to ensure no signal saturation (especially relevant for the monosome peak) is observed in any test condition. This can only be obtained by finding the right combination between the cell and/or RNA amount and the sensitivity parameter on the acquisition module. The baseline parameter is not as critical

nowadays, as most acquisition systems rely on computer software rather than chart paper and recorders. Still, setting up the baseline so that the minimal value recorded aligns with the *x* axis remains good practice.

20. Fraction size will be determined by the injection rate on the syringe pump and the time interval between two fractions set on the fractionator (when using time intervals method for collection) as follows:

$$\text{Fraction size} = \frac{\text{Injection rate}}{\text{Time interval}}$$

For example, to collect 500 μL fractions, it is possible to set the injection rate at 0.5 mL/min with the time interval set at 1 min or set the injection rate at 1 mL/min with the time interval set at 30 s. Smaller fractions will result in better resolution between different ribosome sub-unit alone, monosome and polysome peaks. This is particularly relevant for polysome profiling but can also be of importance for ribo-seq depending on the aim of the study.

21. Ribo-seq is generally performed on RNA molecules bound by at least one ribosome. However, depending on the aim of the study, it is possible to apply this protocol to subsets of this pool, as described in a previous study [5].

22. The RNA amount and incubation time must be strictly observed at this step so as to obtain the typical-sized (around 28–30 nt) ribosome footprints. If the RNA concentration measured after pooling the fractions of interest is too low (i.e., lower than 32 ng/μL, corresponding to 9.5 μg RNA in a total volume of 300 μL), prepare several microcentrifuge tubes with known RNA amount (the sum should be 9.5 μg between the different microcentrifuge tubes) and adjust the RNase I volume to keep the same ratio between RNase I units and RNA amount (1200 U for 9.5 μg RNA). For example, prepare two microcentrifuge tubes with 4.75 μg RNA in each, bring the volume to 300 μL with upH_2O, and add 600 U RNase I before pursuing the protocol. Split samples can be pooled back together at the end of **step 3** (Subheading 3.4) where both RNA pellets can be resuspended with the same 11 μL upH_2O transferred between the different microcentrifuge tubes.

23. Breaking point: samples can be stored at 4 °C overnight or at −20 °C for up to a year.

24. Ensure none of the interphase or the pink, lower organic phase is transferred alongside the clear, aqueous upper phase. Tilting the microcentrifuge tube at a 45 ° angle can help with the pipetting of the aqueous phase, especially when a small amount

is left, as it will form a "bubble" instead of a layer when the tube is held vertically.

25. Alternatively, samples can be precipitated overnight at −20 °C.
26. Run duration might need to be adjusted depending on the generator used. Ideally, run a test sample beforehand to validate the run duration setting: bands of interest should be well separated while not being too close to the bottom of the gel to prevent samples from being damaged or eluted (particularly important for 6% TBE gels).
27. If bands are visible enough using the blue light transilluminator, take a picture through the amber protective screen using a camera. We found the BluPAD dual led blue/white light transilluminator (reference BP001CU from Bio-Helix) to be highly satisfactory in that matter. Alternatively, pictures can be rapidly taken under UV light, but band excision must still be performed using a blue light transilluminator to prevent the introduction of any damage or mutation.
28. The crushed gel should be entirely transferred to the 2 mL microcentrifuge tube. If the gel is not fully disrupted or still partially present in the upper 0.5 mL microcentrifuge tube, there might be a cling film preventing it from sliding down to the 2 mL microcentrifuge tube. Try to remove it with a clean tip and repeat the centrifugation step.
29. When starting a new batch of circligase, run a control reaction in parallel using 5 μL ssDNA control (2 pmol/μL stock concentration) mixed with 6.5 μL upH_2O as template to ensure the enzyme works properly. Deposit the control reaction on the 10% TBE-urea gel with samples after APE 1 re-linearization.
30. In our experience, circularization seems to be a critical step with low yields, and could be optimized. According to the manufacturer's recommendations, conversion to linear form can be improved by increasing the incubation time, increasing the circligase enzyme amount and/or adding 50 μM final concentration ATP in the reaction volume.
31. In our experience, re-linearization by APE 1 under these conditions could be partial, as circular forms could still be observed after migration on a 10% TBE-urea gel. Although not critical, this step could be enhanced by increasing the incubation time and/or increasing the amount of enzyme in the reaction volume.
32. Circular form migrates slower and is found higher on the gel than the linearized one.
33. This migration is part of the final steps of this protocol. Consequently, risks of contamination between samples or between

samples and migration standard and their potential impacts are drastically reduced, enabling the processing of more samples on one gel.

34. For example, we previously performed ribo-seq using an Illumina HiSeq2500 based on the sequence-by-synthesis technology. Runs were 50 bp single-read and sequencing depth was 40 million reads per sample with four samples with different index sequences mixed in the same lane, corresponding to 160 million reads per lane [5].

35. Notable examples of ribo-depletion pipelines include: RiboGalaxy [23], a Galaxy-based platform offering interactive ribo-seq data processing and visualisation; RiboFlow [24], a scalable, Snakemake-based pipeline for processing ribo-seq and RNA-seq data side-by-side; Xtail [25], for differential translation analysis from ribo-seq and matched RNA-seq data.

References

1. Ingolia NT, Ghaemmaghami S, Newman JRS, Weissman JS (2009) Genome-wide analysis in vivo of translation with nucleotide resolution using ribosome profiling. Science 324:218–223. https://doi.org/10.1126/science.1168978
2. Paulet D, David A, Rivals E (2017) Ribo-seq enlightens codon usage bias. DNA Res 24:303–210. https://doi.org/10.1093/dnares/dsw062
3. Tong G, Martinez TF (2025) Ribosome profiling reveals hidden world of small proteins. Trends Genet 41:101–103. https://doi.org/10.1016/j.tig.2024.12.010
4. Lee S, Liu B, Lee S et al (2012) Global mapping of translation initiation sites in mammalian cells at single-nucleotide resolution. Proc Natl Acad Sci USA 109. https://doi.org/10.1073/pnas.1207846109
5. Chevalier C, Dorignac J, Ibrahim Y et al (2023) Physical modeling of ribosomes along messenger RNA: estimating kinetic parameters from ribosome profiling experiments using a ballistic model. PLoS Comput Biol 19:e1011522. https://doi.org/10.1371/journal.pcbi.1011522
6. Xiong Z, Xu K, Lin Z et al (2022) Ultrasensitive Ribo-seq reveals translational landscapes during mammalian oocyte-to-embryo transition and pre-implantation development. Nat Cell Biol 24:968–980. https://doi.org/10.1038/s41556-022-00928-6
7. Su D, Ding C, Qiu J et al (2024) Ribosome profiling: a powerful tool in oncological research. Biomark Res 12:11. https://doi.org/10.1186/s40364-024-00562-4
8. Fremin BJ, Sberro H, Bhatt AS (2020) MetaRibo-Seq measures translation in microbiomes. Nat Commun 11:3268. https://doi.org/10.1038/s41467-020-17081-z
9. Mohammad F, Green R, Buskirk AR (2019) A systematically-revised ribosome profiling method for bacteria reveals pauses at single-codon resolution. elife 8:e42591. https://doi.org/10.7554/eLife.42591
10. Froschauer K, Svensson SL, Gelhausen R et al (2025) Complementary Ribo-seq approaches map the translatome and provide a small protein census in the foodborne pathogen *Campylobacter jejuni*. Nat Commun 16:3078. https://doi.org/10.1038/s41467-025-58329-w
11. Philippe N, Salson M, Commes T, Rivals E (2013) CRAC: an integrated approach to the analysis of RNA-seq reads. Genome Biol 14:R30. https://doi.org/10.1186/gb-2013-14-3-r30
12. Lareau LF, Hite DH, Hogan GJ, Brown PO (2014) Distinct stages of the translation elongation cycle revealed by sequencing ribosome-protected mRNA fragments. elife 3:e01257. https://doi.org/10.7554/eLife.01257
13. Ingolia NT, Lareau LF, Weissman JS (2011) Ribosome profiling of mouse embryonic stem cells reveals the complexity and dynamics of mammalian proteomes. Cell 147:789–802. https://doi.org/10.1016/j.cell.2011.10.002
14. Fresno M, Jiménez A, Vázquez D (1977) Inhibition of translation in eukaryotic systems by

harringtonine. Eur J Biochem 72:323–330. https://doi.org/10.1111/j.1432-1033.1977.tb11256.x

15. David A, Dolan BP, Hickman HD et al (2012) Nuclear translation visualized by ribosome-bound nascent chain puromycylation. J Cell Biol 197:45–57. https://doi.org/10.1083/jcb.201112145
16. Ingolia NT, Brar GA, Rouskin S et al (2013) Genome-wide annotation and quantitation of translation by ribosome profiling. CP Mol Biol 103. https://doi.org/10.1002/0471142727.mb0418s103
17. Garreau De Loubresse N, Prokhorova I, Holtkamp W et al (2014) Structural basis for the inhibition of the eukaryotic ribosome. Nature 513:517–522. https://doi.org/10.1038/nature13737
18. Jimenez A, Carrasco L, Vazquez D (1977) Enzymic and nonenzymic translocation by yeast polysomes. Site of action of a number of inhibitors. Biochemistry 16:4727–4730. https://doi.org/10.1021/bi00640a030
19. Wu CC-C, Zinshteyn B, Wehner KA, Green R (2019) High-resolution ribosome profiling defines discrete ribosome elongation states and translational regulation during cellular stress. Mol Cell 73:959–970.e5. https://doi.org/10.1016/j.molcel.2018.12.009
20. Hussmann JA, Patchett S, Johnson A et al (2015) Understanding biases in ribosome profiling experiments reveals signatures of translation dynamics in yeast. PLoS Genet 11: e1005732. https://doi.org/10.1371/journal.pgen.1005732
21. Grollman AP (1968) Inhibitors of protein biosynthesis. V. Effects of emetine on protein and nucleic acid biosynthesis in HeLa cells. J Biol Chem 243:4089–4094
22. Pestka S (1971) Inhibitors of ribosome functions. Ann Rev Microbiol 25:487–562. https://doi.org/10.1146/annurev.mi.25.100171.002415
23. Michel AM, Mullan JPA, Velayudhan V et al (2016) RiboGalaxy: a browser based platform for the alignment, analysis and visualization of ribosome profiling data. RNA Biol 13:316–319. https://doi.org/10.1080/15476286.2016.1141862
24. Ozadam H, Geng M, Cenik C (2020) RiboFlow, RiboR and RiboPy: an ecosystem for analyzing ribosome profiling data at read length resolution. Bioinformatics 36:2929–2931. https://doi.org/10.1093/bioinformatics/btaa028
25. Xiao Z, Zou Q, Liu Y, Yang X (2016) Genome-wide assessment of differential translations with ribosome profiling data. Nat Commun 7: 11194. https://doi.org/10.1038/ncomms11194

INDEX

Rami Bechara (ed.), *Autoimmunity: Methods and Protocols*, Methods in Molecular Biology, vol. 2990,
https://doi.org/10.1007/978-1-0716-4997-8,

MIX
Papier aus verantwortungsvollen Quellen
Paper from responsible sources
FSC® C105338

If you have any concerns about our products,
you can contact us on
ProductSafety@springernature.com

In case Publisher is established outside the EU,
the EU authorized representative is:
Springer Nature Customer Service Center GmbH
Europaplatz 3, 69115 Heidelberg, Germany

Printed by Libri Plureos GmbH
in Hamburg, Germany